Basic laws of electromagnetism

И. Е. ИРОДОВ

ОСНОВНЫЕ ЗАКОНЫ ЭЛЕКТРОМАГНЕТИЗМА

Издательство «Высшая школа»
Москва

I.E.Irodov

Basic laws of electromagnetism

Translated from Russian
by
Natasha Deineko and Ram Wadhwa

Mir Publishers Moscow

CBS Publishers & Distributors Pvt. Ltd.

New Delhi • Bengaluru • Chennai • Kochi • Kolkata • Mumbai
Hyderabad • Nagpur • Patna • Pune • Vijayawada

First published 1986
Revised from the 1983 Russian edition

На английском языке

ISBN: 81-239-0306-5

First Indian Edition: 1994
Reprint: 1996, 1998, 1999, 2001, 2004

Published by:
Satish Kumar Jain for CBS Publishers & Distributors Pvt. Ltd.,
4819/XI Prahlad Street, 24 Ansari Road, Daryaganj, New Delhi - 110002
delhi@cbspd.com, cbspubs@airtelmail.in • www.cbspd.com
Ph.: 23289259, 23266861, 23266867 • Fax: 011-23243014

Corporate Office: 204 FIE, Industrial Area, Patparganj, Delhi - 110 092
Ph: 49344934 • Fax: 011-49344935
E-mail: publishing@cbspd.com • publicity@cbspd.com

Branches:

- ***Bengaluru:*** 2975, 17th Cross, K.R. Road, Bansankari 2nd Stage, Bengaluru - 70 • Ph: +91-80-26771678/79 • Fax: +91-80-26771680 E-mail: cbsbng@gmail.com, bangalore@cbspd.com
- ***Chennai:*** No. 7, Subbaraya Street, Shenoy Nagar, Chennai - 600030 Ph: +91-44-26681266, 26680620 • Fax: +91-44-42032115 E-mail: chennai@cbspd.com
- ***Kochi:*** Ashana House, 39/1904, A.M. Thomas Road, Valanjambalam, Ernakulum, Kochi • Ph: +91-484-4059061-65 Fax: +91-484-4059065 • E-mail: cochin@cbspd.com
- ***Kolkata:*** 6-B, Ground Floor, Rameshwar Shaw Road, Kolkata - 700014 Ph: +91-33-22891126/7/8 • E-mail: kolkata@cbspd.com
- ***Mumbai:*** 83-C, Dr. E. Moses Road, Worli, Mumbai - 400018 Ph: +91-9833017933, 022-24902340/41 • E-mail: mumbai@cbspd.com

Printed at: Neekunj Print Process, Delhi (India)

Preface

The main idea behind this book is to amalgamate the description of the basic concepts of the theory and the practical methods of solving problems in one book. Therefore, each chapter contains first a description of the theory of the subject being considered (illustrated by concrete examples) and then a set of selected problems with solutions. The problems are closely related to the text and often complement it. Hence they should be analysed together with the text. In author's opinion, the selected problems should enable the reader to attain a deeper understanding of many important topics and to visualize (even without solving the problems but just by going through them) the wide range of applications of the ideas presented in this book.

In order to emphasize the most important laws of electromagnetism, and especially to clarify the most difficult topics, the author has endeavoured to exclude the less important topics. In an attempt to describe the main ideas concisely, clearly and at the same time correctly, the text has been kept free from superfluous mathematical formulas, and the main stress has been laid on the physical aspects of the phenomena. With the same end in view, various model representations, simplifying factors, special cases, symmetry considerations, etc. have been employed wherever possible.

SI units of measurements are used throughout the book. However, considering that the Gaussian system of units is still widely used, we have included in Appendices 3 and 4 the tables of conversion of the most important quantities and formulas from SI to Gaussian units.

The most important statements and terms are given in italics. More complicated material and problems involving cumbersome mathematical calculations are set in brevier type. This material can be omitted on first reading without any loss of continuity. The brevier type is also used for problems and examples.

The book is intended as a textbook for undergraduate students specializing in physics (in the framework of the course on general physics). It can also be used by university teachers.

The author is grateful to Prof. A.A. Detlaf and Reader L.N. Kaptsov who reviewed the manuscript and made a number of valuable comments and suggestions.

I. Irodov

Contents

List of Notations

Vectors are denoted by the bold-face upright letters (e.g. $\mathbf{r}$, $\mathbf{E}$). The same letter in the standard-type print (r, E) denotes the magnitude of the vector.

Average quantities are denoted by angle brackets $\langle\ \rangle$, e.g. $\langle \mathbf{v} \rangle$, $\langle P \rangle$.

The **symbols** in front of the quantities denote:
Δ is the finite increment of the quantity, viz. the difference between its final and initial values, e.g. $\Delta\mathbf{E} = \mathbf{E}_2 - \mathbf{E}_1$, $\Delta\varphi = \varphi_2 - \varphi_1$;
d is the differential (infinitesimal increment), e.g. $d\mathbf{E}$, $d\varphi$;
δ is the incremental value of a quantity, e.g. δA is the elementary work.

Unit vectors:
$\mathbf{e}_x$, $\mathbf{e}_y$, $\mathbf{e}_z$ (or $\mathbf{i}$, $\mathbf{j}$, $\mathbf{k}$) are the unit vectors of Cartesian coordinates x, y, z;
$\mathbf{e}_\rho$, $\mathbf{e}_\varphi$, $\mathbf{e}_z$ are the unit vectors of cylindrical coordinates ρ, φ, z;
$\mathbf{n}$ is the unit vector of the normal to a surface element;
$\boldsymbol{\tau}$ is the unit vector of the tangent to the contour or to an interface.

Time derivative of an arbitrary function f is denoted by $\partial f/\partial t$ or by the dot above the letter denoting the function, $\dot{f}$.

Integrals of any multiplicity are denoted by a single symbol $\int$ and differ only in the notation of the element of integration: dV is the volume element, dS is the surface element and $d\mathbf{l}$ is the element of length. Symbol $\oint$ denotes the integration over a closed contour or a closed surface.

Vector operator ∇ (nabla). The operations involving this operator are denoted as follows: $\nabla\varphi$ is the gradient of φ (grad φ), $\nabla\cdot\mathbf{E}$ is the divergence of $\mathbf{E}$ (div $\mathbf{E}$), and $\nabla\times\mathbf{E}$ is the curl of $\mathbf{E}$ (curl $\mathbf{E}$).

Vector product is denoted $[\mathbf{a}\times\mathbf{b}]$, where $\mathbf{a}$ and $\mathbf{b}$ are vectors.

1. Electrostatic Field in a Vacuum

1.1. Electric Field

Electric Charge. At present it is known that diverse phenomena in nature are based on four fundamental interactions among elementary particles, viz. strong, electromagnetic, weak, and gravitational interactions. Each type of interaction is associated with a certain characteristic of a particle. For example, the gravitational interaction depends on the mass of particles, while the electromagnetic interaction is determined by electric charges.

The electric charge of a particle is one of its basic characteristics. It has the following fundamental properties:

(1) electric charge exists in two forms, i.e. it can be positive or negative;

(2) the algebraic sum of charges in any electrically insulated system does not change (this statement expresses the *law of conservation of electric charge*);

(3) electric charge is a relativistic invariant: its magnitude does not depend on the reference system, in other words, it does not depend on whether the charge moves or is fixed.

It will be shown later that these fundamental properties have far-reaching consequences.

Electric Field. In accordance with modern theory, interaction among charges is accomplished through a field. Any electric charge q alters in a certain way the properties of the space surrounding it, i.e. creates an *electric field*. This means that another, "test" charge placed at some point of the field experiences the action of a force.

Experiments show that the force $\mathbf{F}$ acting on a fixed test point charge q' can always be represented in the form

$$\mathbf{F} = q'\mathbf{E}, \tag{1.1}$$

where vector $\mathbf{E}$ is called the *intensity* of the electric field at a given point. Equation (1.1) shows that vector $\mathbf{E}$ can be defined as the force acting on a positive fixed unit charge. Here we assume that the test charge q' is sufficiently small so that its introduction does not noticeably distort the field under investigation (as a result of possible redistribution of charges creating the field).

The Field of a Point Charge. It follows directly from experiment (Coulomb's law) that the intensity of the field of a fixed point charge q at a distance r from it can be represented in the form

$$\boxed{\mathbf{E}=\frac{1}{4\pi\varepsilon_0}\frac{q}{r^2}\mathbf{e}_r,} \tag{1.2}$$

where ε_0 is the electric constant and $\mathbf{e}_r$ is the unit vector of the radius vector $\mathbf{r}$ drawn from the centre of the field, where the charge q is located, to the point we are interested in. Formula (1.2) is written in SI. Here the coefficient

$$1/4\pi\varepsilon_0 = 9 \times 10^9 \text{ m/F},$$

the charge q is measured in *coulombs* (C) and the field intensity $\mathbf{E}$ in *volts per metre* (V/m). Vector $\mathbf{E}$ is directed along $\mathbf{r}$ or oppositely to it depending on the sign of the charge q.

Formula (1.2) expresses *Coulomb's law* in the "field" form. It is important that the intensity $\mathbf{E}$ of the field created by a point charge is inversely proportional to the square of the distance r. All experimental results indicate that this law is valid for distances from 10^{-13} cm to several kilometres, and there are no grounds to expect that this law will be violated for larger distances.

It should also be noted that the force acting on a test charge in the field created by a fixed point charge does not depend on whether the test charge is at rest or moves. This refers to a system of fixed charges as well.

Principle of Superposition. Besides the law expressed by (1.2), it also follows from experiments that the intensity of the field of a system of fixed point charges is equal to the vector sum of the intensities of the fields that would be created by each charge separately:

$$\boxed{\mathbf{E}=\sum \mathbf{E}_i=\frac{1}{4\pi\varepsilon_0}\sum \frac{q_i}{r_i^2}\mathbf{e}_{ri},} \tag{1.3}$$

where r_i is the distance between the charge q_i and the point under consideration.

This statement is called the *principle of superposition* of electric fields. It expresses one of the most remarkable properties of fields and allows us to calculate field intensity of any system of charges by representing this system as an aggregate of point charges whose individual contributions are given by formula (1.2).

Charge Distribution. In order to simplify mathematical calculations, it is often convenient to ignore the fact that charges have a discrete structure (electrons, nuclei) and assume that they are "smeared" in a certain way in space. In other words, it is convenient to replace the actual distribution of discrete point charges by a fictitious continuous distribution. This makes it possible to simplify calculations considerably without introducing any significant error.

While going over to a continuous distribution, the concept of charge density is introduced, viz. the volume density ρ, surface density σ, and linear density λ. By definition,

$$\rho=\frac{dq}{dV}, \quad \sigma=\frac{dq}{dS}, \quad \lambda=\frac{dq}{dl}, \tag{1.4}$$

where dq is the charge contained in the volume dV, on the surface dS, and in the length dl respectively.

Taking these distributions into consideration we can represent formula (1.3) in a different form. For example, if the charge is distributed over the volume, we must replace q_i by $dq = \rho\, dV$ and summation by integration. This gives

$$\mathbf{E}=\frac{1}{4\pi\varepsilon_0}\int\frac{\rho\mathbf{e}\, dV}{r^2}=\frac{1}{4\pi\varepsilon_0}\int\frac{\rho\mathbf{r}\, dV}{r^3}, \tag{1.5}$$

where the integration is performed over the entire space with nonzero values of ρ.

Thus, knowing the distribution of charges, we can completely solve the problem of finding the electric field intensity by formula (1.3) if the distribution is discrete or by formula (1.5) or by a similar formula if it is continuous. In the general case, the calculation involves certain difficulties (although they are not of principle nature). Indeed, in order to find vector **E**, we must first calculate its projections E_x, E_y, and E_z, which means that we must take three integrals of the type (1.5). As a rule, the problem becomes much simpler in cases when a system of charges has a certain symmetry. Let us consider two examples.

Example 1. The field on the axis of a thin uniformly charged ring. A charge $q > 0$ is uniformly distributed over a thin ring of radius a. Find the electric field intensity E on the axis of the ring as a function of the distance z from its centre.

It can be easily shown that vector **E** in this case must be directed along the axis of the ring (Fig. 1.1). Let us isolate element dl on the

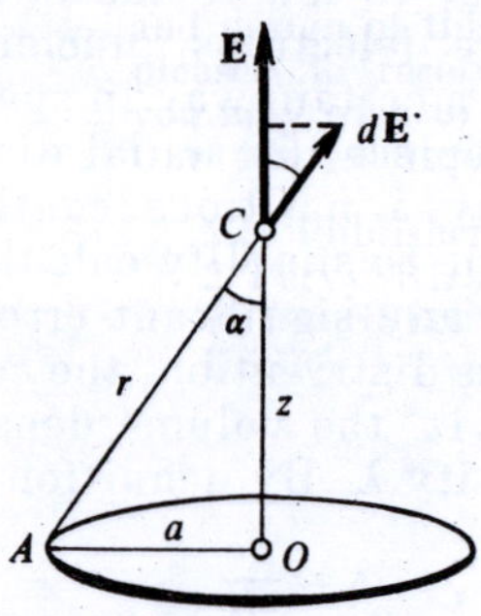

Fig. 1.1

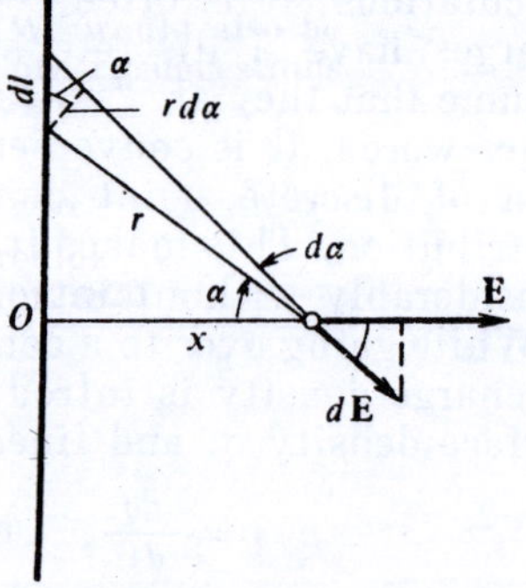

Fig. 1.2

ring in the vicinity of point A. We write the expression for the component dE_z of the field created by this element at a point C:

$$dE_z = \frac{1}{4\pi\varepsilon_0}\frac{\lambda\, dl}{r^2}\cos\alpha,$$

where $\lambda = q/2\pi a$. The values of r and α will be the same for all the elements of the ring, and hence the integration of this equation is simply reduced to the replacement of $\lambda\, dl$ by q. As a result, we obtain

$$E = \frac{q}{4\pi\varepsilon_0}\frac{z}{(a^2+z^2)^{3/2}}.$$

It can be seen that for $z \gg a$ the field $E \approx q/4\pi\varepsilon_0 z^2$, i.e. at large distances the system behaves as a point charge.

Example 2. The field of a uniformly charged straight filament. A thin straight filament of length $2l$ is uniformly charged by a charge q. Find the field intensity E at a point separated by a distance x from the midpoint of the filament and located symmetrically with respect to its ends.

It is clear from symmetry considerations that vector **E** must be directed as shown in Fig. 1.2. This shows the way of solving this problem: we must find the component dE_x of the field created by the element dl of the filament, having the charge dq, and then integrate the result over all the elements of the filament. In this case

$$dE_x = dE \cos\alpha = \frac{1}{4\pi\varepsilon_0}\frac{\lambda\, dl}{r^2}\cos\alpha,$$

where $\lambda = q/2l$ is the linear charge density. Let us reduce this equation to the form convenient for integration. Figure 1.2 shows that $dl \cos\alpha = r d\alpha$ and $r = x/\cos\alpha$; consequently,

$$dE_x = \frac{1}{4\pi\varepsilon_0}\frac{\lambda r\, d\alpha}{r^2} = \frac{\lambda}{4\pi\varepsilon_0 x}\cos\alpha\, d\alpha.$$

This expressio : can be easily integrated:

$$E = \frac{\lambda}{4\pi\varepsilon_0 x}\, 2\int_0^{\alpha_0}\cos\alpha\, d\alpha = \frac{\lambda}{4\pi\varepsilon_0 x}\, 2\sin\alpha_0,$$

where α_0 is the maximum value of the angle α, $\sin\alpha_0 = l/\sqrt{l^2+x^2}$. Thus,

$$E = \frac{q/2l}{4\pi\varepsilon_0 x}\, 2\frac{l}{\sqrt{l^2+x^2}} = \frac{q}{4\pi\varepsilon_0 x\sqrt{l^2+x^2}}.$$

In this case also $E \simeq q/4\pi\varepsilon_0 x^2$ for $x \gg l$ as the field of a point charge.

Geometrical Description of Electric Field. If we know vector **E** at each point, the electric field can be visually represented with the help of field lines, or lines of **E**. Such a line is drawn so that a tangent to it at each point coincides with the direction of vector **E**. The density of the lines, i.e. the number of lines per unit area normal to the lines is proportional to the magnitude of vector **E**. Besides, the lines are directed like vector **E**. This pattern gives the idea about the configuration of a given electric field, i.e. about the direction and magnitude of vector **E** at each point of the field.

On the General Properties of Field E. The field **E** defined above has two very important properties. The knowledge of these properties helped to deeper understand the very concept of the field and formulate its laws, and also made it possible to solve a number of problems in a simple and elegant way. These properties, viz. the Gauss theorem and the theorem on circulation of vector **E**, are associated with two most important mathematical characteristics of all vector fields: the *flux* and the *circulation*. It will be shown below that in terms of these two concepts not only all the laws of electricity but also all the laws of magnetism can be described. Let us go over to a systematic description of these properties.

1.2. The Gauss Theorem

Flux of E. For the sake of clarity, we shall use the geometrical description of electric field (with the help of lines of **E**). Moreover, to simplify the analysis, we shall assume that the density of lines of field **E** is *equal* to the magnitude of vector **E**. Then the number of lines piercing the area element dS, the normal **n** to which forms angle α with vector **E**, is determined as $E \cdot dS \cos \alpha$ (see Fig. 1.3). This quantity is just the flux $d\Phi$ of **E** through the area element dS. In a more compact form, this can be written as

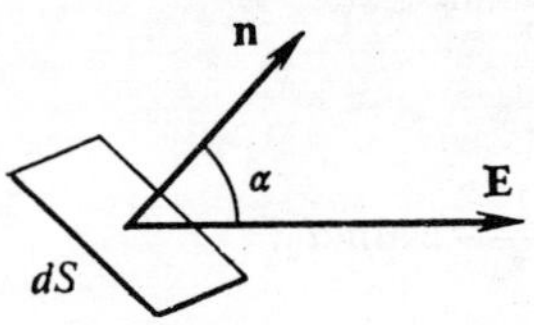

Fig. 1.3

$$d\Phi = E_n \, dS = \mathbf{E} \cdot d\mathbf{S},$$

where E_n is the projection of vector **E** onto the normal **n** to the area element dS, and $d\mathbf{S}$ is the vector whose magnitude is equal to dS and the direction coincides with the direction of the normal. It should be noted that the choice of the direction of **n** (and hence of $d\mathbf{S}$) is arbitrary. This vector could be directed oppositely.

If we have an arbitrary surface S, the flux of **E** through it can be expressed as

$$\Phi = \int_S \mathbf{E} \, d\mathbf{S}. \tag{1.6}$$

This is an algebraic quantity, since it depends not only on the configuration of the field **E** but also on the choice of the normal. If a surface is closed it is customary to direct the normal **n** *outside* the region enveloped by this surface, i.e. to choose the *outward* normal. Henceforth we shall always assume that this is the case.

Although we considered here the flux of **E**, the concept of flux is applicable to any vector field as well.

The Gauss Theorem. The flux of **E** through an arbitrary closed surface S has a remarkable property: it depends only on the algebraic sum of the charges embraced by this surface, i.e.

$$\boxed{\oint \mathbf{E}\, d\mathbf{S} = \frac{1}{\varepsilon_0} q_{\text{in}},} \tag{1.7}$$

where the circle on the integral symbol indicates that the integration is performed over a closed surface.

This expression is essentially the *Gauss theorem: the flux of*

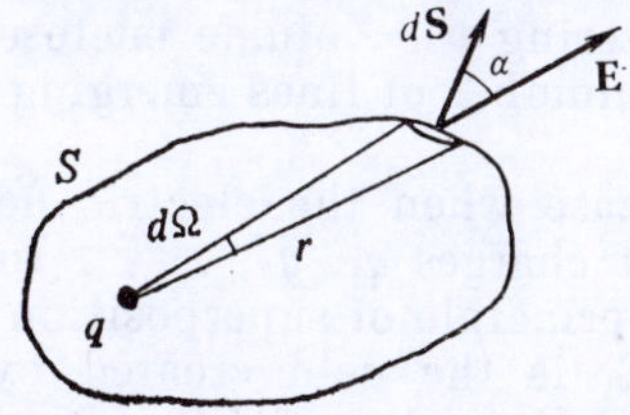

Fig. 1.4

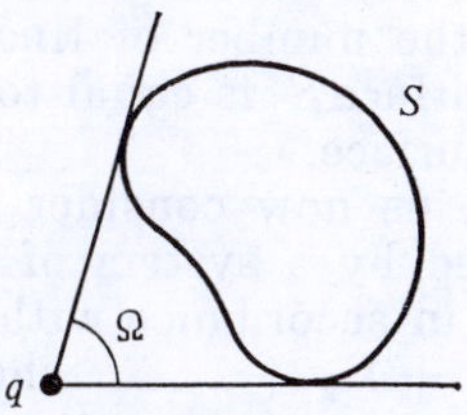

Fig. 1.5

E *through a closed surface is equal to the algebraic sum of the charges enclosed by this surface, divided by* ε_0.

Proof. Let us first consider the field of a single point charge q. We enclose this charge by an arbitrary closed surface S (Fig. 1.4) and find the flux of **E** through the area element $d\mathbf{S}$:

$$d\Phi = E\, dS \cos\alpha = \frac{1}{4\pi\varepsilon_0}\frac{q}{r^2}\, dS \cos\alpha = \frac{q}{4\pi\varepsilon_0}\, d\Omega, \tag{1.8}$$

where $d\Omega$ is the solid angle resting on the area element dS and having the vertex at the point where the charge q is located. The integration of this expression over the entire surface S is equivalent to the integration over the entire solid angle, i.e. to the replacement of $d\Omega$ by 4π. Thus we obtain $\Phi = q/\varepsilon_0$, as is defined by formula (1.7).

It should be noted that for a more complicated shape of a closed surface, the angles α may be greater than $\pi/2$, and hence $\cos\alpha$ and $d\Omega$ in (1.8) generally assume either positive or negative values. Thus, $d\Omega$ is an algebraic quantity: if $d\Omega$ rests on the inner side of the surface S, $d\Omega > 0$, while if it rests on the outer side, $d\Omega < 0$.

In particular, this leads to the following conclusion: if the charge q is located outside a closed surface S, the flux of $\mathbf{E}$ through this surface is equal to zero. In order to prove this, it is sufficient to draw through the charge q a conical surface tangent to the closed surface S. Then the integration of Eq. (1.8) over the surface S is equivalent to the integration over Ω (Fig. 1.5): the outer side of the surface S will be seen from the point q at an angle $\Omega > 0$, while the inner side, at an angle $-\Omega$ (the two angles being equal in magnitude). The sum is equal to zero, and $\Phi = 0$, which also agrees with (1.7). In terms of field lines or lines of $\mathbf{E}$, this means that the number of lines entering the volume enclosed by the surface S is equal to the number of lines emerging from this surface.

Let us now consider the case when the electric field is created by a system of point charges $q_1, q_2, \ldots$. In this case, in accordance with the principle of superposition $\mathbf{E} = \mathbf{E}_1 + \mathbf{E}_2 + \ldots$, where $\mathbf{E}_1$ is the field created by the charge q_1, etc. Then the flux of $\mathbf{E}$ can be written in the form

$$\oint \mathbf{E}\, d\mathbf{S} = \oint (\mathbf{E}_1 + \mathbf{E}_2 + \ldots)\, d\mathbf{S}$$

$$= \oint \mathbf{E}_1\, d\mathbf{S} + \oint \mathbf{E}_2\, d\mathbf{S} + \ldots = \Phi_1 + \Phi_2 + \ldots$$

In accordance with what was said above, each integral on the right-hand side is equal to q_i/ε_0 if the charge q_i is *inside* the closed surface S and is equal to zero if it is *outside* the surface S. Thus, the right-hand side will contain the algebraic sum of *only* those charges that lie inside the surface S.

To complete the proof of the theorem, it remains for us to consider the case when the charges are distributed continuously with the volume density depending on coordinates. In this case, we may assume that each volume element dV contains a "point" charge $\rho\, dV$. Then on the right-hand side of (1.7) we have

$$q_{\text{in}} = \int \rho\, dV, \tag{1.9}$$

where the integration is performed only over the volume contained within the closed surface S.

We must pay attention to the following important circumstance: while the field **E** itself depends on the mutual configuration of *all* the charges, the flux of **E** through an arbitrary closed surface S is determined by the algebraic sum of the charges *inside* the surface S. This means that if we displace the charges, the field **E** will be changed everywhere, and in particular on the surface S. Generally, the flux of **E** through the surface S will also change. However, if the displacement of charges did not involve their crossing of the surface S, the flux of **E** through this surface would remain *unchanged*, although, we stress again, the field **E** itself may change considerably. What a remarkable property of electric field!

1.3. Applications of the Gauss Theorem

Since the field **E** depends on the configuration of *all* charges, the Gauss theorem generally does not allow us to determine this field. However, in certain cases the Gauss theorem proves to be a very effective analytical instrument since it gives answers to certain principle questions without solving the problem and allows us to determine the field **E** in a very simple way. Let us consider some examples and then formulate several general conclusions about the cases when application of the Gauss theorem is the most expedient.

Example 1. On the impossibility of stable equilibrium of a charge in an electric field. Suppose that we have in vacuum a system of fixed point charges in equilibrium. Let us consider one of these charges, e.g. a charge q. Can its equilibrium be stable?

In order to answer this question, let us envelop the charge q by a small closed surface S (Fig. 1.6). For the sake of definiteness, we assume that $q > 0$. For the equilibrium of this charge to be stable, it is necessary that the field **E** created by all the *remaining* charges of the system at all the points of the surface S be directed towards the charge q. Only in this case any small displacement of the charge q from the equilibrium position will give rise to a *restoring* force, and the equilibrium state will actually be stable. But such a configuration of the field **E** around the charge q is in contradiction to the Gauss theorem: the flux of **E** through the surface S is negative, while in accordance with the Gauss theorem it must be equal to zero since it is created by charges lying *outside* the surface S. On the other hand, the fact that **E** is equal to zero indicates that at some points of the surface S vector **E** is directed inside it and at some other points it is directed outside.

Hence it follows that in any electrostatic field a charge cannot be in stable equilibrium.

Example 2. The field of a uniformly charged plane. Suppose that the surface charge density is σ. It is clear from the symmetry of the problem that vector **E** can only be normal to the charged plane. Moreover, at points symmetric with respect to this plane, vectors **E** obviously have the same magnitude but opposite directions. Such a configuration of the field indicates that a right cylinder should be chosen for the closed surface as shown in Fig. 1.7, where we assume that $\sigma > 0$.

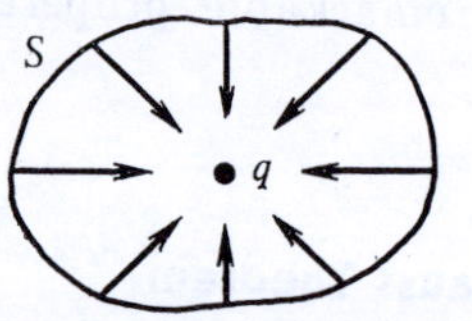

Fig. 1.6

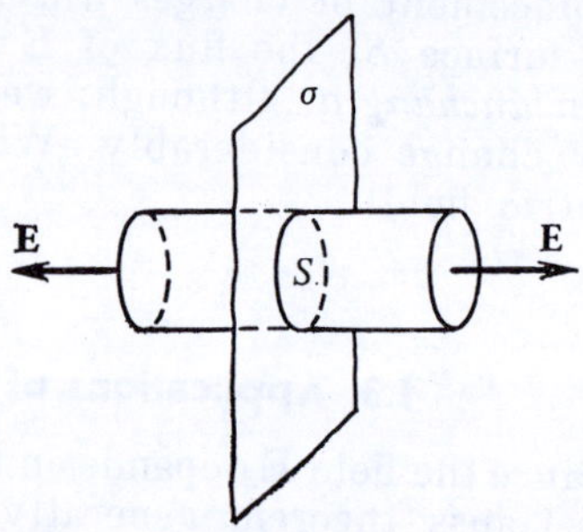

Fig. 1.7

The flux through the lateral surface of this cylinder is equal to zero, and hence the total flux through the entire cylindrical surface is $2E\ \Delta S$, where ΔS is the area of each endface. A charge $\sigma\ \Delta S$ is enclosed within the cylinder. According to the Gauss theorem, $2E\ \Delta S = \sigma\ \Delta S/\varepsilon_0$, whence $E = \sigma/2\varepsilon_0$. In a more exact form, this expression must be written as

$$E_n = \sigma/2\varepsilon_0, \tag{1.10}$$

where E_n is the projection of vector **E** onto the normal **n** to the charged plane, the normal **n** being directed away from this plane. If $\sigma > 0$ then $E_n > 0$, and hence vector **E** is directed away from the charged plane, as shown in Fig. 1.7. On the other hand, if $\sigma < 0$ then $E_n < 0$, and vector **E** is directed towards the charged plane. The fact that **E** is the same at any distance from the plane indicates that the corresponding electric field is uniform (both on the right and on the left of the plane).

The obtained result is valid only for an infinite plane surface, since only in this case we can use the symmetry considerations discussed above. However, this result is approximately valid for the region near the middle of a finite uniformly charged plane surface far from its ends.

Example 3. The field of two parallel planes charged uniformly with densities σ and $-\sigma$ by unlike charges.

This field can be easily found as superposition of the fields created by each plane separately (Fig. 1.8). Here the upper arrows correspond to the field from the positively charged plane, while the lower arrows, to that from the negatively charged plane. In the space between the planes the intensities of the fields being added have the same direction, hence the result (1.10) will be doubled, and the resultant field intensity will be

$$E = \sigma/\varepsilon_0, \tag{1.11}$$

where σ stands for the magnitude of the surface charge density. It can be easily seen that outside this space the field is equal to zero. Thus,

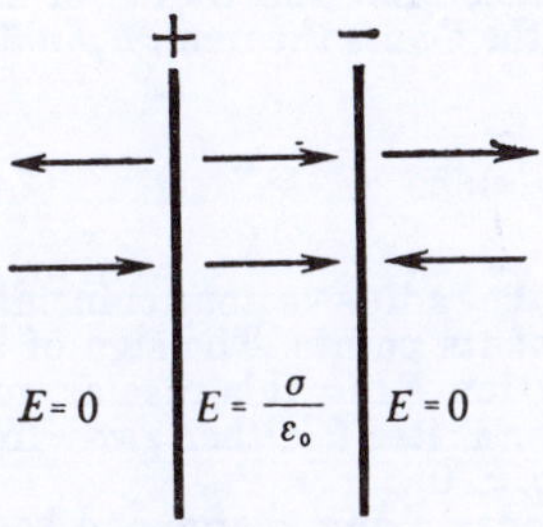

Fig. 1.8

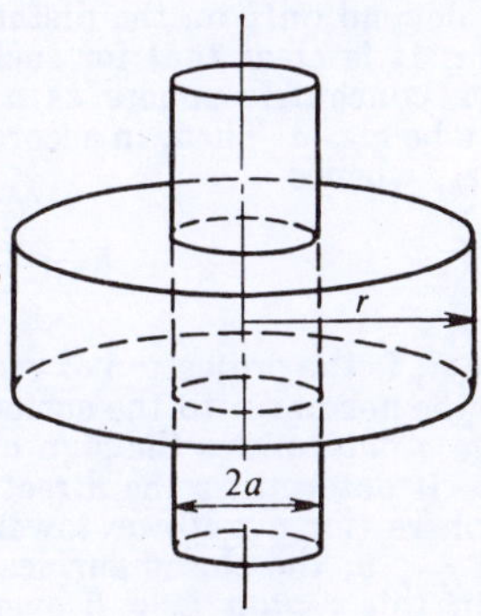

Fig. 1.9

in the given case the field is located between the planes and is uniform.

This result is approximately valid for the plates of finite dimensions as well, if only the separation between the plates is considerably smaller than their linear dimensions (parallel-plate capacitor). In this case, noticeable deviations of the field from uniformity are observed only near the edges of the plates (these distortions are often ignored in calculations).

Example 4. The field of an infinite circular cylinder uniformly charged over the surface so that the charge λ corresponds to its unit length.

In this case, as follows from symmetry considerations, the field is of a radial nature, i.e. vector **E** at each point is perpendicular to the cylinder axis, and its magnitude depends only on the distance r from the cylinder axis to the point. This indicates that a closed surface here should be taken in the form of a coaxial right cylinder (Fig. 1.9). Then the flux of **E** through the endfaces of the cylinder is equal to zero, while the flux through the lateral surface is $E_r\, 2\pi rh$, where E_r is the projection of vector **E** onto the radius vector **r** coinciding with the normal **n** to the lateral surface of the cylinder of radius r and height h. According to the Gauss theorem, $E_r 2\pi rh = \lambda h/\varepsilon_0$ for $r > a$, whence

$$E_r = \frac{\lambda}{2\pi\varepsilon_0 r} \quad (r > a). \tag{1.12}$$

For $\lambda > 0$, $E_r > 0$ as well, i.e. vector **E** is directed away from the charged cylinder, and vice versa.

If $r < a$, the closed surface does not contain any charge since in this region $E = 0$ irrespective of r. Thus, inside a circular infinite cylinder uniformly charged over the surface the field is absent.

Example 5. The field of a spherical surface uniformly charged by the charge q.

This field is obviously centrally symmetric: vector **E** from any point passes through the centre of the sphere, while its magnitude must depend only on the distance r from the point to the centre of the sphere. It is clear that for such a configuration of the field we should take a concentric sphere as a closed surface. Let the radius of this sphere be $r > a$. Then, in accordance with the Gauss theorem, $E_r 4\pi r^2 = q/\varepsilon_0$, whence

$$E_r = \frac{q}{4\pi\varepsilon_0 r^2} \quad (r > a), \tag{1.13}$$

where E_r is the projection of vector **E** onto the radius vector **r** coinciding with the normal **n** to the surface at each of its points. The sign of the charge q determines the sign of the projection E_r in this case as well. Hence it determines the direction of vector **E** itself: either away from the sphere (for $q > 0$) or towards it (for $q < 0$).

If $r < a$, the closed surface does not contain any charge and hence within this region $E = 0$ everywhere. In other words, inside a uniformly charged spherical surface the electric field is absent. Outside this surface the field decreases with the distance r in accordance with the same law as for a point charge.

Example 6. The field of a uniformly charged sphere. Suppose that a charge q is uniformly distributed over a sphere of radius a. Obviously, the field of such a system is centrally symmetric, and hence for determining the field we must take a concentric sphere as a closed surface. It can be easily seen that for the field outside the sphere we obtain the same result as in the previous example [see (1.13)]. However, inside the sphere the expression for the field will be different. The sphere of radius $r < a$ encloses the charge $q' = q\,(r/a)^3$ since in our case the ratio of charges is equal to the ratio of volumes and is proportional to the radii to the third power. Hence, in accordance with the Gauss theorem we have

$$E_r \cdot 4\pi r^2 = \frac{1}{\varepsilon_0}\, q \left(\frac{r}{a}\right)^3,$$

whence

$$E_r = \frac{1}{4\pi\varepsilon_0}\,\frac{q}{a^3}\, r \quad (r \leqslant a), \tag{1.14}$$

i.e. inside a uniformly charged sphere the field intensity grows linearly with the distance r from its centre. The curve representing the dependence of E on r is shown in Fig. 1.10.

General Conclusions. The results obtained in the above

examples could be found by direct integration (1.5) as well. However, it is clear that these problems can be solved in a much simpler way by using the Gauss theorem.

The simple solution of the problems considered above may create an illusive impression about the strength of the method based on the application of the Gauss theorem and about the possibility of solving many other problems by using this theorem. Unfortunately, it is not the case. The number of problems that can be easily solved with the help of the Gauss theorem is limited. We cannot use it even to solve the problem of finding the field of a symmetric charge distribution on a uniformly charged disc. In this case, the field configuration is rather complicated, and a closed surface for a simple calculation of the flux of **E** cannot be found.

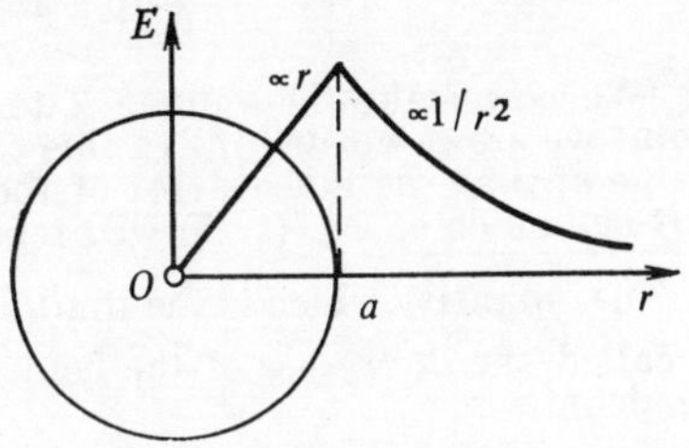

Fig. 1.10

The Gauss theorem can be effectively applied to calculation of fields only when a field has a special symmetry (in most cases plane, cylindrical, or circular). The symmetry, and hence the field configuration, must be such that, firstly, a sufficiently simple closed surface S can be found and, secondly, the calculation of the flux of **E** can be reduced to a simple multiplication of E (or E_r) by the area of the surface S or its part. If these conditions are not satisfied, the problem of finding the field should be solved either directly by formula (1.5) or by using other methods which will be discussed below.

1.4. Differential Form of the Gauss Theorem

A remarkable property of electric field expressed by the Gauss theorem suggests that this theorem be represented in a different form which would broaden its possibilities as an instrument for analysis and calculation.

In contrast to (1.7) which is called the *integral* form we shall seek the *differential* form of the Gauss theorem, which establishes the relation between the volume charge density ρ and the *changes* in the field intensity **E** in the vicinity of a given point in space.

For this purpose, we first represent the charge q in the volume V enveloped by a closed surface S in the form $q_{\text{int}} = \langle\rho\rangle V$, where $\langle\rho\rangle$ is the volume charge density, averaged over the volume V. Then we substitute this expression into Eq. (1.7) and divide both its sides by V, which gives

$$\frac{1}{V}\oint \mathbf{E}\,d\mathbf{S} = \langle\rho\rangle/\varepsilon_0. \tag{1.15}$$

We now make the volume V tend to zero by contracting it to the point we are interested in. In this case, $\langle\rho\rangle$ will obviously tend to the value of ρ at the given point of the field, and hence the ratio on the left-hand side of Eq. (1.15) will tend to ρ/ε_0.

The quantity which is the limit of the ratio of $\oint \mathbf{E}\,d\mathbf{S}$ to V as $V \to 0$ is called the *divergence* of the field **E** and is denoted by div **E**. Thus, by definition,

$$\operatorname{div} \mathbf{E} = \lim_{V\to 0} \frac{1}{V}\oint \mathbf{E}\,d\mathbf{S}. \tag{1.16}$$

The divergence of any other vector field is determined in a similar way. It follows from definition (1.16) that divergence is a scalar function of coordinates.

In order to obtain the expression for the divergence of the field **E**, we must, in accordance with (1.16), take an infinitely small volume V, determine the flux of **E** through the closed surface enveloping this volume, and find the ratio of this flux to the volume. The expression obtained for the divergence will depend on the choice of the coordinate system (in different systems of coordinates it turns out to be different). For example, in Cartesian coordinates it is given by

$$\operatorname{div} \mathbf{E} = \frac{\partial E_x}{\partial x} + \frac{\partial E_y}{\partial y} + \frac{\partial E_z}{\partial z}. \tag{1.17}$$

Thus, we have found that as $V \to 0$ in (1.15), its right-hand side tends to ρ/ε_0, while the left-hand side tends to div **E**. Consequently, the divergence of the field **E** is related to the charge density at the same point through the equation

$$\boxed{\operatorname{div} \mathbf{E} = \rho/\varepsilon_0.} \tag{1.18}$$

This equation expresses the Gauss theorem in the differential form.

The form of many expressions and their applications can be considerably simplified if we introduce the vector differential operator ∇. In Cartesian coordinates, the operator ∇ has the form

$$\nabla = \mathbf{i}\frac{\partial}{\partial x} + \mathbf{j}\frac{\partial}{\partial y} + \mathbf{k}\frac{\partial}{\partial z}, \tag{1.19}$$

where **i**, **j**, and **k** are the unit vectors of the X-, Y-, and Z-axes. The operator ∇ itself does not have any meaning. It becomes meaningful only in combination with a scalar or vector function by which it is symbolically multiplied. For example, if we form the scalar product of vector ∇ and vector **E**, we obtain

$$\nabla \cdot \mathbf{E} = \nabla_x E_x + \nabla_y E_y + \nabla_z E_z = \frac{\partial}{\partial x} E_x + \frac{\partial}{\partial y} E_y + \frac{\partial}{\partial z} E_z.$$

It follows from (1.17) that this is just the divergence of **E**.

Thus, the divergence of the field **E** can be written as div **E** or $\nabla \cdot \mathbf{E}$ (in both cases it is read as "the divergence of **E**"). We shall be using the latter, more convenient notation. Then, for example, the Gauss theorem (1.18) will have the form

$$\boxed{\nabla \cdot \mathbf{E} = \rho/\varepsilon_0.} \qquad (1.20)$$

The Gauss theorem in the differential form is a local theorem: the divergence of the field **E** at a given point depends only on the electric charge density ρ at this point. This is one of the remarkable properties of electric field. For example, the field **E** of a point charge is different at different points. Generally, this refers to the spatial derivatives $\partial E_x/\partial x$, $\partial E_y/\partial y$, and $\partial E_z/\partial z$ as well. However, the Gauss theorem states that the sum of these derivatives, which determines the divergence of **E**, turns out to be equal to zero at all points of the field (outside the charge itself).

At the points of the field where the divergence of **E** is positive, we have the *sources* of the field (positive charges), while at the points where it is negative, we have *sinks* (negative charges). The field lines emerge from the field sources and terminate at the sinks.

1.5. Circulation of Vector E. Potential

Theorem on Circulation of Vector E. It is known from mechanics that any stationary field of central forces is conservative, i.e. the work done by the forces of this field is independent of the path and depends only on the position of the initial and final points. This property is inherent in the electrostatic field, viz. the field created by a system of fixed charges. If we take a unit *positive* charge for the test charge and carry it from point *1* of a given field **E** to point *2*, the elementary work of the forces of the field done over the distance $d\mathbf{l}$ is equal to $\mathbf{E} \cdot d\mathbf{l}$, and the total work of the field forces over the distance between points *1* and *2* is defined as

$$\int_1^2 \mathbf{E}\, d\mathbf{l}. \qquad (1.21)$$

This integral is taken along a certain line (path) and is therefore called the *line* integral.

We shall now show that from the independence of line integral (1.21) of the path between two points it follows that when taken along an arbitrary closed path, this integral is equal to zero. Integral (1.21) over a closed contour is called the *circulation* of vector **E** and is denoted by $\oint$.

Thus we state that circulation of vector **E** in any electrostatic field is equal to zero, i.e.

$$\boxed{\oint \mathbf{E}\, d\mathbf{l} = 0.} \tag{1.22}$$

1248

This statement is called the *theorem on circulation of vector* **E**.

In order to prove this theorem, we break an arbitrary closed path into two parts *1a2* and *2b1* (Fig. 1.11). Since line integral (1.21) $\left(\text{we denote it by } \int_{12}\right)$ does not depend on the path between points *1* and *2*, we have $\int\limits_{12}^{(a)} = \int\limits_{12}^{(b)}$. On the other hand, it is clear that $\int\limits_{12}^{(b)} = -\int\limits_{21}^{(b)}$, where $\int\limits_{21}^{(b)}$ is the integral over the same segment *b* but taken in the opposite direction. Therefore

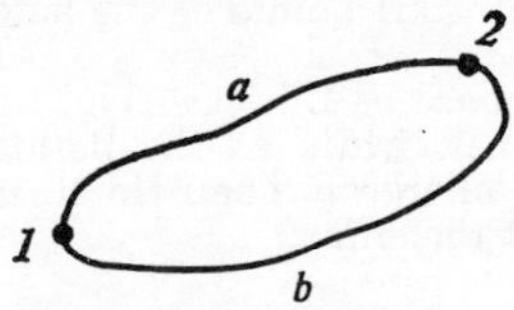

Fig. 1.11

$$\int\limits_{12}^{(a)} + \int\limits_{21}^{(b)} = \int\limits_{11}^{(a)} - \int\limits_{12}^{(b)} = 0,$$

Q.E.D.

A field having property (1.22) is called the *potential field*. Hence, *any electrostatic field is a potential field*.

The theorem on circulation of vector **E** makes it possible to draw a number of important conclusions without resorting to calculations. Let us consider two examples.

Example 1. The field lines of an electrostatic field E cannot be closed.

Indeed, if the opposite were true and some lines of field **E** were closed, then taking the circulation of vector **E** along this line we would immediately come to contradiction with theorem (1.22). This means that actually there are no closed lines of **E** in an electrostatic field: the lines emerge from positive charges and terminate on negative ones (or go to infinity).

Example 2. Is the configuration of an electrostatic field shown in Fig. 1.12 possible?

No, it is not. This immediately becomes clear if we apply the theorem on circulation of vector **E** to the closed contour shown in the figure by the dashed line. The arrows on the contour indicate the direction of circumvention. With such a special choice of the contour, the contribution to the circulation from its vertical parts is equal to zero, since in this case $\mathbf{E} \perp d\mathbf{l}$ and $\mathbf{E} \cdot d\mathbf{l} = 0$. It remains for us to consider the two horizontal segments of equal lengths. The figure shows that the contributions to the circulation from these regions are opposite in sign, and unequal in magnitude (the contribution from the upper segment is larger since the field lines are denser, and hence the value of E is larger). Therefore, the circulation of **E** differs from zero, which contradicts to (1.22).

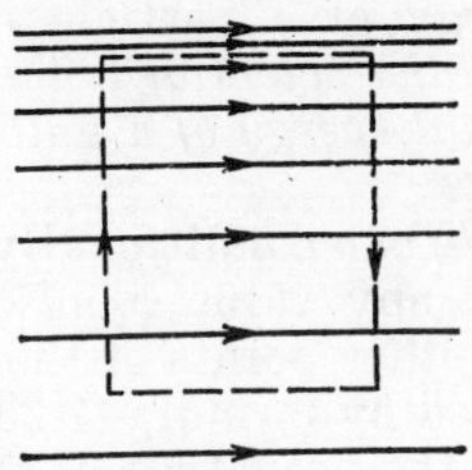

Fig. 1.12

Potential. Till now we considered the description of electric field with the help of vector **E**. However, there exists another adequate way of describing it by using potential φ (it should be noted at the very outset that there is a one-to-one correspondence between the two methods). It will be shown that the second method has a number of significant advantages.

The fact that line integral (1.21) representing the work of the field forces done in the displacement of a unit positive charge from point *1* to point *2* does not depend on the path allows us to state that for electric field there exists a certain scalar function φ (**r**) of coordinates such that its *decrease* is given by

$$\varphi_1 - \varphi_2 = \int_1^2 \mathbf{E}\, d\mathbf{l}, \tag{1.23}$$

where φ_1 and φ_2 are the values of the function φ at the points *1* and *2*. The quantity φ (**r**) defined in this way is called the *field potential*. A comparison of (1.23) with the expression for the work done by the forces of the potential field (the work being equal to the decrease in the potential energy of a particle in the field) leads to the conclusion that *the potential is the quantity numerically equal to the potential energy of a unit positive charge at a given point of the field*.

We can conditionally ascribe to an arbitrary point O of the field any value φ_0 of the potential. Then the potentials of all other points of the field will be unambiguously determined by formula (1.23). If we change φ_0 by a certain value $\Delta\varphi$, the potentials of all other points of the field will change by the same value.

Thus, potential φ is determined to within an arbitrary additive constant. The value of this constant does not play any role, since all electric phenomena depend only on the electric field strength. It is determined, as will be shown later, not by the potential at a given point but by the potential difference between neighbouring points of the field.

The unit of potential is the *volt* (V).

Potential of the Field of a Point Charge. Formula (1.23) contains, in addition to the definition of potential φ, the method of finding this function. For this purpose, it is sufficient to evaluate the integral $\int \mathbf{E}\, d\mathbf{l}$ over any path between two points and then represent the obtained result as a decrease in a certain function which is just $\varphi(\mathbf{r})$. We can make it even simpler. Let us use the fact that formula (1.23) is valid not only for finite displacements but for elementary displacements $d\mathbf{l}$ as well. Then, in accordance with this formula, the elementary *decrease* in the potential over this displacement is

$$-d\varphi = \mathbf{E}\cdot d\mathbf{l}. \tag{1.24}$$

In other words, if we know the field $\mathbf{E}(\mathbf{r})$, then to find φ we must represent $\mathbf{E}\cdot d\mathbf{l}$ (with the help of appropriate transformations) as a decrease in a certain function. This function will be the potential φ.

Let us apply this method for finding the potential of the

field of a fixed point charge:

$$\mathbf{E}\cdot d\mathbf{l} = \frac{1}{4\pi\varepsilon_0}\frac{q}{r^2}\mathbf{e}_r\cdot d\mathbf{l} = \frac{q}{4\pi\varepsilon_0}\frac{dr}{r^2}$$

$$= -d\left(\frac{1}{4\pi\varepsilon_0}\frac{q}{r} + \text{const}\right),$$

where we took into account that $\mathbf{e}_r\, d\mathbf{l} = 1\cdot(d\mathbf{l})_r = dr$, since the projection of $d\mathbf{l}$ onto $\mathbf{e}_r$, and hence on $\mathbf{r}$, is equal to the increment of the magnitude of vector $\mathbf{r}$, i.e. dr. The quantity appearing in the parentheses under the differential is exactly $\varphi(\mathbf{r})$. Since the additive constant contained in the formula does not play any physical role, it is usually omitted in order to simplify the expression for φ. Thus, the potential of the field of a point charge is given by

$$\boxed{\varphi = \frac{1}{4\pi\varepsilon_0}\frac{q}{r}.} \qquad (1.25)$$

The absence of an additive constant in this expression indicates that we conventionally assume that the potential is equal to zero at infinity (for $r \to \infty$).

Potential of the Field of a System of Charges. Let a system consist of fixed point charges $q_1, q_2, \ldots$. In accordance with the principle of superposition, the field intensity at any point of the field is given by $\mathbf{E} = \mathbf{E}_1 + \mathbf{E}_2 + \ldots$, where $\mathbf{E}_1$ is the field intensity from the charge q_1, etc. By using formula (1.24), we can then write

$$\mathbf{E}\cdot d\mathbf{l} = (\mathbf{E}_1 + \mathbf{E}_2 + \ldots)\, d\mathbf{l} = \mathbf{E}_1\cdot d\mathbf{l} + E_2\cdot d\mathbf{l} + \ldots$$

$$= -d\varphi_1 - d\varphi_2 - \ldots = -d\varphi,$$

where $\varphi = \sum \varphi_i$, i.e. the principle of superposition turns out to be valid for potential as well. Thus, the potential of a system of fixed point charges is given by

$$\boxed{\varphi = \frac{1}{4\pi\varepsilon_0}\sum\frac{q_i}{r_i},} \qquad (1.26)$$

where r_i is the distance from the point charge q_i to the point under consideration. Here we also omitted an arbitrary constant. This is in complete agreement with the fact that any real system of charges is bounded in space, and hence its potential can be taken equal to zero at infinity.

If the charges forming the system are distributed continuously, then, as before, we assume that each volume element dV contains a "point" charge $\rho\, dV$, where ρ is the charge density in the volume dV. Taking this into consideration, we can write formula (1.26) in a different form:

$$\boxed{\varphi = \frac{1}{4\pi\varepsilon_0}\int \frac{\rho\, dV}{r},} \tag{1.27}$$

where the integration is performed either over the entire space or over its part containing the charges. If the charges are located only on the surface S, we can write

$$\varphi = \frac{1}{4\pi\varepsilon_0}\int \frac{\sigma\, dS}{r}, \tag{1.28}$$

where σ is the surface charge density and dS is the element of the surface S. A similar expression corresponds to the case when the charges have a linear distribution.

Thus, if we know the charge distribution (discrete or continuous), we can, in principle, find the potential of any system.

1.6. Relation Between Potential and Vector E

It is known that electric field is completely described by vector function $\mathbf{E}(\mathbf{r})$. Knowing this function, we can find the force acting on a charge under investigation at any point of the field, calculate the work of field forces for any displacement of the charge, and so on. And what do we get by introducing potential? First of all, it turns out that if we know the potential $\varphi(\mathbf{r})$ of a given electric field, we can reconstruct the field $\mathbf{E}(\mathbf{r})$ quite easily. Let us consider this question in greater detail.

The relation between φ and $\mathbf{E}$ can be established with the help of Eq. (1.24). Let the displacement $d\mathbf{l}$ be parallel to the X-axis; then $d\mathbf{l} = \mathbf{i}\, dx$, where $\mathbf{i}$ is the unit vector along the X-axis and dx is the increment of the coordinate x. In this case

$$\mathbf{E}\cdot d\mathbf{l} = \mathbf{E}\cdot\mathbf{i}\, dx = E_x\, dx,$$

where E_x is the projection of vector $\mathbf{E}$ onto the unit vector $\mathbf{i}$ (and not on the displacement $d\mathbf{l}$!). A comparison of this expression with formula (1.24) gives

$$E_x = -\partial\varphi/\partial x, \tag{1.29}$$

where the symbol of partial derivative emphasizes that the function $\varphi(x, y, z)$ must be differentiated only with respect to x, assuming that y and z are constant in this case.

In a similar way, we can obtain the corresponding expressions for the projections E_y and E_z. Having determined E_x, E_y, and E_z, we can easily find vector **E** itself:

$$\mathbf{E} = -\left(\frac{\partial\varphi}{\partial x}\mathbf{i} + \frac{\partial\varphi}{\partial y}\mathbf{j} + \frac{\partial\varphi}{\partial z}\mathbf{k}\right). \tag{1.30}$$

The quantity in the parentheses is the *gradient of the potential* φ (grad φ or $\nabla\varphi$). We shall be using the latter, more convenient notation and will formally consider $\nabla\varphi$ as the product of a symbolic vector ∇ and the scalar φ. Then Eq. (1.30) can be represented in the form

$$\boxed{\mathbf{E} = -\nabla\varphi,} \tag{1.31}$$

i.e. the field intensity **E** is equal to the potential gradient with the minus sign. This is exactly the formula that can be used for reconstructing the field **E** if we know the function $\varphi(\mathbf{r})$.

Example. Find the field intensity **E** if the field potential has the form: (1) $\varphi(x, y) = -axy$, where a is a constant; (2) $\varphi(\mathbf{r}) = -\mathbf{a}\cdot\mathbf{r}$, where **a** is a constant vector and **r** is the radius vector of a point under consideration.

(1) By using formula (1.30), we obtain $\mathbf{E} = a(y\mathbf{i} + x\mathbf{j})$.

(2) Let us first represent the function φ as $\varphi = -a_x x - a_y y - a_z z$, where a_x, a_y and a_z are constants. Then with the help of formula (1.30) we find $\mathbf{E} = a_x\mathbf{i} + a_y\mathbf{j} + a_z\mathbf{k} = \mathbf{a}$. It can be seen that in this case the field **E** is uniform.

Let us derive one more useful formula. We write the right-hand side of (1.24) in the form $\mathbf{E}\cdot d\mathbf{l} = E_l\,dl$, where $dl = |d\mathbf{l}|$ is an elementary displacement and E_l is the projection of vector **E** onto the displacement $d\mathbf{l}$. Hence

$$\boxed{E_l = -\partial\varphi/\partial l,} \tag{1.32}$$

i.e. the projection of vector **E** onto the direction of the displacement dl is equal to the directional derivative of the potential (this is emphasized by the symbol of partial derivative).

Equipotential Surfaces. Let us introduce the concept of *equipotential surface*, viz. the surface at all points of which potential φ has the same value. We shall show that vector **E** at each point of the surface is directed along the normal to the equipotential surface and towards the decrease in the potential. Indeed, it follows from formula (1.32) that the projection of vector **E** onto any direction tangent to the equipotential surface at a given point is equal to zero. This means that vector **E** is normal to the given surface. Further, let us take a displacement dl along the normal to the surface, towards decreasing φ. Then $\partial\varphi < 0$, and according to (1.32), $E_l > 0$, i.e. vector **E** is directed towards decreasing φ, or in the direction opposite to that of the vector $\nabla\varphi$.

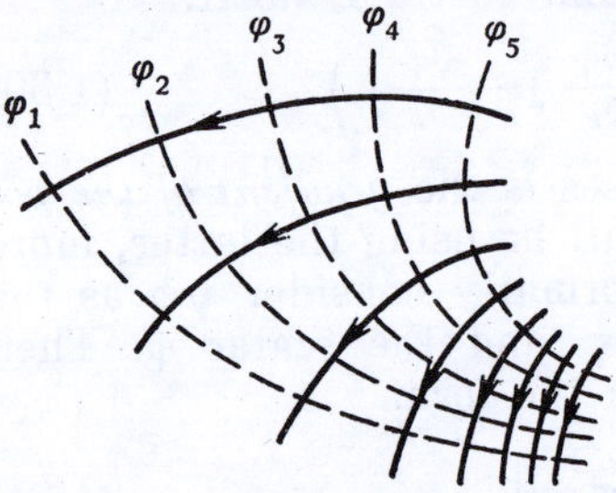

Fig. 1.13

It is expedient to draw equipotential surfaces in such a way that the potential difference between two neighbouring surfaces be the same. Then the density of equipotential surfaces will visually indicate the magnitudes of field intensities at different points. Field intensity will be higher in the regions where equipotential surfaces are denser ("the potential relief is steeper").

Since vector **E** is normal to an equipotential surface everywhere, the field lines are orthogonal to these surfaces.

Figure 1.13 shows a two-dimensional pattern of an electric field. The dashed lines correspond to equipotential surfaces, while the solid lines to the lines of E. Such a representation can be easily visualized. It immediately shows the direction of vector E, the regions where field intensity is higher and where it is lower, as well as the regions with greater steepness of the potential relief. Such a pattern can be used to obtain qualitative answers to a number of questions, such as "In what direction will a charge placed at a certain point move? Where is the magnitude of the potential gradient higher? At which point will the force acting on the charge be greater?" etc.

On the Advantages of Potential. It was noted earlier that electrostatic field is completely characterized by vector function **E** (**r**). Then what is the use of introducing potential? There are several sound reasons for doing that. The concept of potential is indeed very useful, and it is not by chance that this concept is widely used not only in physics but in engineering as well.

1. If we know the potential φ (**r**), we can easily calculate the work of field forces done in the displacement of a point charge q' from point *1* to point *2*:

$$A_{12} = q' (\varphi_1 - \varphi_2), \tag{1.33}$$

where φ_1 and φ_2 are the potentials at points *1* and *2*. This means that the required work is equal to the *decrease* in the potential energy of the charge q' upon its displacement from point *1* to *2*. Calculation of the work of the field forces with the help of formula (1.33) is not just very simple, but is in some cases the only possible resort.

Example. A charge q is distributed over a thin ring of radius a. Find the work of the field forces done in the displacement of a point charge q' from the centre of the ring to infinity.

Since the distribution of the charge q over the ring is unknown, we cannot say anything definite about the intensity **E** of the field created by this charge. This means that we cannot calculate the work by evaluating the integral $\int q'\mathbf{E}\, d\mathbf{l}$ in this case. This problem can be easily solved with the help of potential. Indeed, since all elements of the ring are at the same distance a from the centre of the ring, the potential of this point is $\varphi_0 = q/4\pi\varepsilon_0 a$. And we know that $\varphi = 0$ at infinity. Consequently, the work $A = q'\varphi_0 = q'q/4\pi\varepsilon_0 a$.

2. It turns out in many cases that in order to find electric field intensity **E**, it is easier first to calculate the potential φ and then take its gradient than to calculate the value of **E** directly. This is a considerable advantage of potential. Indeed, for calculating φ, we must evaluate only *one* integral, while for calculating **E** we must take *three* integrals (since it is a vector). Moreover, the integrals for calculating φ are usually simpler than those for E_x, E_y, and E_z.

Let us note here that this does not apply to a comparatively small number of problems with high symmetry, in which the calculation of the field **E** directly or with the help of the Gauss theorem turns out to be much simpler.

There are some other advantages in using potential which will be discussed later.

1.7. Electric Dipole

The Field of a Dipole. The *electric dipole* is a system of two equal in magnitude unlike charges $+q$ and $-q$, separated by a certain distance l. When the dipole field is considered,

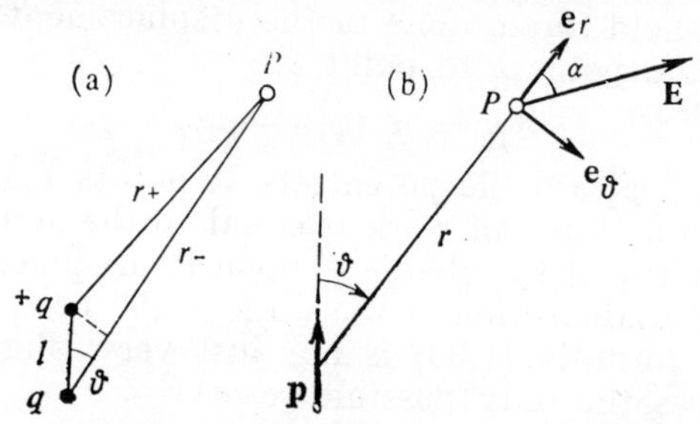

Fig. 1.14

it is assumed that the dipole itself is pointlike, i.e. the distance r from the dipole to the points under consideration is assumed to be much greater than l.

The dipole field is axisymmetric. Therefore, in any plane passing through the dipole axis the pattern of the field is the same, vector **E** lying in this plane.

Let us first find the potential of the dipole field and then its intensity. According to (1.25), the potential of the dipole field at the point P (Fig. 1.14*a*) is defined as

$$\varphi = \frac{1}{4\pi\varepsilon_0}\left(\frac{q}{r_+} - \frac{q}{r_-}\right) = \frac{1}{4\pi\varepsilon_0}\,\frac{q\,(r_- - r_+)}{r_+ r_-}.$$

Since $r \gg l$, it can be seen from Fig. 1.14*a* that $r_- - r_+ = l\cos\vartheta$ and $r_+ r_- = r^2$, where r is the distance from the point P to the dipole (it is pointlike!). Taking this into account, we get

$$\varphi = \frac{1}{4\pi\varepsilon_0}\,\frac{p\cos\vartheta}{r^2}, \tag{1.34}$$

where $p = ql$ is the *electric moment of the dipole*. This quantity corresponds to a vector directed along the dipole axis from the negative to the positive charge:

$$\mathbf{p} = q\mathbf{l}, \tag{1.35}$$

where $q > 0$ and $\mathbf{l}$ is the vector directed as $\mathbf{p}$.

It can be seen from formula (1.34) that the dipole field depends on its electric moment $\mathbf{p}$. It will be shown below that the behaviour of the dipole in an external field also depends on $\mathbf{p}$. Consequently, $\mathbf{p}$ is an important characteristic of the dipole.

It should also be noted that the potential of the dipole field decreases with the distance r faster than the potential of the field of a point charge (in proportion to $1/r^2$ instead of $1/r$).

In order to find the dipole field, we shall use formula (1.32) and calculate the projections of vector $\mathbf{E}$ onto two mutually perpendicular directions along the unit vectors $\mathbf{e}_r$ and $\mathbf{e}_\vartheta$ (Fig. 1.14*b*):

$$E_r = -\frac{\partial \varphi}{\partial r} = \frac{1}{4\pi\varepsilon_0} \frac{2p \cos\vartheta}{r^3}, \quad E_\vartheta = -\frac{\partial \varphi}{r\partial \vartheta} = \frac{1}{4\pi\varepsilon_0} \frac{p \sin\vartheta}{r^3}. \tag{1.36}$$

Hence, the modulus of vector $\mathbf{E}$ will be

$$E = \sqrt{E_r^2 + E_\vartheta^2} = \frac{1}{4\pi\varepsilon_0} \frac{p}{r^3} \sqrt{1 + 3\cos^2\vartheta}. \tag{1.37}$$

In particular, for $\vartheta = 0$ and $\vartheta = \pi/2$ we obtain the expressions for the field intensity on the dipole axis ($E_\parallel$) and on the normal to it ($E_\perp$):

$$E_\parallel = \frac{1}{4\pi\varepsilon_0} \frac{2p}{r^3}, \quad E_\perp = \frac{1}{4\pi\varepsilon_0} \frac{p}{r^3}, \tag{1.38}$$

i.e. for the same r the intensity $E_\parallel$ is twice as high as $E_\perp$.

The Force Acting on a Dipole. Let us place a dipole into a nonuniform electric field. Suppose that $\mathbf{E}_+$ and $\mathbf{E}_-$ are the intensities of the external field at the points where the positive and negative dipole charges are located. Then the resultant force $\mathbf{F}$ acting on the dipole is (Fig. 1.15*a*):

$$\mathbf{F} = q\mathbf{E}_+ - q\mathbf{E}_- = q(\mathbf{E}_+ - \mathbf{E}_-).$$

The difference $\mathbf{E}_+ - \mathbf{E}_-$ is the increment $\Delta\mathbf{E}$ of vector $\mathbf{E}$ on the segment equal to the dipole length l in the direction of vector **l**. Since the length of this segment is small, we can write

$$\Delta \mathbf{E} = \mathbf{E}_+ - \mathbf{E}_- = \frac{\Delta \mathbf{E}}{l} l = \frac{\partial \mathbf{E}}{\partial l} l.$$

Substituting this expression into the formula for **F**, we find that the force acting on the dipole is equal to

$$\boxed{\mathbf{F} = p \frac{\partial \mathbf{E}}{\partial l},} \tag{1.39}$$

where $p = ql$ is the dipole electric moment. The derivative appearing in this expression is called the directional derivative of the vector. The symbol of partial derivative indicates that it is taken with respect to a certain direction, viz. the direction coinciding with vector **l** or **p**.

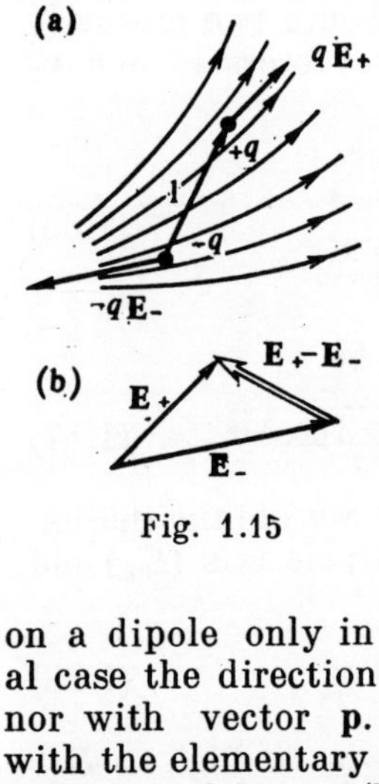

Fig. 1.15

Unfortunately, the simplicity of formula (1.39) is delusive: taking the derivative $\partial \mathbf{E}/\partial l$ is a rather complicated mathematical operation. We shall not discuss this question in detail here but pay attention to the essence of the obtained result. First of all, note that in a uniform field $\partial \mathbf{E}/\partial l = 0$ and $\mathbf{F} = 0$. This means that generally the force acts on a dipole only in a nonuniform field. Next, in the general case the direction of **F** coincides neither with vector **E** nor with vector **p**. Vector **F** coincides in direction only with the elementary increment of vector **E**, taken along the direction of **l** or **p** (Fig. 1.15*b*).

Figure 1.16 shows the directions of the force **F** acting on a dipole in the field of a point charge q for three different dipole orientations. We suggest that the reader prove independently that it is really so.

If we are interested in the projection of force **F** onto a certain direction X, it is sufficient to write equation (1.39) in terms of the projections onto this direction, and we get

$$F_x = p \frac{\partial E_x}{\partial l}, \tag{1.40}$$

where $\partial E_x/\partial l$ is the derivative of the corresponding projection of vector **E** again onto the direction of vector **l** or **p**.

Let a dipole with moment **p** be oriented along the symmetry axis of a certain nonuniform field **E**. We take the positive direction of the X-axis, for example, as shown in Fig. 1.17. Since the increment of the projection E_x in the direction of vector **p** will be negative, $F_x < 0$, and hence vector **F** is directed to the left, i.e. towards increasing field intensity. If we rotate vector **p** shown in the figure through 90° so that the dipole centre coincides with the symmetry axis of the field, it can be easily seen that in this position $F_x = 0$.

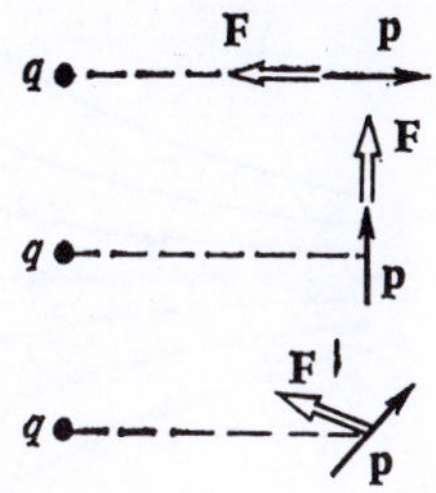

Fig. 1.16

The Moment of Forces Acting on a Dipole. Let us consider behaviour of a dipole in an external electric field in its centre-of-mass system and find out whether the dipole will rotate or not. For this purpose, we must find the moment of external forces with respect to the dipole centre of mass*.

By definition, the moment of forces $\mathbf{F}_+ = q\mathbf{E}_+$ and $\mathbf{F}_- = -q\mathbf{E}_-$ with respect to the centre of mass C (Fig. 1.18) is equal to

$$\mathbf{M} = [\mathbf{r}_+ \times \mathbf{F}_+] + [\mathbf{r}_- \times \mathbf{F}_-] = [\mathbf{r}_+ \times q\mathbf{E}_+] - [\mathbf{r}_- \times q\mathbf{E}_-],$$

where $\mathbf{r}_+$ and $\mathbf{r}_-$ are the radius vectors of the charges $+q$ and $-q$ relative to the point C. For a sufficiently small dipole length, $\mathbf{E}_+ \approx \mathbf{E}_-$ and $\mathbf{M} = [(\mathbf{r}_+ - \mathbf{r}_-) \times q\mathbf{E}]$. It remains for us to take into account that $\mathbf{r}_+ - \mathbf{r}_- = \mathbf{l}$ and $q\mathbf{l} = \mathbf{p}$, which gives

$$\boxed{\mathbf{M} = [\mathbf{p} \times \mathbf{E}].} \tag{1.41}$$

This moment of force tends to rotate the dipole so that its electric moment **p** is oriented along the external field **E**. Such a position of the dipole is stable.

Thus, in a nonuniform electric field a dipole behaves as

---

* We take the moment with respect to the centre of mass in order to eliminate the moment of inertial forces.

follows: under the action of the moment of force (1.41), the dipole tends to get oriented along the field ($\mathbf{p} \uparrow\uparrow \mathbf{E}$), while under the action of the resultant force (1.39) it is

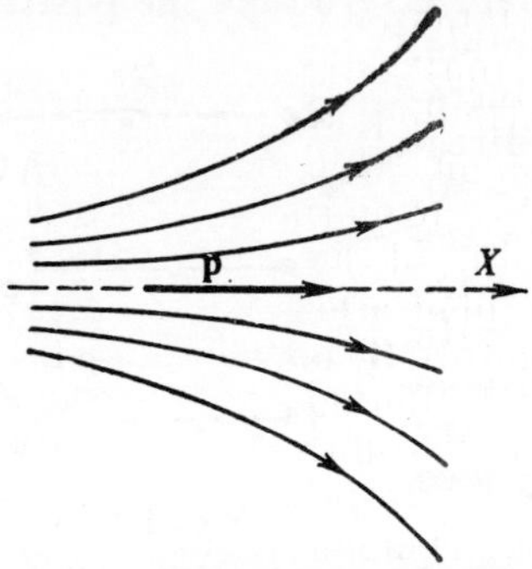

Fig. 1.17

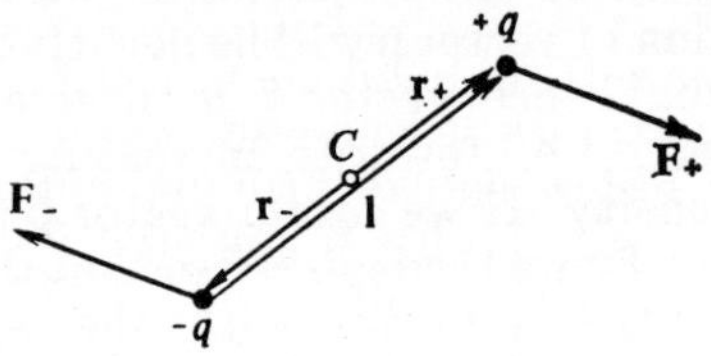

Fig. 1.18

displaced towards the region where the field **E** has larger magnitude. Both these motions are simultaneous.

The Energy of a Dipole in an External Field. We know that the energy of a point charge q in an external field is $W = q\varphi$, where φ is the field potential at the point of location of the charge q. A dipole is the system of two charges, and hence its energy in an external field is

$$W = q_+\varphi_+ + q_-\varphi_- = q\,(\varphi_+ - \varphi_-),$$

where φ_+ and φ_- are the potentials of the external field at the points of location of the charges $+q$ and $-q$. To within a quantity of the second order of smallness, we can write

$$\varphi_+ - \varphi_- = \frac{\partial \varphi}{\partial l}\, l,$$

where $\partial\varphi/\partial l$ is the derivative of the potential in the direction of the vector $\mathbf{l}$. According to (1.32), $\partial\varphi/\partial l = -E_l$, and hence $\varphi_+ - \varphi_- = -E_l l = -\mathbf{E}\cdot\mathbf{l}$, from which we get

$$\boxed{W = -\mathbf{p}\cdot\mathbf{E}.} \qquad (1.42)$$

It follows from this formula that the dipole has the minimum energy ($W_{min} = -pE$) in the position $\mathbf{p} \uparrow\uparrow \mathbf{E}$ (the position of stable equilibrium). If it is displaced from this position, the moment of external forces will return the dipole to the equilibrium position.

Problems

● **1.1.** A very thin disc is uniformly charged with surface charge density $\sigma > 0$. Find the electric field intensity E on the axis of this disc at the point from which the disc is seen at an angle Ω.

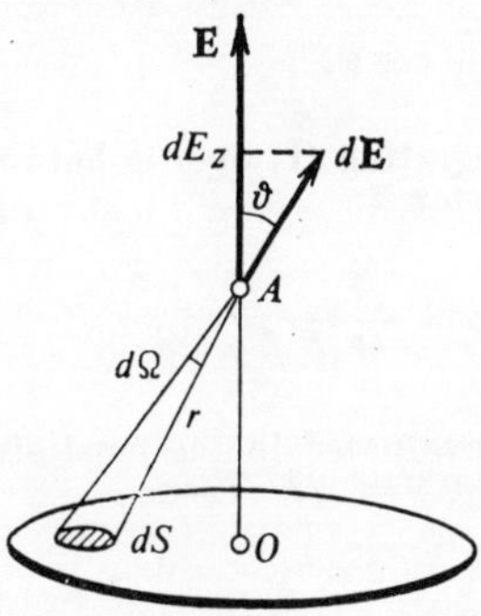

Fig. 1.19

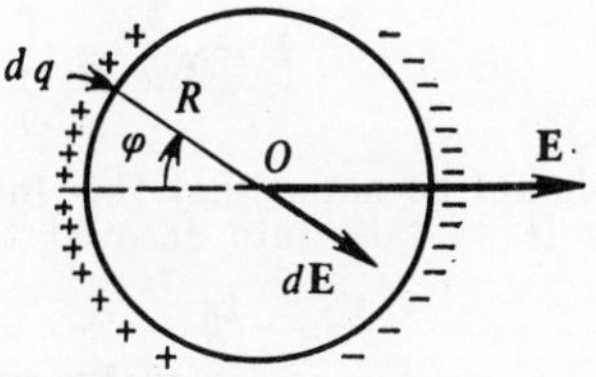

Fig. 1.20

Solution. It is clear from symmetry considerations that on the disc axis vector **E** must coincide with the direction of this axis (Fig. 1.19). Hence, it is sufficient to find the component dE_z from the charge of the area element dS at the point A and then integrate the obtained expression over the entire surface of the disc. It can be easily seen that

$$dE_z = \frac{1}{4\pi\varepsilon_0} \frac{\sigma\, dS}{r^2} \cos\vartheta. \tag{1}$$

In our case $(dS \cos\vartheta)/r^2 = d\Omega$ is the solid angle at which the area element dS is seen from the point A, and expression (1) can be written as

$$dE_z = \frac{1}{4\pi\varepsilon_0} \sigma\, d\Omega.$$

Hence, the required quantity is

$$E = \frac{1}{4\pi\varepsilon_0} \sigma\Omega.$$

It should be noted that at large distances from the disc, $\Omega = S/r^2$, where S is the area of the disc and $E = q/4\pi\varepsilon_0 r^2$ just as the field of the point charge $q = \sigma S$. In the immediate vicinity of the point O, the solid angle $\Omega = 2\pi$ and $E = \sigma/2\varepsilon_0$.

● 1.2. A thin nonconducting ring of radius R is charged with a linear density $\lambda = \lambda_0 \cos\varphi$, where λ_0 is a positive constant and φ is the azimuth angle. Find the electric field intensity E at the centre of the ring.

Solution. The given charge distribution is shown in Fig. 1.20. The symmetry of this distribution implies that vector **E** at the point O is directed to the right, and its magnitude is equal to the sum of the projections onto the direction of **E** of vectors $d\mathbf{E}$ from elementary charges dq. The projection of vector $d\mathbf{E}$ onto vector **E** is

$$dE\cos\varphi = \frac{1}{4\pi\varepsilon_0}\,\frac{dq}{R^2}\cos\varphi, \tag{1}$$

where $dq = \lambda R\,d\varphi = \lambda_0 R\cos\varphi\,d\varphi$. Integrating (1) over φ between 0 and 2π, we find the magnitude of the vector **E**:

$$E = \frac{\lambda_0}{4\pi\varepsilon_0 R}\int_0^{2\pi}\cos^2\varphi\,d\varphi = \frac{\lambda_0}{4\varepsilon_0 R}.$$

It should be noted that this integral is evaluated in the most simple way if we take into account that $\langle\cos^2\varphi\rangle = 1/2$. Then

$$\int_0^{2\pi}\cos^2\varphi\,d\varphi = \langle\cos^2\varphi\rangle\,2\pi = \pi.$$

● 1.3. A semi-infinite straight uniformly charged filament has a charge λ per unit length. Find the magnitude and the direction of the field intensity at the point separated from the filament by a distance y and lying on the normal to the filament, passing through its end.

Solution. The problem is reduced to finding E_x and E_y, viz. the projections of vector **E** (Fig. 1.21, where it is assumed that $\lambda > 0$). Let us start with E_x. The contribution to E_x from the charge element of the segment dx is

$$dE_x = \frac{1}{4\pi\varepsilon_0}\,\frac{\lambda\,dx}{r^2}\sin\alpha. \tag{1}$$

Let us reduce this expression to the form convenient for integration. In our case, $dx = r\,d\alpha/\cos\alpha$, $r = y/\cos\alpha$. Then

$$dE_x = \frac{\lambda}{4\pi\varepsilon_0 y}\sin\alpha\,d\alpha.$$

Integrating this expression over α between 0 and $\pi/2$, we find

$$E_x = \lambda/4\pi\varepsilon_0 y.$$

In order to find the projection E_y, it is sufficient to recall that dE_y differs from dE_x in that $\sin\alpha$ in (1) is simply replaced by $\cos\alpha$.

This gives

$$dE_y = (\lambda \cos \alpha \, d\alpha)/4\pi\varepsilon_0 y \text{ and } E_y = \lambda/4\pi\varepsilon_0 y.$$

We have obtained an interesting result: $E_x = E_y$ independently of y,

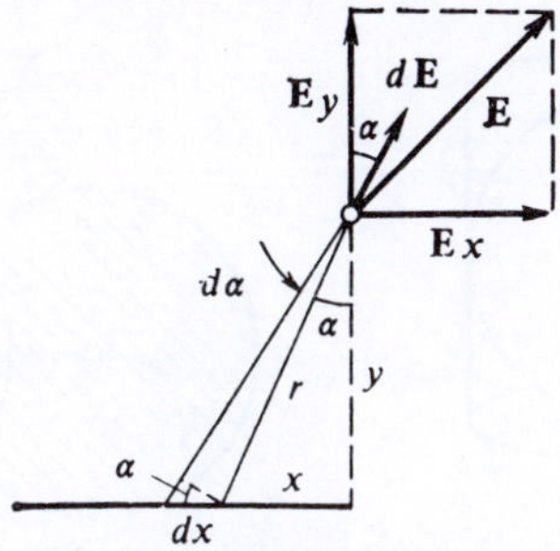

Fig. 1.21

i.e. vector **E** is oriented at the angle of 45° to the filament. The modulus of vector **E** is

$$E = \sqrt{E_x^2 + E_y^2} = \lambda \sqrt{2}/4\pi\varepsilon_0 y.$$

● **1.4. The Gauss theorem.** The intensity of an electric field depends only on the coordinates x and y as follows:

$$\mathbf{E} = a\,(x\mathbf{i} + y\mathbf{j})/(x^2 + y^2),$$

where a is a constant, and **i** and **j** are the unit vectors of the X- and Y-axes. Find the charge within a sphere of radius R with the centre at the origin.

Solution. In accordance with the Gauss theorem, the required charge is equal to the flux of **E** through this sphere, divided by ε_0. In our case, we can determine the flux as follows. Since the field **E** is axisymmetric (as the field of a uniformly charged filament), we arrive at the conclusion that the flux through the sphere of radius R is equal to the flux through the lateral surface of a cylinder having the same radius and the height $2R$, and arranged as shown in Fig. 1.22. Then

$$q = \varepsilon_0 \oint \mathbf{E}\, d\mathbf{S} = \varepsilon_0 E_r S,$$

where $E_r = a/R$ and $S = 2\pi R \cdot 2R = 4\pi R^2$. Finally, we get

$$q = 4\pi\varepsilon_0 a R.$$

● 1.5. A system consists of a uniformly charged sphere of radius R and a surrounding medium filled by a charge with the volume density $\rho = \alpha/r$, where α is a positive constant and r is the distance from the centre of the sphere. Find the charge of the sphere for which the electric field intensity E outside the sphere is independent of r. Find the value of E.

Solution. Let the sought charge of the sphere be q. Then, using the

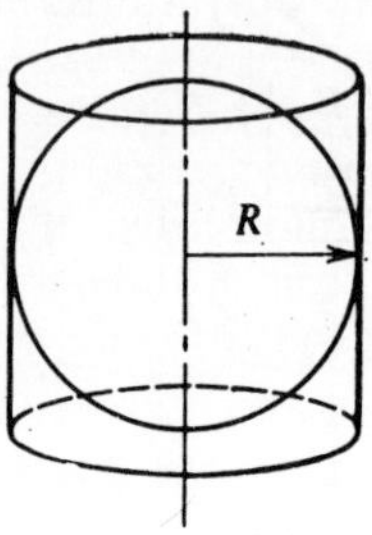

Fig. 1.22

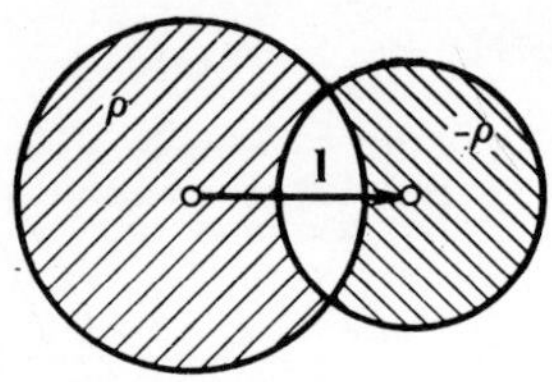

Fig. 1.23

Gauss theorem, we can write the following expression for a spherical surface of radius r (outside the sphere with the charge q):

$$E \cdot 4\pi r^2 = \frac{q}{\varepsilon_0} + \frac{1}{\varepsilon_0} \int_R^r \frac{\alpha}{r} 4\pi r^2\, dr.$$

After integration, we transform this equation to

$$E \cdot 4\pi r^2 = (q - 2\pi\alpha R^2)/\varepsilon_0 + 4\pi\alpha r^2/2\varepsilon_0.$$

The intensity E does not depend on r when the expression in the parentheses is equal to zero. Hence

$$q = 2\pi\alpha R^2 \text{ and } E = \alpha/2\varepsilon_0.$$

● **1.6.** Find the electric field intensity $\mathbf{E}$ in the region of intersection of two spheres uniformly charged by unlike charges with the volume densities ρ and $-\rho$, if the distance between the centres of the spheres is determined by vector $\mathbf{l}$ (Fig. 1.23).

Solution. Using the Gauss theorem, we can easily show that the electric field intensity within a uniformly charged sphere is

$$\mathbf{E} = (\rho/3\varepsilon_0)\, \mathbf{r},$$

where $\mathbf{r}$ is the radius vector relative to the centre of the sphere. We can consider the field in the region of intersection of the spheres as the superposition of the fields of two uniforlmy charged spheres. Then at an arbitrary point A (Fig. 1.24) of this region we have

$$\mathbf{E} = \mathbf{E}_+ + \mathbf{E}_- = \rho\,(\mathbf{r}_+ - \mathbf{r}_-)/3\varepsilon_0 = \rho\mathbf{l}/3\varepsilon_0.$$

Thus, in the region of intersection of these spheres the field is uniform. This conclusion is valid regardless of the ratio between the radii of the spheres and of the distance between their centres. In particular, it is valid when one sphere is completely within the other or, in other words, when there is a spherical cavity in a sphere (Fig. 1.25).

● **1.7.** Using the solution of the previous problem, find the field intensity **E** inside the sphere over which a charge is distributed with

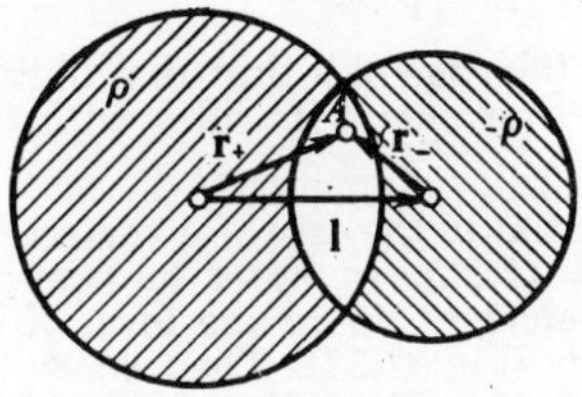

Fig. 1.24

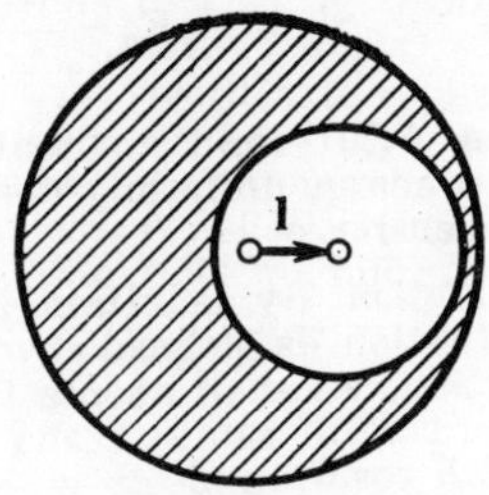

Fig. 1.25

the surface density $\sigma = \sigma_0 \cos\vartheta$, where σ_0 is a constant and ϑ is the polar angle.

Solution. Let us consider two spheres of the same radius, having uniformly distributed volume charges with the densities ρ and $-\rho$. Suppose that the centres of the spheres are separated by the distance **l** (Fig. 1.26). Then, in accordance with the solution of the previous problem, the field in the region of intersection of these spheres will be uniform:

$$\mathbf{E} = (\rho/3\varepsilon_0)\,\mathbf{l}. \quad (1)$$

In our case, the volume charge differs from zero only in the surface layer. For a very small l, we shall arrive at the concept of the surface charge density on the sphere. The thickness of the charged layer at the points determined by angle ϑ (Fig. 1.26) is equal to $l\cos\vartheta$. Hence, in this region the charge per unit area is $\sigma = \rho l \cos\vartheta = \sigma_0 \cos\vartheta$, where $\sigma_0 = \rho l$, and expression (1) can be represented in the form

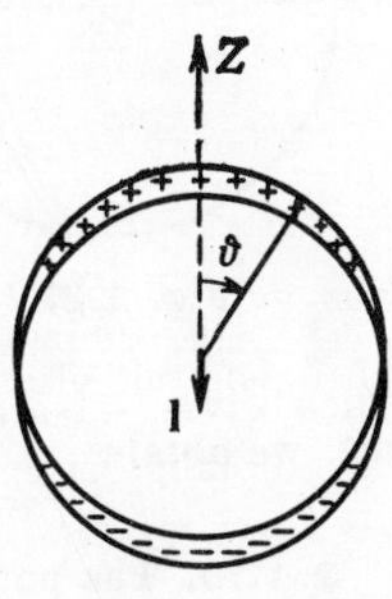

Fig. 1.26

$$\mathbf{E} = -(\sigma_0/3\varepsilon_0)\,\mathbf{k},$$

where **k** is the unit vector of the Z-axis from which the angle ϑ is measured.

● **1.8. Potential.** The potential of a certain electric field has the form $\varphi = \alpha\,(xy - z^2)$. Find the projection of vector **E** onto the direction of the vector $\mathbf{a} = \mathbf{i} + 3\mathbf{k}$ at the point M (2, 1, −3).

Solution. Let us first find vector **E**:

$$\mathbf{E} = -\nabla\varphi = -\alpha\,(y\mathbf{i} + x\mathbf{j} - 2z\mathbf{k}).$$

The sought projection is

$$E_a = \mathbf{E}\cdot\frac{\mathbf{a}}{a} = \frac{-\alpha(y\mathbf{i}-x\mathbf{j}-2z\mathbf{k})(\mathbf{i}+3\mathbf{k})}{\sqrt{1+3^2}} = \frac{-\alpha(y-6z)}{\sqrt{10}}.$$

At the point M we have

$$E_a = \frac{-\alpha(1+18)}{\sqrt{10}} = -\frac{19}{\sqrt{10}}\alpha.$$

● **1.9.** Find the potential φ at the edge of a thin disc of radius R with a charge uniformly distributed over one of its sides with the surface density σ.

Solution. By definition, the potential in the case of a surface charge distribution is defined by integral (1.28). In order to simplify integration, we shall choose the area element dS in the form of a part of the ring of radius r and width dr (Fig. 1.27). Then $dS = 2\vartheta r\, dr$, $r = 2R\cos\vartheta$, and $dr = -2R\sin\vartheta\, d\vartheta$. After substituting these expressions into integral (1.28), we obtain the expression for φ at the point O:

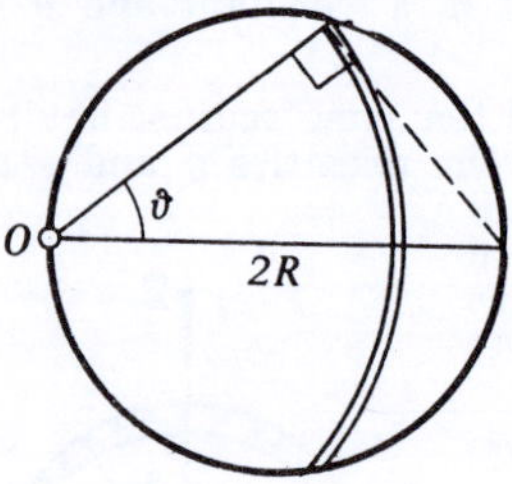

Fig. 1.27

$$\varphi = -\frac{\sigma R}{\pi\varepsilon_0}\int_{\pi/2}^{0}\vartheta\sin\vartheta\, d\vartheta.$$

We integrate by parts, denoting $\vartheta = u$ and $\sin\vartheta\, d\vartheta = dv$:

$$\int\vartheta\sin\vartheta\, d\vartheta = -\vartheta\cos\vartheta + \int\cos\vartheta\, d\vartheta = -\vartheta\cos\vartheta + \sin\vartheta,$$

which gives -1 after substituting the limits of integration. As a result, we obtain

$$\varphi = \sigma R/\pi\varepsilon_0.$$

● **1.10.** The potential of the field inside a charged sphere depends only on the distance r from its centre to the point under consideration in the following way: $\varphi = ar^2 + b$, where a and b are constants. Find the distribution of the volume charge $\rho(r)$ within the sphere.

Solution. Let us first find the field intensity. According to (1.32), we have

$$E_r = -\partial\varphi/\partial r = -2ar. \tag{1}$$

Then we use the Gauss theorem: $4\pi r^2 E_r = q/\varepsilon_0$. The differential of this expression is

$$4\pi\, d(r^2E_r) = \frac{1}{\varepsilon}\, dq = \frac{1}{\varepsilon_0}\rho\cdot 4\pi r^2\, dr,$$

where dq is the charge contained between the spheres of radii r and $r + dr$. Hence

$$r^2\,dE_r + 2rE_r\,dr = \frac{1}{\varepsilon_0}\rho r^2\,dr, \qquad \frac{\partial E_r}{\partial r} + \frac{2}{r}E_r = \frac{\rho}{\varepsilon_0}.$$

Substituting (1) into the last equation, we obtain

$$\rho = -6\varepsilon_0 a,$$

i.e. the charge is distributed uniformly within the sphere.

● **1.11. Dipole.** Find the force of interaction between two point dipoles with moments $\mathbf{p}_1$ and $\mathbf{p}_2$, if the vectors $\mathbf{p}_1$ and $\mathbf{p}_2$ are directed along the straight line connecting the dipoles and the distance between the dipoles is l.

Solution. According to (1.39), we have

$$F = p_1\,|\,\partial E/\partial l\,|,$$

where E is the field intensity from the dipole $\mathbf{p}_2$, determined by the first of formulas (1.38):

$$E = \frac{1}{4\pi\varepsilon_0}\,\frac{2p_2}{l^3}.$$

Taking the derivative of this expression with respect to l and substituting it into the formula for F, we obtain

$$F = \frac{1}{4\pi\varepsilon_0}\,\frac{6p_1p_2}{l^4}.$$

It should be noted that the dipoles will be attracted when $\mathbf{p}_1 \uparrow\uparrow \mathbf{p}_2$ and repulsed when $\mathbf{p}_1 \uparrow\downarrow \mathbf{p}_2$.

2. A Conductor in an Electrostatic Field

2.1. Field in a Substance

Micro- and Macroscopic Fields. The real electric field in any substance (which is called the *microscopic field*) varies abruptly both in space and in time. It is different at different points of atoms and in the interstices. In order to find the intensity $\mathbf{E}$ of a real field at a certain point at a given instant, we should sum up the intensities of the fields of all individual charged particles of the substance, viz. electrons and nuclei. The solution of this problem is obviously not feasible. In any case, the result would be so complicated that it would be impossible to use it. Moreover, the knowledge of this field is not required for the solution of macroscopic problems. In many cases it is sufficient to have a simpler and rougher description which we shall be using henceforth.

Under the electric field **E** in a substance (which is called the *macroscopic field*) we shall understand the microscopic field averaged over space (in this case time averaging is superfluous). This averaging is performed over what is called a *physically infinitesimal volume*, viz. the volume containing a large number of atoms and having the dimensions that are many times smaller than the distances over which the macroscopic field noticeably changes. The averaging over such volumes smoothens all irregular and rapidly varying fluctuations of the microscopic field over the distances of the order of atomic ones, but retains smooth variations of the macroscopic field over macroscopic distances.

Thus, the field in the substance is

$$\mathbf{E} = \mathbf{E}_{\mathrm{macro}} = \langle \mathbf{E}_{\mathrm{micro}} \rangle. \tag{2.1}$$

The Influence of a Substance on a Field. If any substance is introduced into an electric field, the positive and negative charges (nuclei and electrons) are displaced, which in turn leads to a partial separation of these charges. In certain regions of the substance, uncompensated charges of different signs appear. This phenomenon is called the *electrostatic induction*, while the charges appearing as a result of separation are called *induced charges*.

Induced charges create an additional electric field which in combination with the initial (external) field forms the resultant field. Knowing the external field and the distribution of induced charges, we can forget about the presence of the substance itself while calculating the resultant field, since the role of the substance has already been taken into account with the help of induced charges.

Thus, the resultant field in the presence of a substance is determined simply as the superposition of the external field and the field of induced charges. However, in many cases the situation is complicated by the fact that we do not know beforehand how all these charges are distributed in space, and the problem turns out to be not as simple as it could seem at first sight. It will be shown later that the distribution of induced charges is mainly determined by the properties of the substance, i.e. its physical nature and the shape of the bodies. We shall have to consider these questions in greater detail.

2.2. Fields Inside and Outside a Conductor

Inside a Conductor E = 0. Let us place a metallic conductor into an external electrostatic field or impart a certain charge to it. In both cases, the electric field will act on all the charges of the conductor, and as a result all the negative charges (electrons) will be displaced in the direction against the field. This displacement (current) will continue until (this practically takes a small fraction of a second) a certain charge distribution sets in, at which the electric field at all the points inside the conductor vanishes. Thus, in the static case the electric field inside a conductor is absent ($\mathbf{E} = 0$).

Further, since $\mathbf{E} = 0$ everywhere in the conductor, the density of excess (uncompensated) charges inside the conductor is also equal to zero at all points ($\rho = 0$). This can be easily explained with the help of the Gauss theorem. Indeed, since inside the conductor $\mathbf{E} = 0$, the flux of $\mathbf{E}$ through any closed surface inside the conductor is also equal to zero. And this means that there are no excess charges inside the conductor.

Excess charges appear only on the conductor surface with a certain density σ which is generally different for different points of the surface. It should be noted that the excess surface charge is located in a very thin surface layer (whose thickness amounts to one or two interatomic distances).

The absence of a field inside a conductor indicates, in accordance with (1.31), that potential φ in the conductor has the same value for all its points, i.e. any conductor in an electrostatic field is an *equipotential region*, its surface being an *equipotential surface*.

The fact that the surface of a conductor is equipotential implies that in the immediate vicinity of this surface the field $\mathbf{E}$ at each point is directed along the normal to the surface. If the opposite were true, the tangential component of $\mathbf{E}$ would make the charges move over the surface of the conductor, i.e. charge equilibrium would be impossible.

Example. Find the potential of an uncharged conducting sphere provided that a point charge q is located at a distance r from its centre (Fig. 2.1).

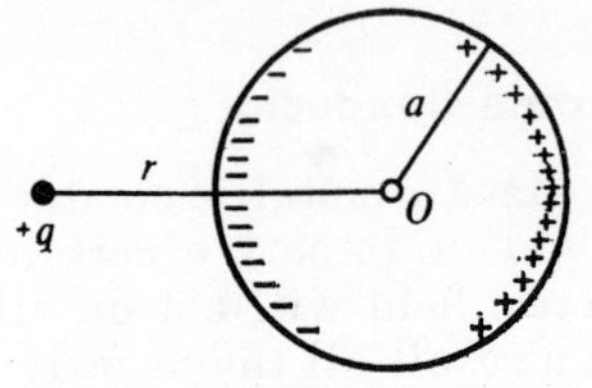

Fig. 2.1

Potential φ is the same for all points of the sphere. Thus we can calculate its value at the centre O of the sphere, because only for this point it can be calculated in the most simple way:

$$\varphi = \frac{1}{4\pi\varepsilon_0} \frac{q}{r} + \varphi', \qquad (1)$$

where the first term is the potential of the charge q, while the second is the potential of the charges induced on the surface of the sphere. But since all induced charges are at the same distance a from the point O and the total induced charge is equal to zero, $\varphi' = 0$ as well. Thus, in this case the potential of the sphere will be determined only by the first term in (1).

Figure 2.2 shows the field and the charge distributions for a system consisting of two conducting spheres one of which (left) is charged. As a result of electric induction, the charges of the opposite sign appear on the surface of the right uncharged sphere. The field of these charges will in turn cause a redistribution of charges on the surface of the left sphere, and their surface distribution will become nonuniform. The solid lines in the figure are the lines of **E**, while the dashed lines show the intersection of equipotential surfaces with the plane of the figure. As we move away from this system, the equipotential surfaces become closer and closer to spherical, and the field lines become closer to radial. The field itself in this case resembles more and more the field of a point charge q, viz. the total charge of the given system.

The Field Near a Conductor Surface. We shall show that the electric field intensity in the immediate vicinity of the surface of a conductor is connected with the local charge density at the conductor surface through a simple relation. This relation can be established with the help of the Gauss theorem.

Suppose that the region of the conductor surface we are interested in borders on a vacuum. The field lines are normal to the conductor surface. Hence for a closed surface we shall take a small cylinder and arrange it as is shown in Fig. 2.3. Then the flux of **E** through this surface will be equal only to the flux through the "outer" endface of the cylinder (the fluxes through the lateral surface and the inner endface are equal to zero). Thus we obtain $E_n \Delta S =$

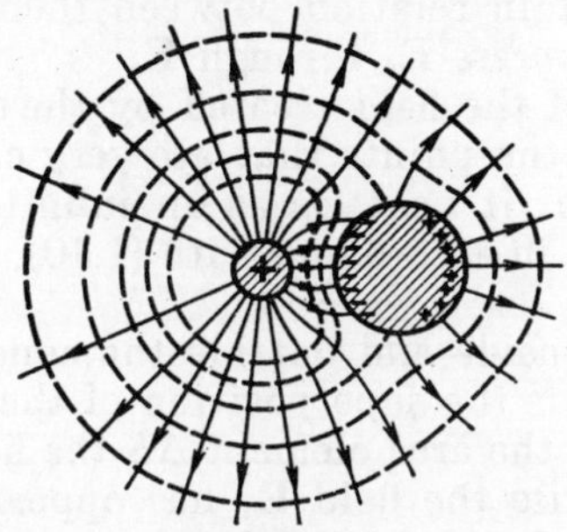

Fig. 2.2

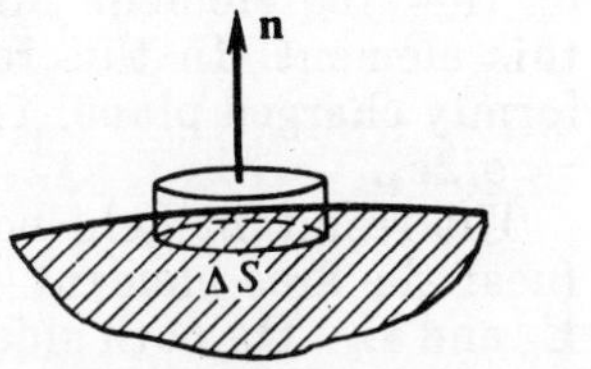

Fig. 2.3

$= \sigma\Delta S/\varepsilon_0$, where E_n is the projection of vector **E** onto the outward normal **n** (with respect to the conductor), ΔS is the cross-sectional area of the cylinder and σ is the local surface charge density of the conductor. Cancelling both sides of this expression by ΔS, we get

$$\boxed{E_n = \sigma/\varepsilon_0.} \tag{2.2}$$

If $\sigma > 0$, then $E_n > 0$, i.e. vector **E** is directed from the conductor surface (coincides in direction with the normal **n**). If $\sigma < 0$, then $E_n < 0$, and vector **E** is directed towards the conductor surface.

Relation (2.2) may lead to the erroneous conclusion that the field **E** in the vicinity of a conductor depends only on the local charge density σ. This is not so. The intensity **E** is determined by *all* the charges of the system under consideration as well as the value of σ itself.

2.3. Forces Acting on the Surface of a Conductor

Let us consider the case when a charged region of the surface of a conductor borders on a vacuum. The force acting on a small area ΔS of the conductor surface is

$$\Delta \mathbf{F} = \sigma\Delta S \cdot \mathbf{E}_0, \tag{2.3}$$

where $\sigma\,\Delta S$ is the charge of this element and $\mathbf{E}_0$ is the field created by all the *other* charges of the system in the region where the charge $\sigma\,\Delta S$ is located. It should be noted at the very outset that $\mathbf{E}_0$ is not equal to the field intensity $\mathbf{E}$ in the vicinity of the given surface element of the conductor, although there exists a certain relation between them. Let us find this relation, i.e. express $\mathbf{E}_0$ through $\mathbf{E}$.

Let $\mathbf{E}_\sigma$ be the intensity of the field created by the charge on the area element ΔS at the points that are very close to this element. In this region, it behaves as an infinite uniformly charged plane. Then, in accordance with (1.10), $E_\sigma = \sigma/2\varepsilon_0$.

The resultant field both inside and outside the conductor (near the area element ΔS) is the superposition of the fields $\mathbf{E}_0$ and $\mathbf{E}_\sigma$. On both sides of the area element ΔS the field $\mathbf{E}_0$ is practically the same, while the field $\mathbf{E}_\sigma$ has opposite directions (see Fig. 2.4 where for the sake of definiteness it is assumed that $\sigma > 0$). From the condition $\mathbf{E} = 0$ inside the conductor, it follows that $E_\sigma = E_0$, and then outside the conductor, near its surface, $E = E_0 + E_\sigma = 2E_0$. Thus,

$$\mathbf{E}_0 = \mathbf{E}/2, \tag{2.4}$$

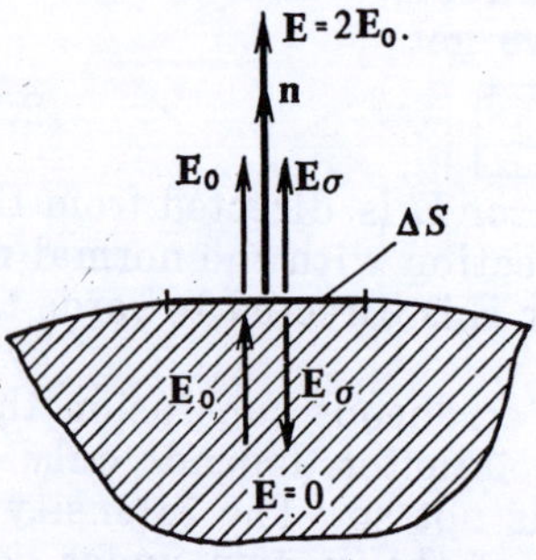

Fig. 2.4

and Eq. (2.3) becomes

$$\Delta\mathbf{F} = \frac{1}{2}\,\sigma\Delta S\cdot\mathbf{E}. \tag{2.5}$$

Dividing both sides of this equation by ΔS, we obtain the expression for the force acting on unit surface of a conductor:

$$\mathbf{F}_u = \frac{1}{2}\,\sigma\mathbf{E}. \tag{2.6}$$

We can write this expression in a different form since the quantities σ and $\mathbf{E}$ appearing in it are interconnected. Indeed, in accordance with (2.2), $E_n = \sigma/\varepsilon_0$, or $\mathbf{E} = (\sigma/\varepsilon_0)\,\mathbf{n}$, where $\mathbf{n}$ is the outward normal to the surface element at a given point of the conductor. Hence

$$\mathbf{F}_u = \frac{\sigma^2}{2\varepsilon_0}\,\mathbf{n} = \frac{\varepsilon_0 E^2}{2}\,\mathbf{n}, \tag{2.7}$$

where we took into account that $\sigma = \varepsilon_0 E_n$ and $E_n^2 = E^2$. The quantity $\mathbf{F}_u$ is called the *surface density of force*. Equation (2.7) shows that regardless of the sign of σ, and hence of the direction of **E**, the force $\mathbf{F}_u$ is always directed outside the conductor, tending to stretch it.

Example. Find the expression for the electric force acting in a vacuum on a conductor as a whole, assuming that the field intensities **E** are known at all points in the vicinity of the conductor surface.

Multiplying (2.7) by dS, we obtain the expression for the force $d\mathbf{F}$ acting on the surface element dS:

$$d\mathbf{F} = \frac{1}{2} \varepsilon_0 E^2 \, d\mathbf{S},$$

where $d\mathbf{S} = \mathbf{n}\, dS$. The resultant force acting on the entire conductor can be found by integrating this equation over the entire conductor surface:

$$\mathbf{F} = \frac{\varepsilon_0}{2} \oint E^2 \, d\mathbf{S}.$$

2.4. Properties of a Closed Conducting Shell

It was shown that in equilibrium there are no excess charges inside a conductor, viz. the material inside the conductor is electrically neutral. Consequently, if the substance is removed from a certain volume inside a conductor (a closed cavity is created), this does not change the field anywhere, i.e. does not affect the equilibrium distribution of charges. This means that the excess charge is distributed on a conductor with a cavity in the same way as on a uniform conductor, viz. on its outer surface.

Thus, in the absence of electric charges within the cavity the electric field is equal to zero in it. *External charges, including the charges on the outer surface of the conductor, do not create any electric field in the cavity inside the conductor.* This forms the basis of *electrostatic shielding*, i.e. the screening of bodies, e.g. measuring instruments, from the influence of external electrostatic fields. In practice, a solid conducting shell can be replaced by a sufficiently dense metallic grating.

That there is no electric field inside an empty cavity can be proved in a different way. Let us take a closed surface S enveloping the cavity and lying completely in the material of the conductor. Since the field **E** is equal to zero inside the conductor, the flux of **E** through the turface S is also equal to zero. Hence, in accordance with the Gauss theorem, the total charge inside S is equal to zero as well. This does not exclude the situation depicted in Fig 2.5, when the surface

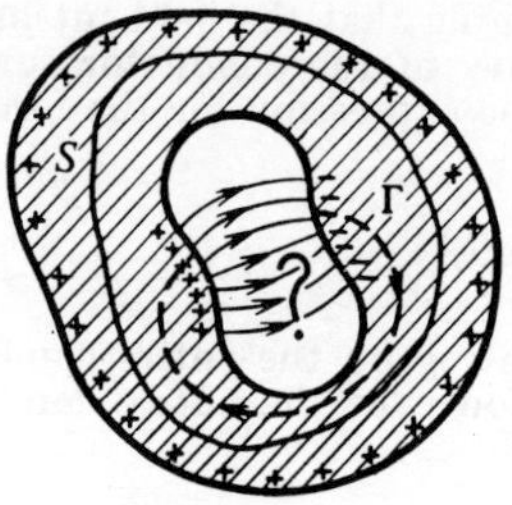

Fig. 2.5

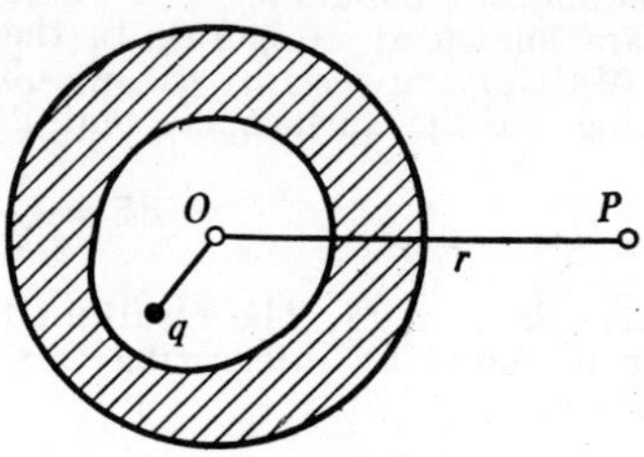

Fig. 2.6

of the cavity itself contains equal quantities of positive and negative charges. However, this assumption is prohibited by another theorem, viz. the theorem on circulation of vector **E**. Indeed, let the contour Γ cross the cavity along one of the lines of **E** and be closed in the conductor material. It is clear that the line integral of vector **E** along this contour differs from zero, which is in contradiction with the theorem on circulation.

Let us now consider the case when the cavity is not empty but contains a certain electric charge q (or several charges). Suppose also that the entire external space is filled by a conducting medium. In equilibrium, the field in this medium is equal to zero, which means that the medium is electrically neutral and contains no excess charges.

Since $\mathbf{E} = 0$ inside the conductor, the field flux through a closed surface surrounding the cavity is also equal to zero. According to the Gauss theorem, this means that the algebraic sum of the charges within this closed surface is equal to zero as well. Thus, the algebraic sum of the charges induced on the cavity surface is equal in magnitude and opposite in sign to the algebraic sum of the charges inside the cavity. In equilibrium the charges induced on the surface of the cavity are arranged so as to compensate completely, in the space outside the cavity, the field created by the charges located inside the cavity.

Since the conducting medium is electrically neutral everywhere, it does not influence the electric field in any way. Therefore, if we remove the medium, leaving only a conducting shell around the cavity, the field will not be changed anywhere, and will remain equal to zero beyond this shell.

Thus, the field of the charges surrounded by a conducting shell and of the charges induced on the surface of the cavity (on the inner surface of the shell) is equal to zero in the entire outer space.

We arrive at the following important conclusion: *a closed conducting shell divides the entire space into the inner and outer parts which are completely independent of one another in respect of electric fields.* This must be interpreted as follows: any arbitrary displacement of charges inside the shell does not introduce any change in the field of the outer space, and hence the charge distribution on the outer surface of the shell remains unchanged. The same refers to the field inside the cavity (if it contains charges) and to the distribution of charges induced on the cavity walls. They will also remain unchanged upon the displacement of charges outside the shell. Naturally, the above arguments are applicable only in the framework of electrostatics.

Example. A point charge q is within an electrically neutral shell whose outer surface has spherical shape (Fig. 2.6). Find the potential φ at the point P lying outside the shell at a distance r from the centre O of the outer surface.

The field at the point P is determined only by charges induced on the outer spherical surface since, as was shown above, the field of the point charge q and of the charges induced on the inner surface of the sphere is equal to zero everywhere outside the cavity. Next, in view of symmetry, the charge on the outer surface of the shell is distributed uniformly, and hence

$$\varphi = \frac{1}{4\pi\varepsilon_0} \frac{q}{r}.$$

An infinite conducting plane is a special case of a closed conducting shell. The space on one side of this plane is electrically independent of the space on its other side.

We shall repeatedly use this property of a closed conducting shell.

2.5. General Problem of Electrostatics. Image Method

Frequently, we must solve problems in which the charge distribution is unknown but the potentials of conductors, their shape and relative arrangement are given. We must find the potential φ (**r**) at any point of the field between the conductors. It should be recalled that if we know the potential φ (**r**), the field **E** (**r**) itself can be easily reconstructed and then its value in the immediate vicinity of the conductor surfaces can be used for determining the surface charge distribution for the conductors.

The Poisson and Laplace equations. Let us derive the differential equation for the function φ (potential). For this purpose, we substitute into the left-hand side of (1.20) the expression for **E** in terms of φ, i.e. $\mathbf{E} = -\nabla\varphi$. As a result, we obtain the general differential equation for potential, which is called the *Poisson equation*:

$$\boxed{\nabla^2\varphi = -\rho/\varepsilon_0,} \tag{2.8}$$

where ∇^2 is the *Laplace operator* (*Laplacian*). In Cartesian coordinates it has the form

$$\nabla^2 = \frac{\partial^2}{\partial x^2} + \frac{\partial^2}{\partial y^2} + \frac{\partial^2}{\partial z^2},$$

i.e. is the scalar product $\nabla \cdot \nabla$ [see (1.19)].

If there are no charges between the conductors ($\rho = 0$), Eq. (2.8) is transformed into a simpler equation, viz. the *Laplace equation*:

$$\boxed{\nabla^2\varphi = 0.} \tag{2.9}$$

To determine potential, we must find a function φ which satisfies Eqs. (2.8) or (2.9) in the entire space between the conductors and acquires the given values φ_1, φ_2, . . . on the surfaces of the conductors.

It can be proved theoretically that this problem has a unique solution. This statement is called the *uniqueness theorem*. From the physical point of view, this conclusion is quite obvious: if there are more than one solution, there will be several potential "reliefs", and hence the field **E** at each point generally has not a single value. Thus we arrive at a physically absurd conclusion.

Using the uniqueness theorem, we can state that in a

static case the charge is distributed over the surface of a conductor in a unique way as well. Indeed, there is a one-to-one correspondence (2.2) between the charges on the conductor and the electric field in the vicinity of its surface: $\sigma = \varepsilon_0 E_n$. Hence it immediately follows that the uniqueness of the field **E** determines the uniqueness of the charge distribution over the conductor surface.

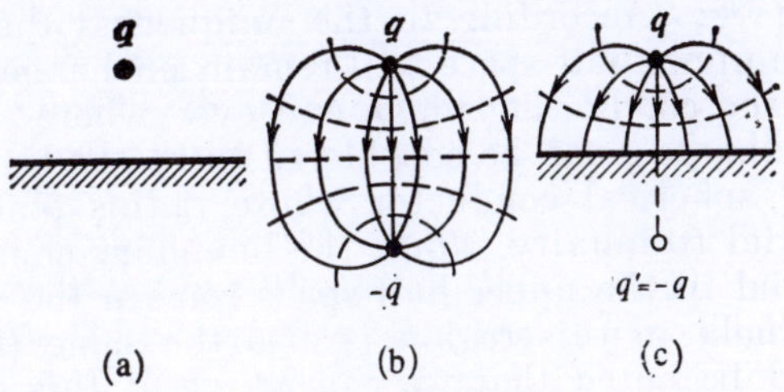

Fig. 2.7

The solution of Eqs. (2.8) and (2.9) in the general case is a complicated and laborious problem. The analytic solutions of these equations were obtained only for a few particular cases. As for the uniqueness theorem, it simplifies the solution of a number of electrostatic problems. If a solution of the problem satisfies the Laplace (or Poisson) equation and the boundary conditions, we can state that it is correct and unique regardless of the methods by which it was obtained (if only by guess).

Example. Prove that in an empty cavity of a conductor the field is absent.

The potential φ must satisfy the Laplace equation (2.9) inside the cavity and acquire a certain value φ_0 at the cavity's walls. The solution of the Laplace equation satisfying this condition can immediately be found. It is $\varphi = \varphi_0$. In accordance with the uniqueness theorem, there can be no other solutions. Hence, $\mathbf{E} = -\nabla\varphi = 0$.

Image Method. This is an artificial method that makes it possible to calculate in a simple way the electric field in some (unfortunately few) cases. Let us consider this method by using a simple example of a point charge q near an infinite conducting plane (Fig. 2.7*a*).

The idea of this method lies in that we must find another problem which can be easily solved and whose solution or a part of it can be used in our problem. In our case such a simple problem is the problem about two charges: q and $-q$.

The field of this system is well known (its equipotential surfaces and field lines are shown in Fig. 2.7b).

Let us make the conducting plane coincide with the middle equipotential surface (its potential $\varphi = 0$) and remove the charge $-q$. According to the uniqueness theorem, the field in the upper half-space will remain unchanged. Indeed, $\varphi = 0$ on the conducting plane and everywhere at infinity. The point charge q can be considered to be the limiting case of a small spherical conductor whose radius tends to zero and potential to infinity. Thus, the boundary conditions for the potential in the upper half-space remain the same, and hence the field in this region is also the same (Fig. 2.7c).

It should be noted that we can arrive at this conclusion proceeding from the properties of a closed conducting shell [see Sec. 2.4], since both half-spaces separated by the conducting plane are electrically independent of one another, and the removal of the charge $-q$ will not affect the field in the upper half-space.

Thus, in the case under consideration the field differs from zero only in the upper half-space. In order to calculate this field, it is sufficient to introduce a fictitious image charge $q' = -q$, opposite in sign to the charge q, by placing it on the other side of the conducting plane at the same distance as the distance from q to the plane. The fictitious charge q' creates in the upper half-space the same field as that of the charges induced on the plane. This is precisely what is meant when we say that the fictitious charge produces the same "effect" as all the induced charges. We must only bear in mind that the "effect" of the fictitious charge extends only to the half-space where the real charge q is located. In another half-space the field is absent.

Summing up, we can say that the image method is essentially based on driving the potential to the boundary conditions, i.e. we strive to find another problem (configuration of charges) in which the field configuration in the region of space we are interested in is the same. The image method proves to be very effective if this can be done with the help of sufficiently simple configurations. Let us consider one more example.

Example. A point charge q is placed between two mutually perpendicular half-planes (Fig. 2.8a). Find the location of fictitious point charges whose action on the charge q is equivalent to the action of all charges induced on these half planes.

We have to find a system of point charges for which the equipotential surfaces with $\varphi = 0$ would coincide with the conducting half-planes. One or two fictitious charges are insufficient in this case; there should be three of them (Fig. 2.8b). Only with such a configuration of the system of four charges we can realize the required "trimming", i.e. ensure that the potential on the conducting half-planes be equal to zero. These three fictitious charges create just the same field within the "right angle" as the field of the charges induced on the conducting planes.

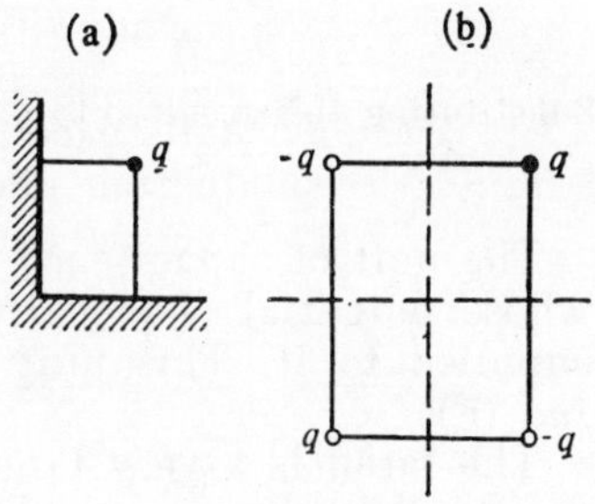

Fig. 2.8

Having found this configuration of point charges (another problem), we can easily answer a number of other questions, for example, find the potential and field intensity on any point within the "right angle" or determine the force acting on the charge q.

2.6. Capacitance. Capacitors

Capacitance of an Isolated Conductor. Let us consider a solitary conductor, i.e. the conductor removed from other conductors, bodies, and charges. Experiments show that the charge q of this conductor is directly proportional to its potential φ (we assumed that at infinity potential is equal to zero): $\varphi \propto q$. Consequently, the ratio q/φ does not depend on the charge q and has a certain value for each solitary conductor. The quantity

$$C = q/\varphi \tag{2.10}$$

is called the *electrostatic capacitance* of an isolated conductor (or simply *capacitance*). It is numerically equal to the charge that must be supplied to the conductor in order to increase its potential by unity. The capacitance depends on the size and shape of the conductor.

Example. Find the capacitance of an isolated conductor which has the shape of a sphere of radius R.

It can be seen from formula (2.10) that for this purpose we must mentally charge the conductor by a charge q and calculate its potential φ. In accordance with (1.23), the potential of a sphere is

$$\varphi=\int_R^\infty E_r\,dr=\frac{1}{4\pi\varepsilon_0}\int_R^\infty \frac{q}{r^2}\,dr=\frac{1}{4\pi\varepsilon_0}\,\frac{q}{R}.$$

Substituting this result into (2.10), we find

$$C=4\pi\varepsilon_0 R. \tag{2.11}$$

The unit of capacitance is the capacitance of a conductor whose potential changes by 1 V when a charge of 1 C is supplied to it. This unit of capacitance is called the *farad* (F).

The farad is a very large quantity. It corresponds to the capacitance of an isolated sphere 9×10^6 km in radius, which is 1500 times the radius of the Earth (the capacitance of the Earth is 0.7 mF). In actual practice, we encounter capacitances between 1 μF and 1 pF.

Capacitors. If a conductor is not isolated, its capacitance will considerably increase as other bodies approach it. This is due to the fact that the field of the given conductor causes a redistribution of charges on the surrounding bodies, i.e. induces charges on them. Let the charge of the conductor be $q>0$. Then negative induced charges will be nearer to the conductor than the positive charges. For this reason, the potential of the conductor, which is the algebraic sum of the potentials of its own charge and of the charges induced on other bodies will decrease when other uncharged bodies approach it. This means that its capacitance increases.

This circumstance made it possible to create the system of conductors, which has a considerably higher capacitance than that of an isolated conductor. Moreover, the capacitance of this system does not depend on surrounding bodies. Such a system is called a *capacitor*. The simplest capacitor consists of two conductors (plates) separated by a small distance.

In order to exclude the effect of external bodies on the capacitance of a capacitor, its plates are arranged with respect to one another in such a way that the field created by the charges accumulated on them is concentrated almost completely inside the capacitor. This means that the lines of **E** emerging on one plate must terminate on the other, i.e. the charges on the plates must be equal in magnitude and opposite in sign (q and $-q$).

The basic characteristic of a capacitor is its capacitance. Unlike the capacitance of an isolated conductor, the capacitance of a capacitor is defined as the ratio of its charge to the potential difference between the plates (this difference is called the voltage):

$$\boxed{C = q/U.} \tag{2.12}$$

The charge q of a capacitor is the charge of its positively charged plate.

Naturally, the capacitance of a capacitor is also measured in farads.

The capacitance of a capacitor depends on its geometry (size and shape of its plates), the gap between the plates, and the material that fills the capacitor. Let us derive the expressions for the capacitances of some capacitors assuming that there is a vacuum between their plates.

Capacitance of a Parallel-plate Capacitor. This capacitor consists of two parallel plates separated by a gap of width h. If the charge of the capacitor is q, then, according to (1.11), the intensity of the field between its plates is $E = \sigma/\varepsilon_0$, where $\sigma = q/S$ and S is the area of each plate. Consequently, the voltage between the plates is

$$U = Eh = qh/\varepsilon_0 S.$$

Substituting this expression into (2.12), we obtain

$$C = \varepsilon_0 S/h. \tag{2.13}$$

This calculation was made without taking into account field distortions near the edges of the plates (edge effects). The capacitance of a real plane capacitor is determined by this formula the more accurately the smaller the gap h in comparison with the linear dimensions of the plates.

Capacitance of a Spherical Capacitor. Let the radii of the inner and outer capacitor plates be a and b respectively. If the charge of the capacitor is q, field intensity between the plates is determined by the Gauss theorem:

$$E_r = \frac{1}{4\pi\varepsilon_0}\frac{q}{r^2}.$$

The voltage of the capacitor is

$$U=\int_a^b E_r\,dr=\frac{q}{4\pi\varepsilon_0}\left(\frac{1}{a}-\frac{1}{b}\right).$$

It can be easily seen that the capacitance of a spherical capacitor is given by

$$C=4\pi\varepsilon_0\,\frac{ab}{b-a}. \tag{2.14}$$

It is interesting that when the gap between the plates is small, i.e. when $b-a\ll a$ (or b), this expression is reduced to (2.13), viz. the expression for the capacitance of a parallel-plate capacitor.

Capacitance of a Cylindrical Capacitor. By using the same line of reasoning as in the case of a spherical capacitor, we obtain

$$C=\frac{2\pi\varepsilon_0 l}{\ln(b/a)}, \tag{2.15}$$

where l is the capacitor's length, a and b are the radii of the inner and outer cylindrical plates. Like in the previous case, the obtained expression is reduced to (2.13) when the gap between the plates is small.

The influence of the medium on the capacitance of a capacitor will be discussed in Sec. 3.7.

Problems

● **2.1. On the determination of potential.** A point charge q is at a distance r from the centre O of an uncharged spherical conducting layer, whose inner and outer radii are equal to a and b respectively. Find the potential at the point O if $r<a$.

Solution. As a result of electrostatic induction, say, negative charges will be induced on the inner surface of the layer and positive charges on its outer surface (Fig. 2.9). According to the principle of superposition, the sought potential at the point O can be represented in the form

$$\varphi=\frac{1}{4\pi\varepsilon_0}\left(\frac{q}{r}+\oint\frac{\sigma_-\,dS}{a}+\oint\frac{\sigma_+\,dS}{b}\right),$$

where the first integral is taken over all the charges induced on the inner surface of the layer, while the second integral, over all the charges on the outer surface. It follows from this expression that

$$\varphi=\frac{q}{4\pi\varepsilon_0}\left(\frac{1}{r}-\frac{1}{a}+\frac{1}{b}\right).$$

It should be noted that the potential can be found in such a simple form only at the point O since all the like induced charges are at the same distance from this point and their distribution (which is unknown to us) does not play any role.

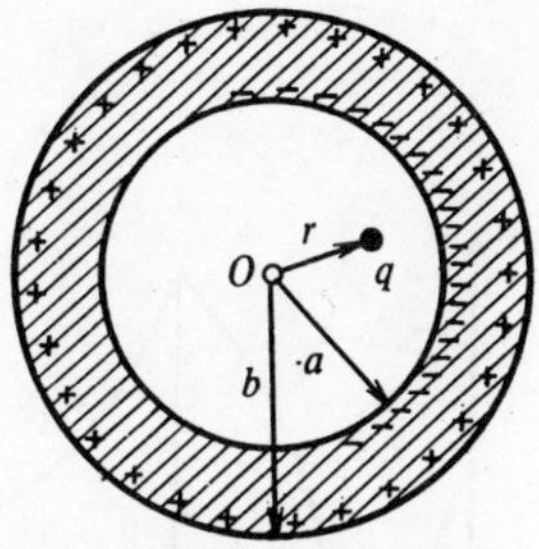

Fig. 2.9

● **2.2.** A system consists of two concentric spheres, the inner sphere of radius R_1 having a charge q_1. What charge q_2 must be placed onto the outer sphere of radius R_2 to make the potential of the inner sphere equal to zero? What will be the dependence of potential φ on the distance r from the centre of the system? Plot schematically the graph of this dependence, assuming that $q_1 < 0$.

Solution. We write the expressions for potentials outside the system (φ_{II}) and in the region between the spheres (φ_I):

$$\varphi_{II}=\frac{1}{4\pi\varepsilon_0}\frac{q_1+q_2}{r},\qquad \varphi_I=\frac{1}{4\pi\varepsilon_0}\frac{q_1}{r}+\varphi_0,$$

where φ_0 is a certain constant. Its value can be easily found from the boundary condition: for $r=R_2$, $\varphi_{II}=\varphi_I$. Hence

$$\varphi_0=q_2/4\pi\varepsilon_0 R_2.$$

From the condition $\varphi_I(R_1)=0$ we find that $q_2=-q_1R_2/R_1$. The $\varphi(r)$ dependence (Fig. 2.10) will have the form:

$$\varphi_{II}=\frac{q_1}{4\pi\varepsilon_0}\frac{1-R_2/R_1}{r},\quad \varphi_I=\frac{q_1}{4\pi\varepsilon_0}\left(\frac{1}{r}-\frac{1}{R_1}\right).$$

● **2.3. The force acting on a surface charge.** An uncharged metallic sphere of radius R is placed into an external uniform field, as a result of which an induced charge appears on the sphere with surface density $\sigma=\sigma_0\cos\vartheta$, where σ_0 is a positive constant and ϑ is a polar angle. Find the magnitude of the resultant electric force acting on like charges.

Solution. According to (2.5), the force acting on the area element dS is

$$d\mathbf{F}=\frac{1}{2}\sigma\mathbf{E}\,dS. \tag{1}$$

It follows from symmetry considerations that the resultant force $\mathbf{F}$ is directed along the Z-axis (Fig. 2.11), and hence it can be represented as the sum (integral) of the projections of elementary forces (1) onto the Z-axis:

$$dF_z = dF \cos \vartheta. \tag{2}$$

It is expedient to take for the area element dS a spherical zone $dS =$

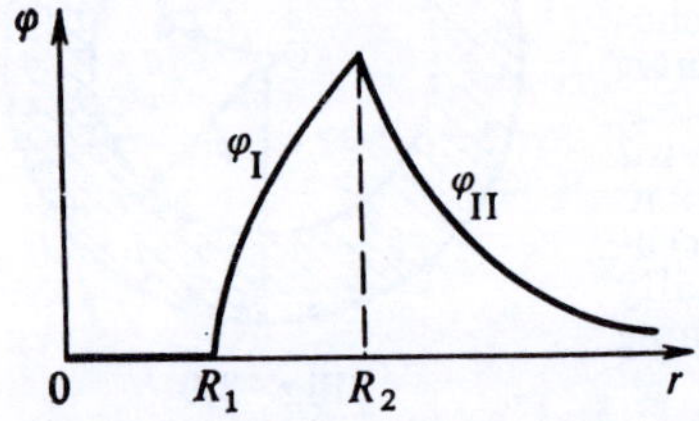

Fig. 2.10

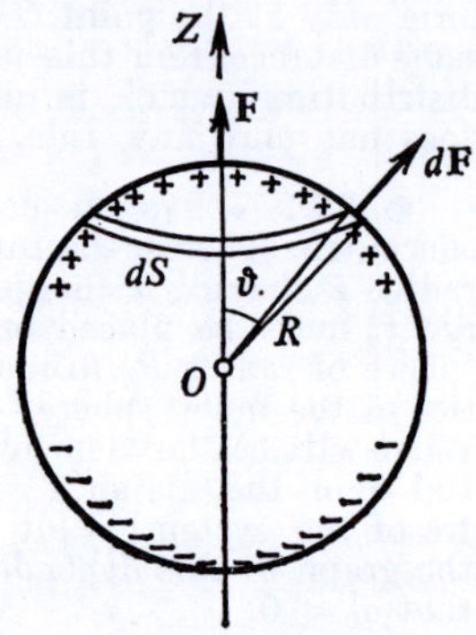

Fig. 2.11

$= 2\pi R \sin \vartheta \cdot R\, d\vartheta$. Considering that $E = \sigma/\varepsilon_0$, we transform (2) as follows:

$$dF_z = (\pi\sigma^2 R^2/\varepsilon_0) \sin \vartheta \cos \vartheta \, d\vartheta = -(\pi\sigma_0^2 R^2/\varepsilon_0) \cos^3 \vartheta \, d(\cos \vartheta).$$

Integrating this expression over the half-sphere (i.e. with respect to $\cos \vartheta$ between 1 and 0), we obtain

$$F = \pi\sigma_0^2 R^2/4\varepsilon_0.$$

● **2.4. Image method.** A point charge q is at a distance l from an infinite conducting plane. Find the density of surface charges induced on the plane as a function of the distance r from the base of the perpendicular dropped from the charge q onto the plane.

Solution. According to (2.2), the surface charge density on a conductor is connected with the electric field near its surface (in vacuum) through the relation $\sigma = \varepsilon_0 E_n$. Consequently, the problem is reduced to determining the field E in the vicinity of the conducting plane.

Using the image method, we find that the field at the point P (Fig. 2.12) which is at a distance r from the point O is

$$E = 2E_q \cos \alpha = 2 \frac{q}{4\pi\varepsilon_0 x^2} \frac{l}{x}.$$

Hence

$$\sigma = -\frac{ql}{2\pi (l^2 + r^2)^{3/2}},$$

where the minus sign indicates that the induced charge is opposite to sign to the point charge q.

● **2.5.** A point charge q is at a distance l from an infinite conducting plane. Find the work of the electric force acting on the charge q done upon its slow removal to a very large distance from the plane.

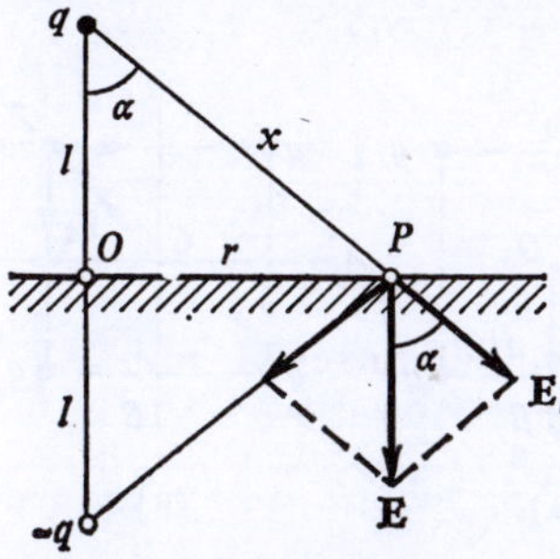

Fig. 2.12

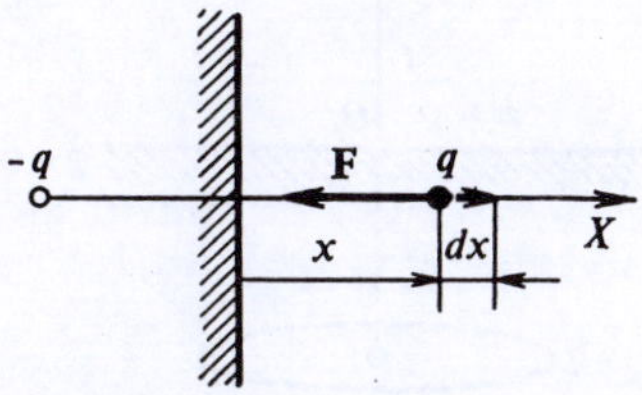

Fig. 2.13

Solution. By definition, the work of this force done upon an elementary displacement dx (Fig. 2.13) is given by

$$\delta A = F_x\, dx = -\frac{q^2}{4\pi\varepsilon_0 (2x)^2}\, dx,$$

where the expression for the force is obtained with the help of the image method. Integrating this equation over x between l and ∞, we find

$$A = -\frac{q^2}{16\pi\varepsilon_0}\int_l^\infty \frac{dx}{x^2} = -\frac{q^2}{16\pi\varepsilon_0 l}.$$

Remark. An attempt to solve this problem in a different way (through potential) leads to an erroneous result which differs from what was obtained by us by a factor of two. This is because the relation $A = q(\varphi_1 - \varphi_2)$ is valid only for potential fields. However, in the reference system fixed to the conducting plane, the electric field of induced charges is not a potential field: a displacement of the charge q leads to a redistribution of the induced charges, and their field turns out to be time-dependent.

● **2.6.** A thin conducting ring of radius R, having a charge q, is arranged so that it is parallel to an infinite conducting plane at a distance l from it. Find (1) the surface charge density at a point of the plane, which is symmetric with respect to the ring and (2) the electric field potential at the centre of the ring.

Solution. It can be easily seen that in accordance with the image method, a fictitious charge $-q$ must be located on a similar ring but on the other side of the conducting plane (Fig. 2.14). Indeed, only in this case the potential of the midplane between these rings is equal to zero, i.e. it coincides with the potential of the conducting plane. Let us now use the formulas we already know.

(1) In order to find σ at the point O, we must, according to (2.2), find the field E at this point (Fig. 2.14). The expression for E on the

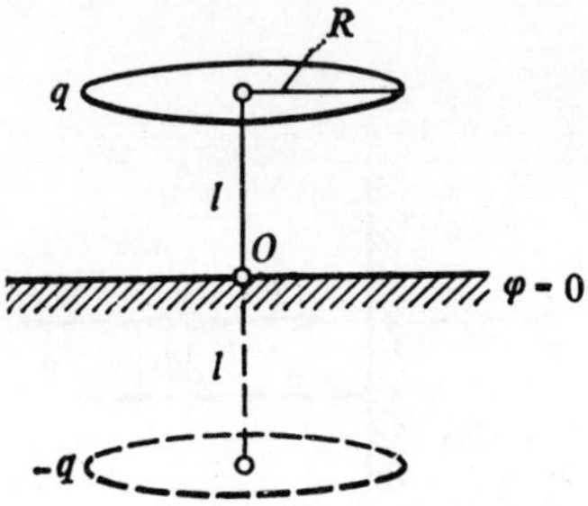

Fig. 2.14

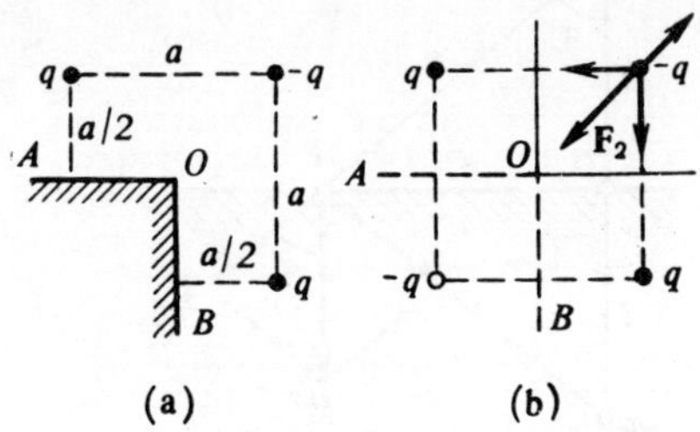

Fig. 2.15

axis of a ring was obtained in Example 1 (see p. 14). In our case, this expression must be doubled. As a result, we obtain

$$\sigma = \frac{ql}{2\pi (R^2 + l^2)^{3/2}}.$$

(2) The potential at the centre of the ring is equal to the algebraic sum of the potentials at this point created by the charges q and $-q$:

$$\varphi = \frac{1}{4\pi\varepsilon_0}\left(\frac{q}{R} - \frac{q}{\sqrt{R^2+4l^2}}\right).$$

● **2.7.** Three unlike point charges are arranged as shown in Fig. 2.15a, where AOB is the right angle formed by two conducting half-planes. The magnitude of each of the charges is $|q|$ and the distances between them are shown in the figure. Find (1) the total charge induced on the conducting half-planes and (2) the force acting on the charge $-q$.

Solution. The half-planes forming the angle AOB go to infinity, and hence their potential $\varphi = 0$. It can be easily seen that a system having equipotential surfaces with $\varphi = 0$ coinciding with the conducting half-planes has the form shown in Fig. 2.15b. Hence the action of the charges induced on the conducting half-planes is equivalent to the action of the fictitious charge $-q$ placed in the lower left corner of the dashed square.

Thus we have already answered the first question: $-q$.

By reducing the system to four point charges, we can easily find the required force (see Fig. 2.15*b*)

$$F = F_2 - F_1 = \frac{2\sqrt{2}-1}{4\pi\varepsilon_0}\,\frac{q^2}{2a^2},$$

and answer the second question.

● **2.8. Capacitance of parallel wires.** Two long straight wires with the same cross section are arranged in air parallel to one another.

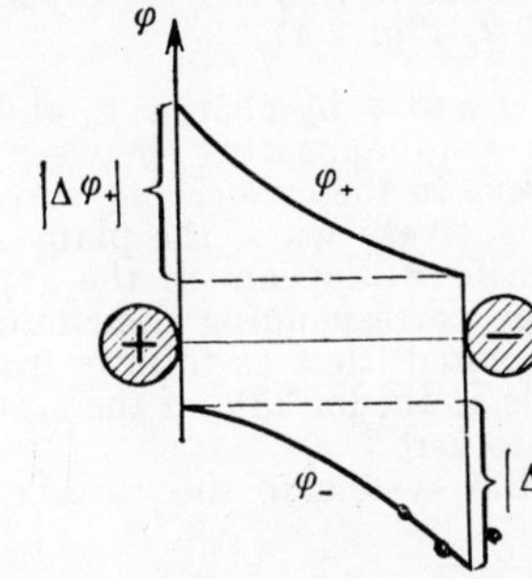

Fig. 2.16

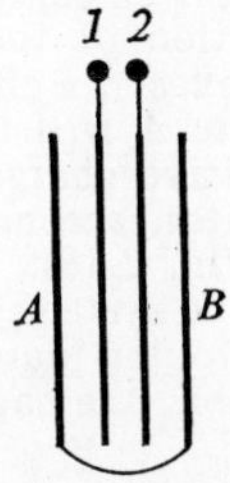

Fig. 2.17

The distance between the wires is η times larger than the radius of the wires' cross section. Find the capacitance of the wires per unit length provided that $\eta \gg 1$.

Solution. Let us mentally charge the two wires by charges of the same magnitude and opposite signs so that the charge per unit length is equal to λ. Then, by definition, the required capacitance is

$$C_u = \lambda / U, \tag{1}$$

and it remains for us to find the potential difference between the wires.

It follows from Fig. 2.16 showing the dependences of the potentials φ_+ and φ_- on the distance between the plates that the sought potential difference is

$$U = |\Delta\varphi_+| + |\Delta\varphi_-| = 2|\Delta\varphi_+|. \tag{2}$$

The intensity of the electric field created by one of the wires at a distance x from its axis can be easily found with the help of the Gauss theorem: $E = \lambda/2\pi\varepsilon_0 x$. Then

$$|\Delta\varphi_+| = \int_a^{b-a} E\,dx = \frac{\lambda}{2\pi\varepsilon_0}\ln\frac{b-a}{a}, \tag{3}$$

where a is the radius of the wires' cross section and b is the separation between the axes of the wires.

It follows from (1), (2) and (3) that

$$C_u = \pi\varepsilon_0/\ln \eta,$$

where we took into account that $b \gg a$.

● **2.9.** Four identical metallic plates are arranged in air at the same distance h from each other. The outer plates are connected by a conductor. The area of each plate is equal to S. Find the capacitance of this system (between points *1* and *2*, Fig. 2.17).

Solution. Let us charge the plates *1* and *2* by charges q_0 and $-q_0$. Under the action of the dissipation field appearing between these plates (edge effect), a charge will move in the connecting wire, after which the plate A will be charged negatively while the plate B will acquire a positive charge. An electric field appears in the gaps between the plates, accompanied by the corresponding distribution of potential φ (Fig. 2.18). It should be noted that as follows from the symmetry of the system, the potentials at the middle of the system as well as on its outer plates are equal to zero.

By definition, the capacitance of the system in this case is

$$C = q_0/U \tag{1}$$

where U is the required potential difference between the points *1* and *2*. Figure 2.18 shows that the potential difference U between the inner plates is twice as large as the potential difference between the outside pair of plates (both on the right and on the left). This also refers to the field intensity:

$$E = 2E'. \tag{2}$$

And since $E \propto \sigma$, we can state that according to (2) the charge q_0 on plate *1* is divided into two parts: $q_0/3$ on the left side of the plate *1* and $2q_0/3$ on its right side. Hence

$$U = Eh = \sigma h/\varepsilon_0 = 2q_0h/3\varepsilon_0 S,$$

and the capacitance of the system (between points *1* and *2*) is

$$C = \frac{3\varepsilon_0 S}{2h}.$$

● **2.10. Distribution of an induced charge.** A point charge q is placed between two large parallel conducting plates *1* and *2* separated by a distance l. Find the total charges q_1 and q_2 induced on each plate, if the plates are connected by a wire and the charge q is located at a distance l_1 from the left plate *1* (Fig. 2.19*a*).

Solution. Let us use the superposition principle. We mentally place somewhere on a plane P the same charge q. Clearly, this will double the surface charge on each plate. If we now distribute uniformly on the surface P a certain charge with surface density σ, the electric field can be easily calculated (Fig. 2.19*b*).

The plates are connected by the wire, and hence the potential difference between them is equal to zero. Consequently,

$$E_{1x}l_1 + E_{2x}(l - l_1) = 0,$$

where E_{1x} and E_{2x} are the projections of vector **E** onto the X-axis to the left and to the right of the plane P (Fig. 2.19*b*).

On the other hand, it is clear that

$$\sigma = -(\sigma_1 + \sigma_2),$$

where, in accordance with (2.2), $\sigma_1 = \varepsilon_0 E_{1n_1} = \varepsilon_0 E_{1x}$ and $\sigma_2 = \varepsilon_0 E_{2n_2} = -\varepsilon_0 E_{2x}$ (the minus sign indicates that the normal $\mathbf{n}_2$ is directed oppositely to the unit vector of the X-axis).

Eliminating E_{1x} and E_{2x} from these equations, we obtain

$$\sigma_1 = -\sigma(l - l_1)/l, \quad \sigma_2 = -\sigma l_1/l.$$

The formulas for charges q_1 and q_2 in terms of q have a similar form.

It would be difficult, however, to solve this problem with the help of the image method, since it would require an infinite series of fictitious charges arranged on both sides of the charge q, and to find the field of such a system is a complicated problem.

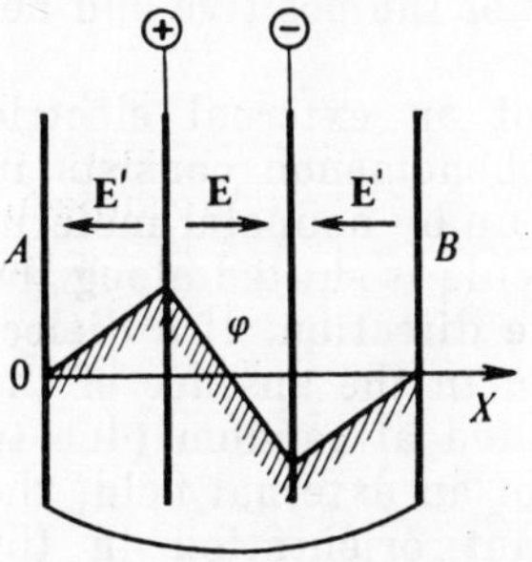

Fig. 2.18

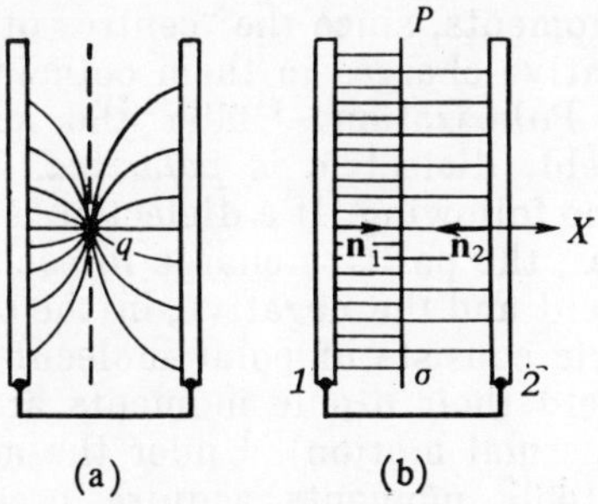

Fig. 2.19

3. Electric Field in Dielectrics

3.1. Polarization of Dielectrics

Dielectrics. *Dielectrics* (or insulators) are substances that practically do not conduct electric current. This means that in contrast, for example, to conductors dielectrics do not contain charges that can move over considerable distances and create electric current.

When even a neutral dielectric is introduced into an external electric field, appreciable changes are observed in the field and in the dielectric itself. This is because the dielectric is acted upon by a force, the capacitance of a capacitor increases when it is filled by a dielectric, and so on.

In order to understand the nature of these phenomena, we must take into consideration that dielectrics consist either of neutral molecules or of charged ions located at the sites of a crystal lattice (ionic crystals, for example, of the NaCl type). The molecules can be either *polar* or *nonpolar*. In a polar molecule, the centre of "mass" of the negative charge is displaced relative to the centre of "mass" of the positive charge. As a result, the molecule acquires an intrinsic dipole moment **p**. Nonpolar molecules do not have intrinsic dipole moments, since the "centres of mass" of the positive and negative charges in them coincide.

Polarization. Under the action of an external electric field, dielectric is *polarized.* This phenomenon consists in the following. If a dielectric is made up by nonpolar molecules, the positive charge in each molecule is shifted along the field and the negative, in the opposite direction. If a dielectric consists of polar molecules, then in the absence of the field their dipole moments are oriented at random (due to thermal motion). Under the action of an external field, the dipole moments acquire predominant orientation in the direction of the external field. Finally, in dielectric crystals of the NaCl type, an external field displaces all the positive ions along the field and the negative ions, against the field.*

Thus, the mechanism of polarization depends on the structure of a dielectric. For further discussion it is only important that regardless of the polarization mechanism, all the positive charges during this process are displaced along the field, while the negative charges, against the field. It should be noted that under normal conditions the displacements of charges are very small even in comparison with the dimensions of the molecules. This is due to the fact that the intensity of the external field acting on the dielectric is consider-

* There exist ionic crystals polarized even in the absence of an external field. This property is inherent in dielectrics which are called *electrets* (they resemble permanent magnets).

ably lower than the intensities of internal electric fields in the molecules.

Bulk and Surface Bound Charges. As a result of polarization, uncompensated charges appear on the dielectric surface as well as in its bulk. To understand better the mechanism

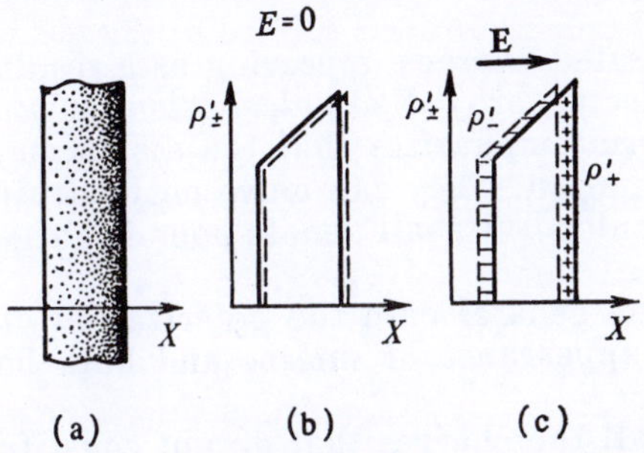

Fig. 3.1

of emergence of these charges (and especially bulk charges), let us consider the following model. Suppose that we have a plate made of a neutral inhomogeneous dielectric (Fig. 3.1*a*) whose density increases with the coordinate x according to a certain law. We denote by ρ'_+ and ρ'_- the magnitudes of the volume densities of the positive and negative charges in the material (these charges are associated with nuclei and electrons).

In the absence of an external field, $\rho'_+ = \rho'_-$ at each point of the dielectric, since the dielectric is electrically neutral. However, ρ'_+ as well as ρ'_- increase with x due to inhomogeneity of the dielectric (Fig. 3.1*b*). This figure shows that in the absence of external field, these two distributions exactly coincide (the distribution of $\rho'_+(x)$ is shown by the solid line, while that of $\rho'_-(x)$, by the dashed line).

Switching on of the external field leads to a displacement of the positive charges along the field and of the negative charges against the field, and the two distributions will be shifted relative to one another (Fig. 3.1*c*). As a result, uncompensated charges will appear on the dielectric surface as well as in the bulk (in Fig. 3.1 an uncompensated negative charge appears in the bulk). It should be noted that the reversion of the field direction changes the sign of all these charges. It can be easily seen that in the case of a plate made of a homogeneous dielectric, the distributions $\rho'_+(x)$ and $\rho'_-(x)$ would be Π-shaped, and only uncompensated surface charges would appear upon their relative displacement in the field **E**.

Uncompensated charges appearing as a result of polarization of a dielectric are called *polarization*, or *bound*, charges. The latter term emphasizes that the displacements of these charges are limited. They can move only within electrically neutral molecules. We shall denote bound charges by a prime (q', ρ', σ').

Thus, in the general case the polarization of a dielectric leads to the appearance of surface and bulk bound charges in it.

We shall call the charges that do not constitute dielectric molecules the *extraneous* charges.* These charges may be located both inside and outside the dielectric.

The Field in a Dielectric. The field **E** in a dielectric is the term applied to the superposition of the field $\mathbf{E}_0$ of extraneous charges and the field $\mathbf{E}'$ of bound charges:

$$\mathbf{E} = \mathbf{E}_0 + \mathbf{E}', \tag{3.1}$$

where $\mathbf{E}_0$ and $\mathbf{E}'$ are macroscopic fields, i.e. the microscopic fields of extraneous and bound charges, averaged over a physically infinitesimal volume. Clearly, the field **E** in the dielectric defined in this way is also a macroscopic field.

3.2. Polarization

Definition. It is natural to describe polarization of a dielectric with the help of the dipole moment of a unit volume. If an external field or a dielectric (or both) are nonuniform, polarization turns out to be different at different points of the dielectric. In order to characterize the polarization at a given point, we must mentally isolate an infinitesimal volume ΔV containing this point and then find the vector sum of the

* Extraneous charges are frequently called *free* charges, but this term is not convenient in some cases since extraneous charges may be not free.

dipole moments of the molecules in this volume and write the ratio

$$\mathbf{P} = \frac{1}{\Delta V} \sum \mathbf{p}_i. \tag{3.2}$$

Vector **P** defined in this way is called the *polarization of a dielectric*. This vector is numerically equal to the dipole moment of a unit volume of the substance.

There are two more useful representations of vector **P**. Let a volume ΔV contain ΔN dipoles. We multiply and divide the right-hand side of (3.2) by ΔN. Then we can write

$$\mathbf{P} = n \langle \mathbf{p} \rangle, \tag{3.3}$$

where $n = \Delta N/\Delta V$ is the concentration of molecules (their number in a unit volume) and $\langle \mathbf{p} \rangle = (\Sigma \mathbf{p}_i)/\Delta N$ is the mean dipole moment of a molecule.

Another expression for **P** corresponds to the model of a dielectric as a mixture of positive and negative "fluids". Let us isolate a very small volume ΔV inside the dielectric. Upon polarization, the positive charge $\rho'_+ \Delta V$ contained in this volume will be displaced relative to the negative charge by a distance **l**, and these charges will acquire the dipole moment $\Delta \mathbf{p} = \rho'_+ \Delta V \cdot \mathbf{l}$. Dividing both sides of this formula by ΔV, we obtain the expression for the dipole moment of a unit volume, i.e. vector **P**:

$$\mathbf{P} = \rho'_+ \mathbf{l}. \tag{3.4}$$

The unit of polarization **P** is the *coulomb per square meter* (C/m^2).

Relation Between P and E. Experiments show that for a large number of dielectrics and a broad class of phenomena, polarization **P** *linearly* depends on the field **E** in a dielectric. For an isotropic dielectric and for not very large **E**, there exists a relation

$$\mathbf{P} = \varkappa \varepsilon_0 \mathbf{E}, \tag{3.5}$$

where $\varkappa$ is a dimensionless quantity called the *dielectric susceptibility* of a substance. This quantity is independent of **E** and characterizes the properties of the dielectric itself. $\varkappa$ is always greater than zero.

Henceforth, if the opposite is not stipulated, we shall

consider only isotropic dielectrics for which relation (3.5) is valid.

However, there exist dielectrics for which (3.5) is not applicable. These are some ionic crystals (see footnote on page 68) and *ferroelectrics*. The relation between **P** and **E** for ferroelectrics is *nonlinear* and depends on the history of the dielectric, i.e. on the previous values of **E** (this phenomenon is called *hysteresis*).

3.3. Properties of the Field of P

The Gauss Theorem for the Field of P. We shall show that the field of **P** has the following remarkable and important property. It turns out that the flux of **P** through an

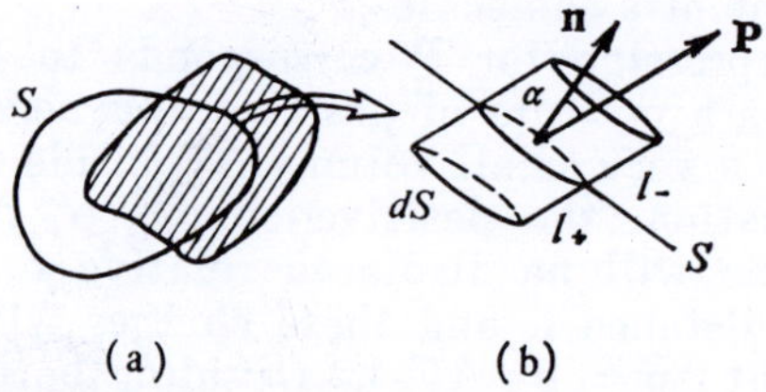

Fig. 3.2

arbitrary closed surface S is equal to the excess bound charge (with the reverse sign) of the dielectric in the volume enclosed by the surface S, i.e.

$$\oint \mathbf{P}\, d\mathbf{S} = -q'_{\text{int}}. \tag{3.6}$$

This equation expresses the *Gauss theorem for vector* **P**.

Proof of the theorem. Let an arbitrary closed surface S envelope a part of a dielectric (Fig. 3.2*a*, the dielectric is hatched). When an external electric field is switched on, the dielectric is polarized—its positive charges are displaced relative to the negative charges. Let us find the charge which passes through an element dS of the closed surface S in the outward direction (Fig. 3.2*b*).

Let $\mathbf{l}_+$ and $\mathbf{l}_-$ be vectors characterizing the displacement of the positive and negative bound charges as a result of polarization. Then it is clear that the positive charge $\rho'_+ l_+ dS \cos\alpha$ inclosed in the "inner" part of the oblique cylinder will pass through the area element dS from the surface S outwards (Fig. 3.2*b*). Besides, the negative charge $\rho'_- l_- dS \cos\alpha$ enclosed in the "outer" part of the oblique cylinder will enter the surface S through the area element dS. But be know that the transport of a negative charge in a certain direction is equivalent to the transport of the positive charge in the opposite direction. Taking this into account, we can write the expression for the total bound charge passing through the area element dS of the surface S in the outward direction:

$$dq' = \rho'_+ l_+ \, dS \cos\alpha + |\rho'_-| l_- \, dS \cos\alpha.$$

Since $|\rho'_-| = \rho'_+$ we have

$$dq' = \rho'_+ (l_+ + l_-) \, dS \cos\alpha = \rho'_+ l \, dS \cos\alpha, \tag{3.7}$$

where $l = l_+ + l_-$ is the relative displacement of positive and negative bound charges in the dielectric during polarization.

Next, according to (3.4), $\rho'_+ l = P$ and $dq' = P \, dS \cos\alpha$, or

$$dq' = P_n \, dS = \mathbf{P} \cdot d\mathbf{S}. \tag{3.8}$$

Integrating this expression over the entire closed surface S, we find the total charge that left the volume enclosed by the surface S upon polarization. This charge is equal to $\oint \mathbf{P} \cdot d\mathbf{S}$. As a result, a certain excess bound charge q' will be left inside the surface S. Clearly, the charge leaving the volume must be equal to the excess bound charge remaining within the surface S, taken with the opposite sign. Thus, we arrive at (3.6).

Differential form of Eq. (3.6). Equation (3.6), viz. the Gauss theorem for the field of vector **P**, can be written in the differential form as follows:

$$\boxed{\nabla \cdot \mathbf{P} = -\rho',} \tag{3.9}$$

i.e. the divergence of the field of vector **P** is equal to the volume density of the excess bound charge at the same point, but taken with the opposite sign. This equation can be obtained from (3.6) in the same manner as the similar expression for vector **E** was obtained (see p. 24). For this purpose, it is sufficient to replace **E** by **P** and ρ by ρ'..

When Is ρ' Equal to Zero in a Dielectric? We shall show that the volume density of excess bound charges in a dielectric is equal to zero if two conditions are simultaneously satisfied: (1) the dielectric is homogeneous and (2) there are no extraneous charges within it ($\rho = 0$).

Indeed, it follows from the main property (3.6) of the field of vector **P** that in the case of a homogeneous dielectric, we can substitute $\varkappa\varepsilon_0\mathbf{E}$ for **P** in accordance with (3.5), take $\varkappa$ out of the integral, and write

$$\varkappa \oint \varepsilon_0 \mathbf{E} \cdot d\mathbf{S} = -q'.$$

The remaining integral is just the algebraic sum of all the charges—extraneous and bound—inside the closed surface S under consideration, i.e. it is equal to $q + q'$. Hence, $\varkappa (q + q') = -q'$, from which we obtain

$$q' = -\frac{\varkappa}{1+\varkappa} q. \tag{3.10}$$

This relation between the excess bound charge q' and the extraneous charge q is valid for any volume inside the dielectric, in particular, for a physically infinitesimal volume, when $q' \to dq' = \rho' dV$ and $q \to dq = \rho dV$. Then, after cancelling out dV, Eq. (3.10) becomes

$$\rho' = -\frac{\varkappa}{1+\varkappa} \rho. \tag{3.11}$$

Hence it follows that in a homogeneous dielectric $\rho' = 0$ when $\rho = 0$.

Thus, if we place a homogeneous isotropic dielectric of any shape into an arbitrary electric field, we can be sure that its polarization will give rise only to the surface bound charge, while the bulk excess bound charge will be zero at all points of such a dielectric.

Boundary Conditions for Vector P. Let us consider the behaviour of vector **P** at the interface between two homogeneous isotropic dielectrics. We have just shown that in such a dielectric there is no excess bound bulk charge and only a surface bound charge appears as a result of polarization.

Let us find the relation between polarization **P** and the surface density σ' of the bound charge at the interface between the dielectrics. For this purpose, let us use property (3.6) of the field of vector **P**. We choose the closed surface in the form of a flat cylinder whose endfaces are on different sides of the interface (Fig. 3.3). We shall assume that the height of the cylinder is negligibly small and the area ΔS of each endface is so small that vector **P** is the same at all points of each endface (this also refers to the surface density σ' of the bound charge). Let **n** be the common normal to the interface at a given point. We shall always draw vector **n** from dielectric *1* to dielectric *2*.

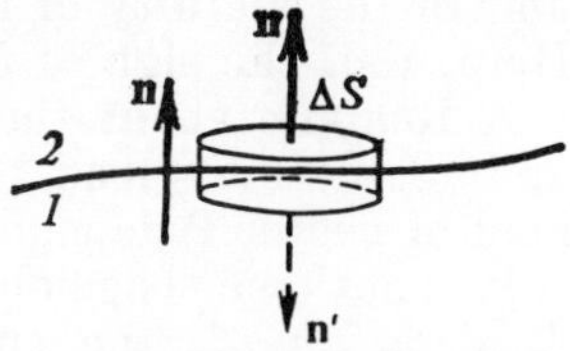

Fig. 3.3

Disregarding the flux of **P** through the lateral surface of the cylinder, we can write, in accordance with (3.6):

$$P_{2n}\,\Delta S + P_{1n'}\,\Delta S = -\sigma'\,\Delta S,$$

where P_{2n} and $P_{1n'}$ are the projections of vector **P** in dielectric *2* onto the normal **n** and in dielectric *1* onto the normal **n**$'$ (Fig. 3.3). Considering that the projection of vector **P** onto the normal **n**$'$ is equal to the projection of this vector onto the opposite (common) normal **n**, taken with the opposite sign, i.e. $P_{1n'} = -P_{1n}$, we can write the previous equation in the following form (after cancelling ΔS):

$$P_{2n} - P_{1n} = -\sigma'. \tag{3.12}$$

This means that at the interface between dielectrics the normal component of vector **P** has a discontinuity, whose magnitude depends on σ'. In particular, if medium *2* is a vacuum, then $P_{2n} = 0$, and condition (3.12) acquires a simpler form:

$$\sigma' = P_n, \tag{3.13}$$

where P_n is the projection of vector **P** onto the outward normal to the surface of a given dielectric. The sign of the projection P_n determines the sign of the surface bound charge σ' at a given point. Formula (3.13) can be written in a different form. In accordance with (3.5), we can write

$$\sigma' = \varkappa\varepsilon_0 E_n, \tag{3.14}$$

where E_n is the projection of vector **E** (inside the dielectric and in the vicinity of its surface) onto the outward normal. Here, too, the sign of E_n determines the sign of σ'.

A Remark about the Field of Vector P. Relations (3.6) and (3.13) may lead to the erroneous conclusion that the field of vector **P** depends only on the bound charge. Actually, this is not true. The field of vector **P**, as well as the field of **E**, depends on *all* the charges, both bound and extraneous. This can be proved if only by the fact that vectors **P** and **E** are connected through the relation $\mathbf{P} = \varkappa\varepsilon_0\mathbf{E}$. The bound charge determines the flux of vector **P** through a closed surface S rather than the field of **P**. Moreover, this flux is determined not by the whole bound charge but by its part enclosed by the surface S.

3.4. Vector D

The Gauss Theorem for Field D. Since the sources of an electric field **E** are all the electric charges—extraneous and bound, we can write the Gauss theorem for the field **E** in the following form:

$$\oint \varepsilon_0 \mathbf{E}\, d\mathbf{S} = (q + q')_{\text{int}}, \tag{3.15}$$

where q and q' are the extraneous and bound charges enclosed by the surface S. The appearance of the bound charge q' complicates the analysis, and formula (3.15) turns out to be of little use for finding the field **E** in a dielectric even in the case of a "sufficiently good" symmetry. Indeed, this formula expresses the properties of unknown field **E** in terms of the bound charge q' which in turn is determined by unknown field **E**.

This difficulty, however, can be overcome by expressing the charge q' in terms of the flux of **P** by formula (3.6). Then expression (3.15) can be transformed as follows:

$$\oint (\varepsilon_0 \mathbf{E} + \mathbf{P})\, d\mathbf{S} = q_{\text{int}}. \tag{3.16}$$

The quantity in the parentheses in the integrand is denoted by **D**. Thus, we have defined an auxiliary vector **D**

$$\boxed{\mathbf{D} = \varepsilon_0 \mathbf{E} + \mathbf{P},} \tag{3.17}$$

whose flux through an arbitrary closed surface is equal to the algebraic sum of extraneous charges enclosed by this surface:

$$\boxed{\oint \mathbf{D}\, d\mathbf{S} = q_{\text{int}}.} \tag{3.18}$$

This statement is called the *Gauss theorem for field* **D**.

It should be noted that vector **D** is the sum of two completely different quantities: $\varepsilon_0 \mathbf{E}$ and **P**. For this reason, it is indeed an auxiliary vector which does not have any deep physical meaning. However, the property of the field of vector **D**, expressed by equation (3.18), justifies the introduction of this vector: in many cases it considerably simplifies the analysis of the field in dielectrics.*

Relations (3.17) and (3.18) are valid for any dielectric, both isotropic and anisotropic.

Expression (3.17) shows that the dimensions of vector **D** are the same as those of vector **P**. The quantity **D** is measured in coulombs per square metre (C/m²).

Differential form of Eq. (3.18) is

$$\boxed{\nabla \cdot \mathbf{D} = \rho,} \tag{3.19}$$

i.e. the divergence of the field **D** is equal to the volume density of an extraneous charge at the same point.

This equation can be obtained from (3.18) in the same way as it was done for the field **E** (see p. 24). It suffices to replace **E** by **D** and take into account only extraneous charges.

---

* The quantity **D** is often called *dielectric displacement*, or *electrostatic induction*. We shall not be using this term, however, in order to emphasize the auxiliary nature of vector **D**.

At the points where the divergence of vector **D** is positive we have the *sources* of the field **D** ($\rho > 0$), while at the points where the divergence is negative, the *sinks* of the field **D** ($\rho < 0$).

Relation Between Vectors D and E. In the case of isotropic dielectrics, polarization $\mathbf{P} = \varkappa\varepsilon_0\mathbf{E}$. Substituting this expression into (3.17), we obtain $\mathbf{D} = \varepsilon_0 (1 + \varkappa) \mathbf{E}$, or

$$\boxed{\mathbf{D} = \varepsilon_0\varepsilon\mathbf{E},} \tag{3.20}$$

where ε is the *dielectric constant* of a substance:

$$\varepsilon = 1 + \varkappa. \tag{3.21}$$

The dielectric constant ε (as well as $\varkappa$) is the basic electric characteristic of a dielectric. For materials $\varepsilon > 1$, while for vacuum $\varepsilon = 1$. The value of ε depends on the nature of the dielectric and varies between the values slightly differing from unity (for gases) and several thousands (for some ceramics). The value of ε for water is rather high ($\varepsilon = 81$).

Formula (3.20) shows that in isotropic dielectrics vector **D** is collinear to vector **E**. For anisotropic dielectrics, these vectors are generally noncollinear.

The field **D** can be graphically represented by the lines of vector **D**, whose direction and density are determined in the same way as for vector **E**. The lines of **E** may emerge and terminate on extraneous as well as bound charges. We say that *any* charges may be the sources and sinks of vector **E**. The sources and sinks of field **D**, however, are only *extraneous* charges, since only on these charges the lines of **D** emerge and terminate. The lines of **D** pass without discontinuities through the regions of the field containing bound charges.

A Remark about the Field of Vector D. The field of vector **D** generally depends on extraneous as well as bound charges (just as the field of vector **E**). This follows if only from the relation $\mathbf{D} = \varepsilon_0\varepsilon\mathbf{E}$. However, in *certain* cases the field of vector **D** is determined only by extraneous charges. It is just the cases for which vector **D** is especially useful. At the same time, this may lead to the erroneous conclusion that vector **D** always depends only on extraneous charges and to an incorrect interpretation of the laws (3.18) and (3.19). These laws express only a certain property of field **D** but do not determine this field proper.

Let us illustrate what was said above by several examples.

Example 1. An extraneous point charge q is located at the centre of a sphere of radius a, made of an isotropic dielectric with a dielectric

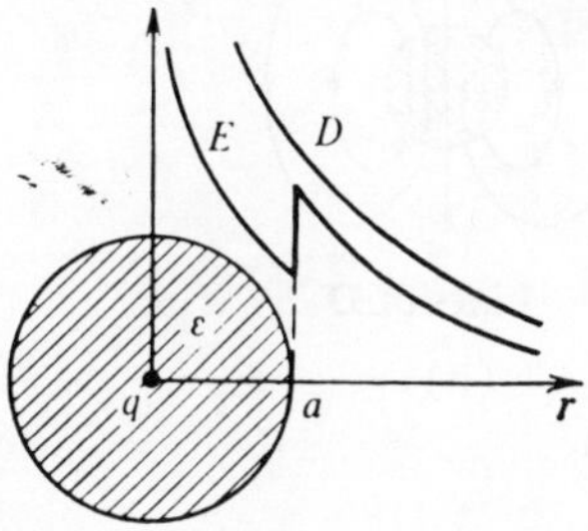

Fig. 3.4

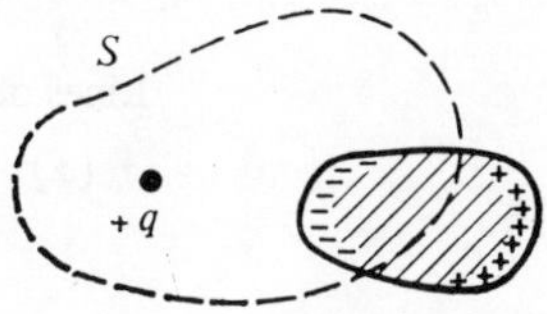

Fig. 3.5

constant ε. Find the projection E_r of field intensity **E** as a function of the distance r from the centre of this sphere.

The symmetry of the system allows us to use the Gauss theorem for vector **D** for solving the problem (we cannot use here the similar theorem for vector **E**, since the bound charge is unknown to us). For a sphere of radius r with the centre at the point of location of the charge q, we can write the following relation: $4\pi r^2 D_r = q$. Hence we can find D_r and then, using formula (3.20), the required quantity E_r:

$$E_r\,(r<a)=\frac{1}{4\pi\varepsilon_0}\,\frac{q}{\varepsilon r^2},\qquad E_r\,(r>a)=\frac{1}{4\pi\varepsilon_0}\,\frac{q}{r^2}.$$

Figure 3.4 shows the curves $D\,(r)$ and $E\,(r)$.

Example 2. Suppose that a system consists of a point charge $q>0$ and an arbitrary sample of a homogeneous isotropic dielectric (Fig. 3.5), where S is a certain closed surface. Find out what will happen to the fields of vectors **E** and **D** (and to their fluxes through the surface S) if the dielectric is removed.

The field **E** at any point of space is determined by the charge q and by bound charges of the polarized dielectric. Since in our case $\mathbf{D}=\varepsilon_0\varepsilon\mathbf{E}$, this refers to the field **D** as well: it is also determined by the extraneous charge q and by the bound charges of the dielectric.

The removal of the dielectric will change the field **E**, and hence the field **D**. The flux of vector **E** through the surface S will also change, since negative bound charges will vanish from inside this surface. However, the flux of vector **D** through the surface S will remain the same in spite of the change in the field **D**.

Example 3. Let us consider a system containing no extraneous charges and having only bound charges. Such a system can, for example, be a sphere made of an electret (see p. 68). Figure 3.6a shows the field **E** of this system. What can we say about the field **D**?

First of all, the absence of extraneous charges means that the field **D** has no sources: the lines of **D** do not emerge or terminate anywhere. However, the field **D** exists and is shown in Fig. 3.6*b*. The lines of **E**

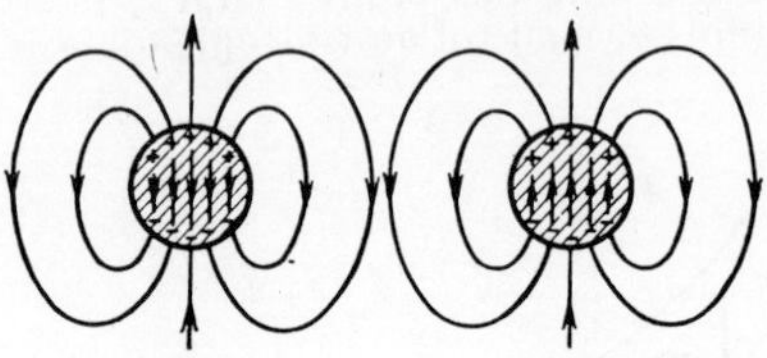

Lines of **E** (a) Lines of **D** (b)

Fig. 3.6

and **D** coincide outside the sphere, but inside the sphere they have opposite directions, since here the relation $\mathbf{D} = \varepsilon_0\varepsilon\mathbf{E}$ is no longer valid and $\mathbf{D} = \varepsilon_0\mathbf{E} + \mathbf{P}$.

3.5. Boundary Conditions

Let us first consider the behaviour of vectors **E** and **D** at the interface between two homogeneous isotropic dielectrics. Suppose that, for greater generality, an extraneous surface charge exists at the interface between these dielectrics. The required conditions can be easily obtained with the help of two theorems: the theorem on circulation of vector **E** and the Gauss theorem for vector **D**:

$$\oint \mathbf{E}\, d\mathbf{l} = 0 \quad \text{and} \quad \oint \mathbf{D}\, d\mathbf{S} = q_{\text{int}}.$$

Boundary Condition for Vector E. Let the field near the interface be $\mathbf{E}_1$ in dielectric *1* and $\mathbf{E}_2$ in dielectric *2*. We choose a small elongated rectangular contour and orient it as shown in Fig. 3.7. The sides of the contour parallel to the interface must have such a length that the field **E** over this length in each dielectric can be assumed constant. The "height" of the contour must be negligibly small. Then, in accordance with the theorem on circulation of vector **E**, we have

$$E_{2\tau}l + E_{1\tau'}l = 0,$$

where the projections of vector **E** are taken on the direction of the circumvention of the contour, shown by arrows in the figure. If in the lower region of the contour we take the

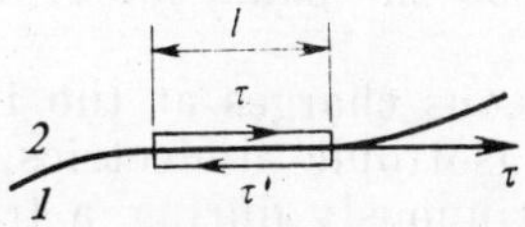

Fig. 3.7

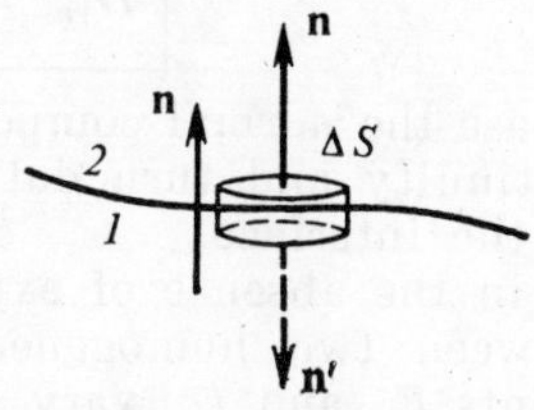

Fig. 3.8

projection of vector **E** not onto the unit vector $\boldsymbol{\tau}'$ but onto the common unit vector $\boldsymbol{\tau}$, then $E_{1\tau'} = -E_{1\tau}$, and it follows from the above equation that

$$\boxed{E_{1\tau} = E_{2\tau},} \tag{3.22}$$

i.e. the tangential component of vector **E** turns out to be the same on both sides of the interface (it does not have a discontinuity).

Boundary Condition for Vector D. Let us take a cylinder of a very small height and arrange it at the interface between two dielectrics (Fig. 3.8). The cross section of the cylinder must be such that vector **D** is the same within each of its endfaces. Then, in accordance with the Gauss theorem for vector **D**, we have

$$D_{2n} \cdot \Delta S + D_{1n'} \cdot \Delta S = \sigma \Delta S,$$

where σ is the surface density of the extraneous charge at the interface. Taking both projections of vector **D** onto the common normal **n** (which is directed from dielectric *1* to dielectric *2*), we obtain $D_{1n'} = -D_{1n}$, and the previous equation can be reduced to the form

$$\boxed{D_{2n} - D_{1n} = \sigma.} \tag{3.23}$$

It follows from this relation that the normal component of vector **D** generally has a discontinuity when passing through the interface. If, however, there are no extraneous charges at the interface ($\sigma = 0$), we obtain

$$\boxed{D_{1n} = D_{2n}.} \tag{3.24}$$

In this case the normal components of vector **D** do not have a discontinuity and turn out to be the same on different sides of the interface.

Thus, in the absence of extraneous charges at the interface between two homogeneous isotropic dielectrics, the components E_τ and D_n vary continuously during a transition through this interface, while the components E_n and D_τ have discontinuities.

Refraction of E and D Lines. The boundary conditions which we obtained for the components of vectors **E** and **D** at the interface between two dielectrics indicate (as will be shown later) that these vectors have a break at this interface, i.e. are refracted (Fig. 3.9). Let us find the relation between the angles α_1 and α_2.

In the absence of extraneous charges at the interface, we have, in accordance with (3.22) and (3.24), $E_{2\tau} = E_{1\tau}$ and $\varepsilon_2 E_{2n} = \varepsilon_1 E_{1n}$. Figure 3.9 shows that

$$\frac{\tan \alpha_2}{\tan \alpha_1} = \frac{E_{2\tau}/E_{2n}}{E_{1\tau}/E_{1n}}.$$

Taking into account the above conditions, we obtain the law of refraction of lines **E**, and hence of lines **D**:

$$\frac{\tan \alpha_2}{\tan \alpha_1} = \frac{\varepsilon_2}{\varepsilon_1}. \tag{3.25}$$

This means that lines of **D** and **E** will form a larger angle with the normal to the interface in the dielectric with a larger value of ε (in Fig. 3.9, $\varepsilon_2 > \varepsilon_1$).

Example. Let us represent graphically the fields **E** and **D** at the interface between two homogeneous dielectrics *1* and *2*, assuming that $\varepsilon_2 > \varepsilon_1$ and there is no extraneous charge on this surface.

Since $\varepsilon_2 > \varepsilon_1$, in accordance with (3.25) $\alpha_2 > \alpha_1$ (Fig. 3.10). Considering that the tangential component of vector **E** remains unchanged and using Fig. 3.9, we can easily show that $E_2 < E_1$ in magnitude, i.e. the lines of **E** in dielectric *1* must be denser than in dielectric *2*, as is shown in Fig. 3.10. The fact that the normal components of vectors **D** are equal leads to the conclusion that $D_2 > D_1$ in magnitude, i.e. the lines of **D** must be denser in dielectric *2*.

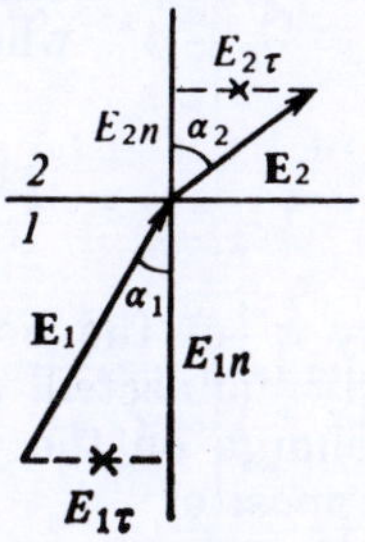

Fig 3.9

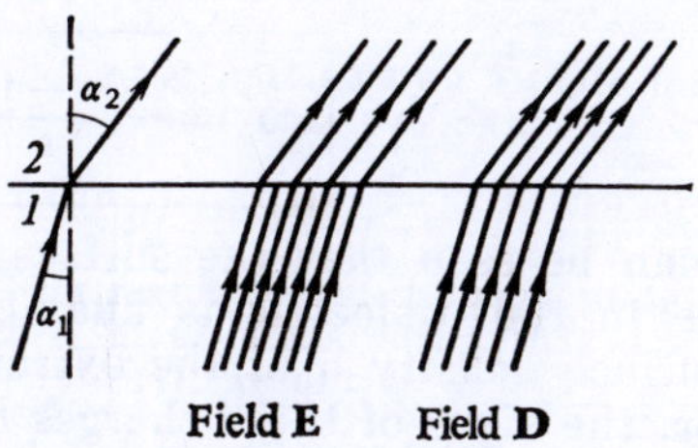

Fig. 3.10

We see that in the case under consideration, the lines of **E** are refracted and undergo discontinuities (due to the presence of bound charges), while the lines of **D** are only refracted, without discontinuities (since there are no extraneous charges at the interface).

Boundary Condition on the Conductor-Dielectric Interface. If medium *1* is a conductor and medium *2* is a dielectric (see Fig. 3.8), it follows from formula (3.23) that

$$\boxed{D_n = \sigma,} \tag{3.26}$$

where **n** is the conductor's *outward* normal (we omitted the subscript *2* since it is inessential in the given case). Let us verify formula (3.26). In equilibrium, the electric field inside a conductor is $\mathbf{E} = 0$, and hence the polarization $\mathbf{P} = 0$. This means, according to (3.17), that vector $\mathbf{D} = 0$ inside the conductor, i.e. in the notations of formula (3.23) $\mathbf{D}_1 = 0$ and $D_{1n} = 0$. Hence $D_{2n} = \sigma$.

Bound Charge at the Conductor Surface. If a homogeneous dielectric adjoins a charged region of the surface of a conductor, bound charges of a certain density σ' appear at the conductor-dielectric interface (recall that the volume density of bound charges $\rho' = 0$ for a homogeneous dielectric). Let us now apply the Gauss theorem to vector **E** in the same way as it was done while deriving formula (2.2). Considering that there are both bound and extraneous charges (σ

and σ') at the conductor-dielectric interface we arrive at the following expression: $E_n = (\sigma + \sigma')/\varepsilon_0$. On the other hand, according to (3.26) $E_n = D_n/\varepsilon\varepsilon_0 = \sigma/\varepsilon\varepsilon_0$. Combining these two equations, we obtain $\sigma/\varepsilon = \sigma + \sigma'$, whence

$$\boxed{\sigma' = -\frac{\varepsilon - 1}{\varepsilon}\sigma.} \qquad (3.27)$$

It can be seen that the surface density σ' of the bound charge in the dielectric is unambiguously connected with the surface density σ of the extraneous charge on the conductor, the signs of these charges being opposite.

3.6. Field in a Homogeneous Dielectric

It was noted in Sec. 2.1 that the determination of the resultant field **E** in a substance is associated with considerable difficulties, since the distribution of induced charges in the substance is not known beforehand. It is only clear that the distribution of these charges depends on the nature and shape of the substance as well as on the configuration of the external field $\mathbf{E}_0$.

Consequently, in the general case, while solving the problem about the resultant field **E** in a dielectric, we encounter serious difficulties: determination of the macroscopic field $\mathbf{E}'$ of bound charges in each specific case is generally a complicated independent problem, since unfortunately there is no universal formula for finding $\mathbf{E}'$.

An exception is the case when the entire space where there is a field $\mathbf{E}_0$ is filled by a homogeneous isotropic dielectric. Let us consider this case in greater detail. Suppose that we have a charged conductor (or several conductors) in a vacuum. Normally, extraneous charges are located on conductors. As we already know, in equilibrium the field **E** inside the conductor is zero, which corresponds to a certain unique distribution of the surface charge σ. Let the field created in the space surrounding the conductor be $\mathbf{E}_0$.

Let us now fill the entire space of the field with a homogeneous dielectric. As a result of polarization, only surface bound charges σ will appear in this dielectric at the interface with the conductor. According to (3.27) the charges

σ' are unambiguously connected with the extraneous charges σ on the surface of the conductor.

As before, there will be no field inside the conductor ($\mathbf{E} = 0$). This means that the distribution of surface charges (extraneous charges σ and bound charges σ') at the conductor-dielectric interface will be *similar* to the previous distribution of extraneous charges (σ), and the configuration of the resultant field $\mathbf{E}$ in the dielectric will remain the same as in the absence of the dielectric. Only the magnitude of the field at each point will be different.

In accordance with the Gauss theorem, $\sigma + \sigma' = \varepsilon_0 E_n$, where $E_n = D_n/\varepsilon\varepsilon_0 = \sigma/\varepsilon\varepsilon_0$, and hence

$$\sigma + \sigma' = \sigma/\varepsilon. \tag{3.28}$$

But if the charges creating the field have decreased by a factor of ε everywhere at the interface, the field $\mathbf{E}$ itself has become less than the field $\mathbf{E}_0$ by the same factor:

$$\mathbf{E} = \mathbf{E}_0/\varepsilon. \tag{3.29}$$

Multiplying both sides of this equation by $\varepsilon\varepsilon_0$, we obtain

$$\mathbf{D} = \mathbf{D}_0, \tag{3.30}$$

i.e. the field of vector $\mathbf{D}$ does not change in this case.

It turns out that formulas (3.29) and (3.30) are also valid in a more general case when a homogeneous dielectric fills the volume enclosed between the equipotential surfaces of the field $\mathbf{E}_0$ of extraneous charges (or of an external field). In this case also $\mathbf{E} = \mathbf{E}_0/\varepsilon$ and $\mathbf{D} = \mathbf{D}_0$ inside the dielectric.

In the cases indicated above, the intensity $\mathbf{E}$ of the field of bound charges is connected by a simple relation with the polarization $\mathbf{P}$ of the dielectric, namely,

$$\mathbf{E}' = -\mathbf{P}/\varepsilon_0. \tag{3.31}$$

This relation can be easily obtained from the formula $\mathbf{E} = \mathbf{E}_0 + \mathbf{E}'$ if we take into account that $\mathbf{E}_0 = \varepsilon\mathbf{E}$ and $\mathbf{P} = \varkappa\varepsilon_0\mathbf{E}$.

As was mentioned above, in other cases the situation is much more complicated, and formulas (3.29)-(3.31) are inapplicable.

Corollaries. Thus, if a homogeneous dielectric fills the entire space occupied by a field, the intensity $\mathbf{E}$ of the field will be lower than the intensity $\mathbf{E}_0$ of the field of the same extraneous charges, but in the absence of dielectric, by a factor of ε. Hence it follows that potential φ at all points will also decrease by a factor of ε:

$$\varphi = \varphi_0/\varepsilon, \tag{3.32}$$

where φ_0 is the field potential in the absence of the dielectric.

The samo applies to the potential difference:

$$U = U_0/\varepsilon, \tag{3.33}$$

where U_0 is the potential difference in a vacuum, in the absence of dielectric.

In the simplest case, when a homogeneous dielectric fills the entire space between the plates of a *capacitor*, the potential difference U between its plates will be by a factor of ε less than that in the absence of dielectric (naturally, at the same magnitude of the charge q on the plates). And since it is so, the capacitance $C = q/U$ of the capacitor filled by dielectric will increase ε times

$$C' = \varepsilon C, \tag{3.34}$$

where C is the capacitance of the capacitor in the absence of dielectric. It should be noted that this formula is valid when the *entire* space between the plates is filled and edge effects are ignored.

Problems

● **3.1. Polarization of a dielectric and the bound charge.** An extraneous point charge q is at the centre of a spherical layer of a heterogeneous isotropic dielectric whose dielectric constant varies only in the radial direction as $\varepsilon = \alpha/r$, where α is a constant and r is the distance from the centre of the system. Find the volume density ρ' of a bound charge as a function of r within the layer.

Solution. We shall use Eq. (3.6), taking a sphere of radius r as the closed surface, the centre of the sphere coinciding with the centre of the system. Then

$$4\pi r^2 \cdot P_r = -q'(r),$$

where $q'(r)$ is the bound charge inside the sphere. Let us take the differential of this expression:

$$4\pi\, d(r^2 \cdot P_r) = -\,dq'. \tag{1}$$

Here dq' is the bound charge in a thin layer between the spheres with radii r and $r + dr$. Considering that $dq' = \rho' 4\pi r^2\, dr$, we transform (1) as follows:

$$r^2\, dP_r + 2rP_r\, dr = -\rho' r^2\, dr,$$

whence

$$\rho' = -\left(\frac{dP_r}{dr} + \frac{2}{r} P_r\right). \qquad (2)$$

In the case under consideration we have

$$P_r = \varkappa\varepsilon_0 E_r = \frac{\varepsilon - 1}{\varepsilon} D_r = \frac{\varepsilon - 1}{\varepsilon} \frac{q}{4\pi r^2}$$

and after certain transformation expression (2) will have the form

$$\rho' = \frac{1}{4\pi\alpha} \frac{q}{r^2}$$

which is just the required result.

● **3.2. The Gauss theorem for vector D.** An infinitely large plate made of a homogeneous dielectric with the dielectric constant ε is uniformly charged by an extraneous charge with volume density $\rho > 0$. The thickness of the plate is $2a$. (1) Find the magnitude of vector **E** and the potential φ as functions of the distance l from the middle of the plate (assume that the potential is zero in the middle of the plate), choosing the X-axis perpendicular to the plate. Plot schematic curves for the projection $E_x(x)$ of vector **E** and the potential $\varphi(x)$. (2) Find the surface and volume densities of the bound charge.

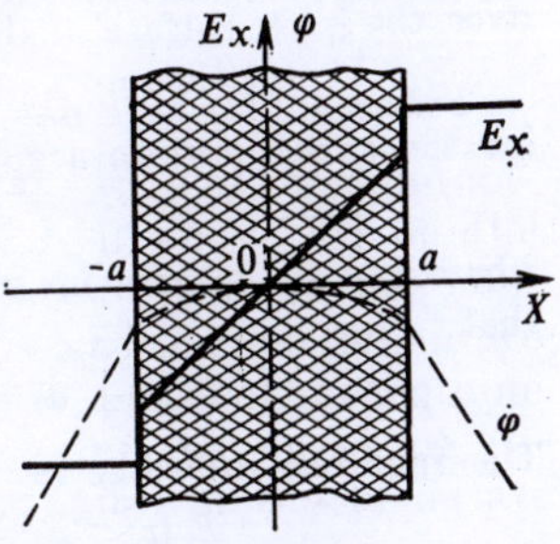

Fig. 3.11

Solution. (1) From symmetry considerations it is clear that $E = 0$ in the middle of the plate, while at all other points vectors **E** are perpendicular to the surface of the plate. In order to determine **E**, we shall use the Gauss theorem for vector **D** (since we know the distribution of only extraneous charges). We take for the closed surface a right cylinder of height l, one of whose endfaces coincides with the midplane ($x = 0$). Let the cross-sectional area of this cylinder be S. Then

$$DS = \rho S l, \quad D = \rho l, \quad E = \rho l/\varepsilon\varepsilon_0 \ (l \leqslant a),$$

$$DS = \rho S a, \quad D = \rho a, \quad E = \rho a/\varepsilon_0 \ (l \geqslant a).$$

The graphs of the funtions $E_x(x)$ and $\varphi(x)$ are shown in Fig. 3.11. It is useful to verify that the graph of $E_x(x)$ corresponds to the derivative $-\partial\varphi/\partial x$.

(2) In accordance with (3.13), the surface density of the bound charge is

$$\sigma' = P_n = \varkappa\varepsilon_0 E_n = (\varepsilon - 1)\,\rho a/\varepsilon = \frac{\varepsilon-1}{\varepsilon}\,\rho a > 0.$$

This result is valid for both sides of the plate. Thus, if the extraneous charge $\rho > 0$, the bound charges appearing on both surfaces of the plate are also positive.

In order to find the volume density of the bound charge, we use Eq. (3.9) which in our case will have a simpler form:

$$\rho' = -\frac{\partial P_x}{\partial x} = -\frac{\partial}{\partial x}\left(\frac{\varepsilon-1}{\varepsilon}\,\rho x\right) = -\frac{\varepsilon-1}{\varepsilon}\,\rho.$$

It can be seen that the bound charge is uniformly distributed over the bulk and has the sign opposite to that of the extraneous charge.

● **3.3.** A homogeneous dielectric has the shape of a spherical layer whose inner and outer radii are a and b. Plot schematically the curves of intensity E and potential φ of the electric field as functions of the distance r from the centre of the system, if the dielectric is charged by a positive extraneous charge distributed uniformly (1) over the inner surface of the layer, (2) over the layer's bulk.

Solution. (1) We use the Gauss theorem for vector **D**, taking for the closed surface a sphere of radius r:

$$4\pi r^2 D = q,$$

where q is the extraneous charge within this sphere. Hence it follows that

$$D\,(r < a) = 0, \quad D\,(r > a) = q/4\pi r^2.$$

The required intensity is

$$E\,(r < a) = 0, \quad E\,(r > a) = D/\varepsilon\varepsilon_0.$$

The curve for $E(r)$ is shown in Fig. 3.12*a*. The curve for $\varphi(r)$ is also shown in this figure. The curve $\varphi(r)$ must have such a shape that the derivative $\partial\varphi/\partial r$ taken with the opposite sign corresponds to the curve of the function $E(r)$. Besides, we must take into account the normalization condition: $\varphi \to 0$ as $r \to \infty$.

It should be noted that the curve corresponding to the function $\varphi(r)$ is *continuous*. At the points where the function $E(r)$ has finite discontinuities, the function $\varphi(r)$ is only broken.

(2) In this case, in accordance with the Gauss theorem, we have

$$4\pi r^2 D = \frac{4}{3}\pi\,(r^3 - a^3)\,\rho,$$

where ρ is the volume density of the extraneous charge. Hence

$$E=\frac{D}{\varepsilon\varepsilon_0}=\frac{\rho}{3\varepsilon_0\varepsilon}\,\frac{r^3-a^3}{r^2}.$$

The corresponding curves for $E(r)$ and $\varphi(r)$ are shown in Fig. 3.12*b*.

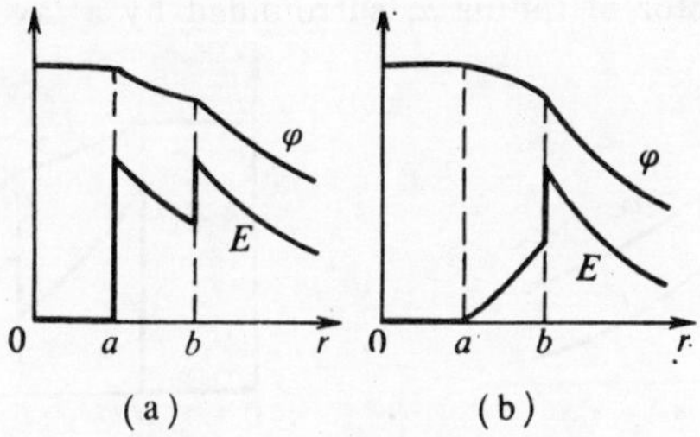

Fig. 3.12

● **3.4.** Extraneous charge is uniformly distributed with the volume density $\rho>0$ over a sphere of radius a, made of a homogeneous dielectric with the permittivity ε. Find (1) the magnitude of vector **E** as a function of the distance r from the centre of the sphere and plot the curves of the functions $E(r)$ and $\varphi(r)$; (2) the surface and volume densities of bound charge.

Solution. (1) In order to determine E, we shall use the Gauss theorem for vector **D** since we know the distribution of only extraneous charge:

$$r\leqslant a,\quad 4\pi r^2 D=\frac{4}{3}\pi r^3\rho,\quad D=\frac{\rho}{3}r,\quad E=\frac{D}{\varepsilon\varepsilon_0}=\frac{\rho}{3\varepsilon\varepsilon_0}r,$$

$$r\geqslant a,\quad 4\pi r^2 D=\frac{4}{3}\pi a^3\rho,\quad D=\frac{\rho a^3}{3r^2},\quad E=\frac{D}{\varepsilon_0}=\frac{\rho a^3}{3\varepsilon_0}\,\frac{1}{r^2}.$$

The curves for the functions $E(r)$ and $\varphi(r)$ are shown in Fig. 3.13.

(2) The surface density of the bound charge is

$$\sigma'=P_n=\frac{\varepsilon-1}{\varepsilon}\,\frac{\rho a}{3}.$$

In order to find the volume density of bound charge, it is sufficient to repeat the reasoning that led us to formula (3.11), and we get

$$\rho'=-\frac{\varepsilon-1}{\varepsilon}\rho. \tag{1}$$

This result can be obtained in a different way, viz. by using Eq. (3.9). Since $\mathbf{P}=\varkappa\varepsilon_0\mathbf{E}$ and $\varkappa$ does not depend on coordinates (inside the sphere), we obtain

$$\rho' = -\nabla \cdot \mathbf{P} = -\varkappa\varepsilon_0 \nabla \cdot \mathbf{E},$$

where $\varepsilon_0 \nabla \cdot \mathbf{E} = \rho + \rho'$. Hence $\rho' = -\varkappa(\rho + \rho')$, which gives formula (1).

● **3.5. Capacitance of a conductor.** Find the capacitance of a spherical conductor of radius a, surrounded by a layer of a homoge-

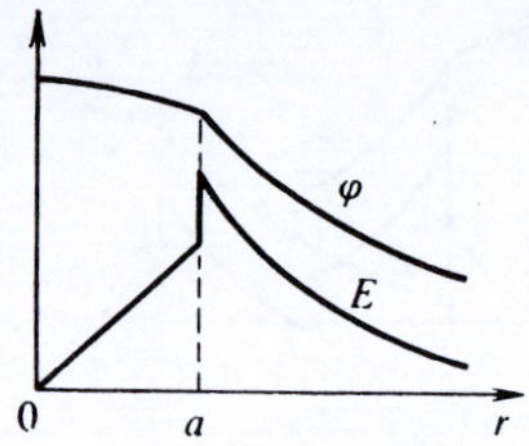

Fig. 3.13

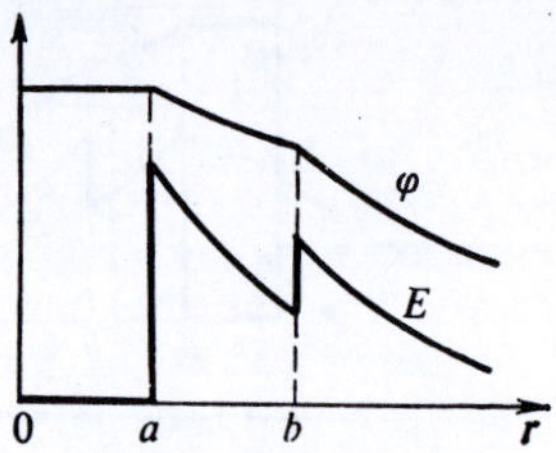

Fig. 3.14

neous dielectric and having the outer radius b and the dielectric constant ε. Plot approximate curves for $E(r)$ and $\varphi(r)$, where r is the distance from the centre of the sphere, if the sphere is charged positively.

Solution. By definition, the capacitance $C = q/\varphi$. Let us find the potential φ of the conductor, supplying mentally a charge q to it:

$$\varphi = \int_a^\infty E_r\,dr = \frac{1}{4\pi\varepsilon_0}\int_a^b \frac{q}{\varepsilon r^2}\,dr + \frac{1}{4\pi\varepsilon_0}\int_b^\infty \frac{q}{r^2}\,dr.$$

Integrating this expression, we obtain

$$\varphi = \frac{q}{4\pi\varepsilon_0\varepsilon}\left(\frac{1}{a} + \frac{\varepsilon - 1}{b}\right), \quad C = \frac{4\pi\varepsilon_0\varepsilon a}{1 + (\varepsilon - 1)\,a/b}.$$

The curves for $E(r)$ and $\varphi(r)$ are shown in Fig. 3.14.

● **3.6. Capacitance of a capacitor.** A spherical capacitor with the radii of the plates a and b, where $a < b$, is filled with an isotropic heterogeneous dielectric whose permittivity depends on the distance r from the centre of the system as $\varepsilon = \alpha/r$, where α is a constant. Find the capacitance of the capacitor.

Solution. In accordance with the definition of the capacitance of a capacitor ($C = q/U$), the problem is reduced to determination of the potential difference U for a given charge q:

$$U = \int_a^b E\,dr, \tag{1}$$

where we assume that the charge of the inner plate is $q > 0$. Let us find E with the help of the Gauss theorem for vector $\mathbf{D}$:

$$4\pi r^2 D = q, \qquad E = \frac{D}{\varepsilon\varepsilon_0} = \frac{1}{4\pi\varepsilon_0} \frac{q}{\varepsilon r^2} = \frac{1}{4\pi\varepsilon_0} \frac{q}{\alpha r}.$$

Substituting the latter expression into (1) and integrating, we find

$$U = \frac{q}{4\pi\varepsilon_0\alpha} \ln \frac{b}{a}, \qquad C = \frac{4\pi\varepsilon_0\alpha}{\ln(b/a)}.$$

● **3.7. The Gauss theorem and the principle of superposition.** Suppose that we have a dielectric sphere which retains polarization

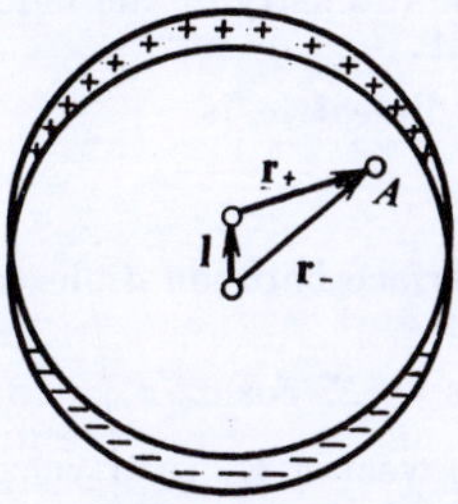

Fig. 3.15

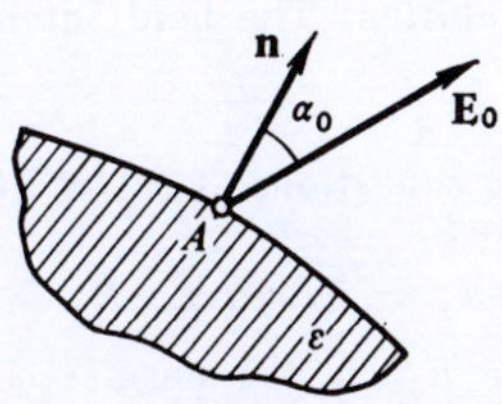

Fig. 3.16

after an external electric field is switched off. If the sphere is polarized uniformly, the field intensity inside it is $\mathbf{E}' = -\mathbf{P}/3\varepsilon_0$, where $\mathbf{P}$ is the polarization. (1) Derive this formula assuming that the sphere is polarized as a result of a small displacement of all positive charges of the dielectric with respect to all its negative charges. (2) Using this formula, find the intensity $\mathbf{E}_0$ of the field in the spherical cavity inside an infinite homogeneous dielectric with the permittivity ε if the field intensity in the dielectric away from the cavity is $\mathbf{E}$.

Solution. (1) Let us represent this sphere as a combination of two spheres of the same radii, bearing uniformly distributed charges with volume densities ρ and $-\rho$. Suppose that as a result of a small shift, the centres of the spheres are displaced relative to one another by a distance $\mathbf{l}$ (Fig. 3.15). Then at an arbitrary point A inside the sphere we have

$$\mathbf{E}' = \mathbf{E}'_+ + \mathbf{E}'_- = \frac{\rho}{3\varepsilon_0}(\mathbf{r}_+ - \mathbf{r}_-) = -\frac{\rho\mathbf{l}}{3\varepsilon_0},$$

where we used the fact that field intensity inside a uniformly charged sphere is $\mathbf{E} = \rho\mathbf{r}/3\varepsilon_0$, which directly follows from the Gauss theorem. It remains for us to take into account that in accordance with (3.4), $\rho\mathbf{l} = \mathbf{P}$.

(2) The creation of a spherical cavity in a dielectric is equivalent to the removal from a sphere of a ball made of a polarized material.

Consequently, in accordance with the principle of superposition, the field $\mathbf{E}$ inside the dielectric can be represented as the sum $\mathbf{E} = \mathbf{E}' + \mathbf{E}_0$. Hence

$$\mathbf{E}_0 = \mathbf{E} - \mathbf{E}' = \mathbf{E} + \mathbf{P}/3\varepsilon_0.$$

Considering that $\mathbf{P} = (\varepsilon - 1)\,\varepsilon_0\mathbf{E}$, we obtain

$$\mathbf{E}_0 = (2 + \varepsilon)\,\mathbf{E}/3.$$

● **3.8. Boundary conditions.** In the vicinity of point A (Fig. 3.16) belonging to a dielectric-vacuum interface, the electric field intensity in vacuum is equal to E_0, and the vector $\mathbf{E}_0$ forms the angle α_0 with the normal to the interface at the given point. The dielectric permittivity is ε. Find the ratio E/E_0, where E is the intensity of the field inside the dielectric in the vicinity of point A.

Solution. The field intensity inside the dielectric is

$$E = \sqrt{E_\tau^2 + E_n^2}\,. \tag{1}$$

Using conditions (3.22) and (3.24) at the interface between dielectrics, we find

$$E_\tau = E_0 \sin\alpha_0,\quad E_n = D_n/\varepsilon\varepsilon_0 = E_{0n}/\varepsilon = E_0 \cos\alpha_0/\varepsilon,$$

where E_{0n} is the normal component of the vector $\mathbf{E}_0$ in a vacuum. Substituting these expressions into (1), we obtain

$$\frac{E}{E_0} = \sqrt{\sin^2\alpha_0 + \frac{\cos^2\alpha_0}{\varepsilon^2}} < 1,$$

i.e. $E < E_0$.

● **3.9.** A point charge q is in a vacuum at a distance l from the plane surface of a homogeneous dielectric filling the half-space below the plane. The dielectric's permittivity is ε. Find (1) the surface density of the bound charge as a function of the distance r from the point charge q and analyse the obtained result; (2) the total bound charge on the surface of the dielectric.

Solution. Let us use the continuity of the normal component of vector $\mathbf{D}$ at the dielectric-vacuum interface (Fig. 3.17):

$$D_{2n} = D_{1n},\quad E_{2n} = \varepsilon E_{1n}$$

or

$$-\frac{1}{4\pi\varepsilon_0}\frac{q}{r^2}\cos\vartheta + \frac{\sigma'}{2\varepsilon_0} = \varepsilon\left(-\frac{1}{4\pi\varepsilon_0}\frac{q}{r^2}\cos\vartheta - \frac{\sigma'}{2\varepsilon_0}\right),$$

where the term $\sigma'/2\varepsilon_0$ is the component of the electric field created near the region of the plane under consideration, where the surface charge density is σ'. From the last equality it follows that

$$\sigma' = -\frac{\varepsilon - 1}{\varepsilon + 1}\frac{ql}{2\pi r^3}\,. \tag{1}$$

Here we took into account that $\cos\vartheta = l/r$. As $l \to 0$ the quantity $\sigma' \to 0$, i.e. if the charge q is just at the interface, there is no surface charge on the plane.

(2) Let us imagine at the interface a thin ring with the centre at the point O (Fig. 3.17). Suppose that the inner and outer radii of this ring are r' and $r' + dr'$. The surface bound charge within this ring is $dq' = \sigma' \cdot 2\pi r'\,dr'$. It can be seen from the figure that $r^2 = l^2 + r'^2$, whence $r\,dr = r'\,dr'$, and the expression for dq' combined with (1) gives

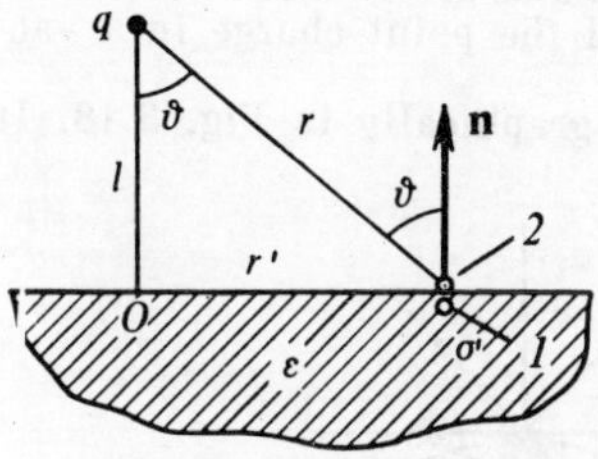

Fig. 3.17

$$dq' = -\frac{\varepsilon-1}{\varepsilon+1}\, ql\,\frac{dr}{r^2}.$$

Integrating this equation over r between l and ∞, we obtain

$$q' = -\frac{\varepsilon-1}{\varepsilon+1}\, q.$$

● **3.10.** A point charge q is on the plane separating a vacuum from an infinite homogeneous dielectric with the dielectric permittivity ε. Find the magnitudes of vectors **D** and **E** in the entire space.

Solution. In this case, it follows from the continuity of the normal component of vector **D** that $E_{2n} = \varepsilon E_{1n}$. Only the surface charge σ' will contribute to the normal component of vector **E** in the vicinity of the point under consideration. Hence the above equality can be written in the form

$$\sigma'/2\varepsilon_0 = \varepsilon\,(-\sigma'/2\varepsilon_0).$$

We immediately find that $\sigma' = 0$.

Thus, in the given case there is no bound surface charge (with the exception of the points in direct contact with the extraneous point charge q). This means that the electric field in the surrounding space is the field of the point charge $q + q'$, and E depends only on the distance r from this charge. But the charge q' is unknown, and hence we must use the Gauss theorem for vector **D**. Taking for the closed surface a sphere of radius r with the centre at the point of location of the charge q, we can write

$$2\pi r^2 D_0 + 2\pi r^2 D = q,$$

where D_0 and D are the magnitudes of vector **D** at the distance r from the charge q in the vacuum and in the dielectric respectively.

Besides, from the continuity of the tangential component of vector **E** it follows that

$$D = \varepsilon D_0.$$

Combining these two conditions, we find

$$D_0 = \frac{q}{2\pi(1+\varepsilon)\,r^2}, \quad D = \frac{\varepsilon q}{2\pi(1+\varepsilon)\,r^2},$$

and the electric field intensity in the entire space is

$$E=\frac{D_0}{\varepsilon_0}=\frac{q}{2\pi(1+\varepsilon)\varepsilon_0 r^2}.$$

It can be seen that for $\varepsilon = 1$ these formulas are reduced to the already familiar expressions for D and E of the point charge in a vacuum.

The obtained results are represented graphically in Fig. 3.18. It

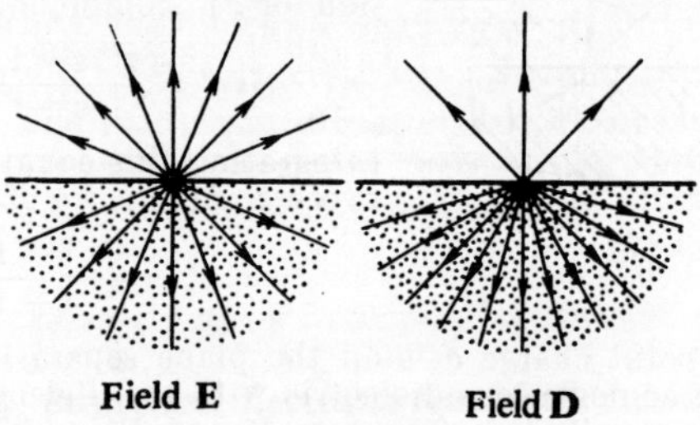

Fig. 3.18

should be noted that the field **D** in this case is not determined by the extraneous charge alone (otherwise it would have the form of the field of a point charge).

4. Energy of Electric Field

4.1. Electric Energy of a System of Charges

Energy Approach to Interaction. The energy approach to interaction between electric charges is, as will be shown, rather fruitful in respect of its applications. Besides, this approach makes it possible to consider the electric field from a different point of view.

First of all, let us find out how we can arrive at the concept of the energy of interaction in a system of charges.

1. Let us first consider a system of *two* point charges *1* and *2*. We shall find the algebraic sum of the elementary works of the forces $\mathbf{F}_1$ and $\mathbf{F}_2$ of interaction between the charges. Suppose that in a certain system of reference K the charges were displaced by $d\mathbf{l}_1$ and $d\mathbf{l}_2$ during the time

dt. Then the corresponding work of these forces is

$$\delta A_{1,2} = \mathbf{F}_1 \cdot d\mathbf{l}_1 + \mathbf{F}_2 \cdot d\mathbf{l}_2.$$

Considering that $\mathbf{F}_2 = -\mathbf{F}_1$ (according to the Newton third law), we can write this expression in the form

$$\delta A_{1,2} = \mathbf{F}_1 (d\mathbf{l}_1 - d\mathbf{l}_2).$$

The quantity in the parentheses is the displacement of charge *1* relative to charge *2*. In other words, it is the displacement of charge *1* in the system of reference K', which is rigidly fixed to charge *2* and accomplishes with it a translational motion relative to the initial reference system K. Indeed, the displacement $d\mathbf{l}_1$ of charge *1* in system K can be represented as the displacement $d\mathbf{l}_2$ of system K' plus the displacement $d\mathbf{l}'_1$ of charge *1* relative to system K': $d\mathbf{l}_1 = d\mathbf{l}_2 + d\mathbf{l}'_1$. Hence, $d\mathbf{l}_1 - d\mathbf{l}_2 = d\mathbf{l}'_1$, and

$$\delta A_{1,2} = \mathbf{F}_1 \cdot d\mathbf{l}'_1.$$

Thus, it turns out that the sum of the elementary works done by two charges in an arbitrary system of reference K is always equal to the elementary work done by the force acting on one charge in another reference system (K') in which the other charge is at rest. In other words, the work $\delta A_{1,2}$ does not depend on the choice of the initial system of reference.

The force $\mathbf{F}_1$ acting on charge *1* from charge *2* is conservative (as a central force). Consequently, the work of the given force in the displacement $d\mathbf{l}'_1$ can be represented as the decrease in the potential energy of interaction between the pair of charges under consideration:

$$\delta A_{1,2} = -dW_{12},$$

where W_{12} is the quantity depending only on the distance between these charges.

2. Let us now go over to a system of *three* point charges (the result obtained for this case can be easily generalized for a system of an arbitrary number of charges). The work performed by all forces of interaction during elementary displacements of all the charges can be represented as the sum of the works of three pairs of interactions, i.e. $\delta A = \delta A_{1,2} + \delta A_{1,3} + \delta A_{2,3}$. But as it has just been shown,

for each pair of interactions $\delta A_{i,k} = -dW_{ik}$, and hence

$$\delta A = -d(W_{12} + W_{13} + W_{23}) = -dW,$$

where W is the *energy of interaction* for the given system of charges:

$$W = W_{12} + W_{13} + W_{23}.$$

Each term of this sum depends on the distance between corresponding charges, and hence the energy W of the given system of charges is a function of its configuration.

Similar arguments are obviously valid for a system of any number of charges. Consequently, we can state that to each configuration of an arbitrary system of charges, there corresponds a certain value of energy W, and the work of all the forces of interaction upon a change in this configuration is equal to the decrease in the energy W

$$\delta A = -dW. \tag{4.1}$$

Energy of Interaction. Let us find the expression for the energy W. We again first consider a system of three point charges, for which we have found that $W = W_{12} + W_{13} + W_{23}$. This sum can be transformed as follows. We represent each term W_{ik} in a symmetric form: $W_{ik} = \frac{1}{2}(W_{ik} + W_{ki})$, since $W_{ik} = W_{ki}$. Then

$$W = \frac{1}{2}(W_{12} + W_{21} + W_{13} + W_{31} + W_{23} + W_{32}).$$

Let us group the terms with similar first indices:

$$W = \frac{1}{2}[(W_{12} + W_{13}) + (W_{21} + W_{23}) + (W_{31} + W_{32})].$$

Each sum in the parentheses is the energy W_i of interaction between the ith charge with all the remaining charges. Hence the latter expression can be written in the form

$$W = \frac{1}{2}(W_1 + W_2 + W_3) = \frac{1}{2}\sum_{i=1}^{3} W_i.$$

This expression can be generalized to a system of an arbitrary number of charges, since the above line of reasoning obviously does not depend on the number of charges constituting the system. Thus, the energy of interaction for a system of point charges is

$$W=\frac{1}{2}\sum W_i. \tag{4.2}$$

Considering that $W_i = q_i\varphi_i$, where q_i is the ith charge of the system and φ_i is the potential created at the point of location of the ith charge by *all the remaining* charges, we obtain the final expression for the energy of interaction of the system of point charges:

$$\boxed{W=\frac{1}{2}\sum q_i\varphi_i.} \tag{4.3}$$

Example. Four similar point charges q are located at the vertices of a tetrahedron with an edge a (Fig. 4.1). Find the energy of interaction of charges in this system.

The energy of interaction for each pair of charges of this system is the same and equal to $W_1 = q^2/4\pi\varepsilon_0 a$. It can be seen from the figure that the total number of interacting pairs is six, and hence the energy of interaction of all point charges of the system is

$$W = 6W_1 = 6q^2/4\pi\varepsilon_0 a.$$

Another approach to the solution of this problem is based on formula (4.3). The potential φ at the point of location of one of the charges, created by the field of all the other charges, is $\varphi = 3q/4\pi\varepsilon_0 a$. Hence

$$W=\frac{1}{2}\sum_{i=1}^{4} q_i\varphi_i=\frac{1}{2}4q\varphi=\frac{1}{4\pi\varepsilon_0}\frac{6q^2}{a}.$$

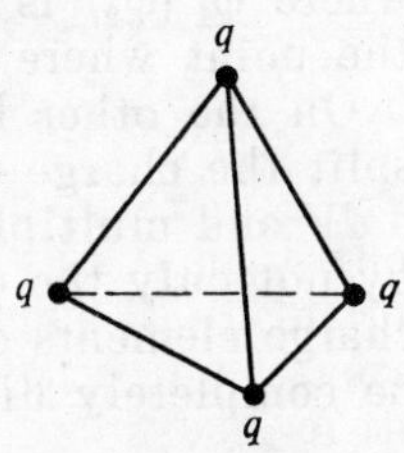

Fig. 4.1

Total Energy of Interaction. If the charges are arranged continuously, then, representing the system of charges as a combination of elementary charges $dq = \rho\, dV$ and going over in (4.3) from summation to integration, we obtain

$$\boxed{W=\frac{1}{2}\int \rho\varphi\, dV,} \tag{4.4}$$

where φ is the potential created by all the charges of the system in the volume element dV. A similar expression can be written, for example, for a surface distribution of charges. For this purpose, we must replace in (4.4) ρ by σ and dV by dS.

Expression (4.4) can be erroneously interpreted (and this often leads to misunderstanding) as a modification of expression (4.3), corresponding to the replacement of the concept of point charges by that of a continuously distributed charge. Actually, this is not so since the two expressions differ *essentially*. The origin of this difference is in different meanings of the potential φ appearing in these expressions. Let us explain this difference with the help of the following example.

Suppose a system consists of two small balls having charges q_1 and q_2. The distance between the balls is considerably larger than their dimensions, hence q_1 and q_2 can be assumed to be point charges. Let us find the energy W of the given system by using both formulas.

According to (4.3), we have

$$W = \frac{1}{2}(q_1\varphi_1 + q_2\varphi_2) = q_1\varphi_1 = q_2\varphi_2,$$

where φ_1 (φ_2) is the potential created by charge q_2 (q_1) at the point where charge q_1 (q_2) is located.

On the other hand, according to formula (4.4) we must split the charge of each ball into infinitely small elements $\rho\, dV$ and multiply each of them by the potential φ created by not only the charge elements of another ball but by the charge elements of *this* ball as well. Clearly, the result will be completely different:

$$W = W_1 + W_2 + W_{12}, \tag{4.5}$$

where W_1 is the energy of interaction of the charge elements of the first ball with each other, W_2 the same but for the second ball, and W_{12} the energy of interaction between the charge elements of the first ball and the charge elements of the second ball. The energies W_1 and W_2 are called the *intrinsic* energies of charges q_1 and q_2, while W_{12} is the energy of *interaction* between charge q_1 and charge q_2.

Thus, we see that the energy W calculated by formula (4.3) corresponds only to the energy W_{12}, while calculation by formula (4.4) gives the *total energy of interaction*: in addition to W_{12}, it gives intrinsic energies W_1 and W_2. Disregard of this circumstance is frequently a cause of gross errors.

We shall return to this question in Sec. 4.4. Now, we shall use formula (4.4) for obtaining several important results.

4.2. Energies of a Charged Conductor and a Charged Capacitor

Energy of an Isolated Conductor. Let a conductor have a charge q and a potential φ. Since the value of φ is the same at all points where charge is located, we can take φ in formula (4.4) out of the integral. The remaining integral is just the charge q on the conductor, and we obtain

$$\boxed{W = \frac{q\varphi}{2} = \frac{C\varphi^2}{2} = \frac{q^2}{2C}.} \tag{4.6}$$

These three expressions are written assuming that $C = q/\varphi$.

Energy of a Capacitor. Let q and φ_+ be the charge and potential of the positively charged plate of a capacitor. According to (4.4), the integral can be split into two parts (for the two plates). Then

$$W = \frac{1}{2}(q_+\varphi_+ + q_-\varphi_-).$$

Since $q_- = -q_+$, we have

$$W = \frac{1}{2}q_+(\varphi_+ - \varphi_-) = \frac{1}{2}qU,$$

where $q = q_+$ is the charge of the capacitor and U is the potential difference across its plates. Considering that $C = q/U$, we obtain the following expression for the energy of the capacitor:

$$\boxed{W = \frac{qU}{2} = \frac{CU^2}{2} = \frac{q^2}{2C}.} \tag{4.7}$$

It should be noted here that these formulas determine the *total* energy of interaction, viz. not only the energy of interaction between the charges of one plate and those of the other plate, but also the energy of interaction of charges within each plate.

And What if We Have a Dielectric? We shall show that *formulas* (4.6) *and* (4.7) *are valid in the presence of a dielectric.* For this purpose, we consider the process of charging a capacitor as a transport of small portions of charge (dq') from one plate to the other.

The elementary work performed against the forces of the field in this case is

$$\delta A = U' \, dq' = (q'/C) \, dq',$$

where U' is the potential difference between the plates at the moment when the next portion of charge dq' is being transferred.

Integrating this expression over q' between 0 and q, we obtain

$$A = q^2/2C,$$

which coincides with the expression for the total energy of a capacitor. Consequently, the work done against the forces of the electric field is completely spent for accumulating the energy W of the charged capacitor. Moreover, the expression obtained for the work A is also valid in the case when there is a dielectric between the plates of a capacitor. Thus, we have proved the validity of (4.7) in the presence of a dielectric.

Obviously, all this applies to (4.6) as well.

4.3. Energy of Electric Field

On Localization of Energy. Formula (4.4) defines electric energy W of any system in terms of charge and potential. It turns out, however, that energy W can also be expressed through another quantity characterizing the field itself, viz. through field intensity **E**. Let us at first demonstrate this by using the simple example of a parallel-plate capacitor, ignoring field distortions near the edges of the plates (edge effect). Substituting into the formula $W = CU^2/2$

the expression $C = \varepsilon\varepsilon_0 S/h$, we obtain

$$W = \frac{CU^2}{2} = \frac{\varepsilon\varepsilon_0 SU^2}{2} = \frac{\varepsilon\varepsilon_0}{2}\left(\frac{U}{h}\right)^2 Sh.$$

And since $U/h = E$ and $Sh = V$ (the volume between the capacitor plates), we get

$$W = \frac{1}{2}\varepsilon_0\varepsilon E^2 V. \tag{4.8}$$

This formula is valid for the uniform field which fills the volume V.

In the general theory, it is proved that the energy W can be expressed in terms of **E** (in the case of an *isotropic* dielectric) through the formula

$$\boxed{W = \int \frac{\varepsilon_0\varepsilon E^2}{2}\,dV = \int \frac{\mathbf{E}\cdot\mathbf{D}}{2}\,dV.} \tag{4.9}$$

The integrand in this equation has the meaning of the energy contained in the volume dV. This leads us to a very important and fruitful physical idea about *localization of energy in the field.* This assumption was confirmed in experiments with fields varying in time. It is the domain where we encounter phenomena that can be explained with the help of the notion of energy localization in the field. These varying fields may exist independently of electric charges which have generated them and may propagate in space in the form of electromagnetic waves. Experiments show that electromagnetic waves carry energy. This circumstance confirms the idea that the field itself is a carrier of energy.

The last two formulas show that the electric energy is distributed in space with the volume density

$$\boxed{w = \frac{\varepsilon_0\varepsilon E^2}{2} = \frac{\mathbf{E}\cdot\mathbf{D}}{2}.} \tag{4.10}$$

It should be noted that this formula is valid only in the case of *isotropic* dielectrics for which the relation $\mathbf{P} = \varkappa\varepsilon_0\mathbf{E}$ holds.

For anisotropic dielectrics the situation is more complicated.

Another substantiation for formula (4.9). It is known that the energy of an isolated charged conductor is $W = q\varphi/2$. Let us show that this formula is correct, proceeding from the idea of localization of energy in the field.

Let us consider an arbitrary positively charged conductor. We mentally isolate a tube of infinitesimal cross section, which is bounded by lines of **E** (Fig. 4.2), and take in it an elementary volume $dV = dS\,dl$. This volume contains the energy

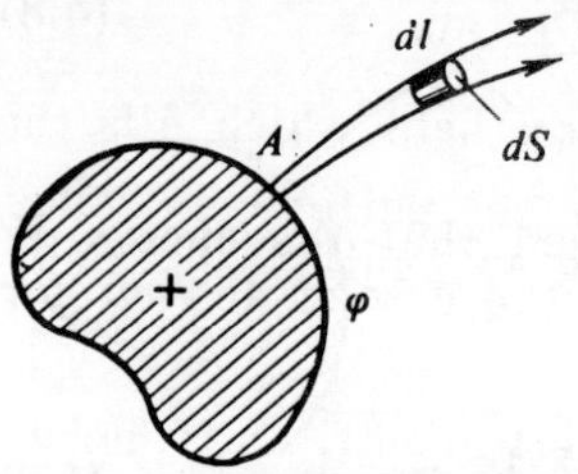

Fig. 4.2

$$\frac{ED}{2}\,dS\,dl = \frac{D\,dS}{2}\,E\,dl.$$

Let us now find the energy localized in the entire isolated tube. For this purpose, we integrate the last expression, considering that the product $D\,dS$ is the same in all cross sections of the tube and hence it can be taken out of the integral:

$$dW = \frac{D\,dS}{2}\int_A^\infty E\,dl = \frac{D\,dS}{2}\,\varphi,$$

where A is a point at the beginning of the imaginary tube.

It remains for us to make the last step, i.e. integrate the obtained expression over all the tubes, and find the energy localized in the entire field. Considering that potential φ is the same at the endfaces of all the tubes (since they originate on the surface of the conductor), we write

$$W = \frac{\varphi}{2}\oint D\,dS,$$

where the integration is performed over a closed surface coinciding with one of the equipotential surfaces. In accordance with the Gauss theorem, this integral is equal to the charge q of the conductor, and we finally get

$$W = q\varphi/2,$$

Q.E.D.

Let us consider two examples illustrating the advantages we get by using the idea of energy localization in the field.

Example 1. A point charge q is at the centre of a spherical layer made of a homogeneous dielectric with the dielectric constant ε. The inner and outer radii of the layer are equal to a and b respectively. Find the electric energy contained in this dielectric layer.

We mentally isolate in the dielectric a very thin concentric spherical layer of radius from r to $r + dr$. The energy localized in this layer is given by

$$dW = \frac{\varepsilon_0 \varepsilon E^2}{2} 4\pi r^2 dr,$$

where $E = q^2/4\pi\varepsilon_0\varepsilon r^2$. Integrating this expression over r between a and b, we obtain

$$W = \frac{q^2}{8\pi\varepsilon_0\varepsilon}\left(\frac{1}{a} - \frac{1}{b}\right).$$

Example 2. Find the work that must be done against the electric forces in order to remove a dielectric plate from a parallel-plate charged capacitor. It is assumed that the charge q of the capacitor remains unchanged and that the dielectric fills the entire space between the capacitor plates. The capacitance of the capacitor in the absence of the dielectric is C.

The work against the electric forces in this system is equal to the increment of the electric energy of the system:

$$A = \Delta W = W_2 - W_1,$$

where W_1 is the energy of the field between the capacitor plates in the presence of the dielectric and W_2 is the same quantity in the absence of the dielectric. Bearing in mind that the magnitude of vector $\mathbf{D}$ will not change as a result of the removal of the plate, i.e. that $D_2 = D_1 = \sigma$, we can write

$$A = W_2 - W_1 = \left(\frac{D^2}{2\varepsilon_0} - \frac{D^2}{2\varepsilon_0\varepsilon}\right) V = \frac{q^2}{2C}\left(1 - \frac{1}{\varepsilon}\right),$$

where $V = Sh$ and $C = \varepsilon_0 S/h$, S being the area of each plate and h the distance between them.

The Work of the Field During Polarization of a Dielectric. An analysis of formula (4.10) for the volume energy density reveals that for the same value of E, the quantity w is ε times greater in the presence of a dielectric than when it is absent. At first glance this may seem strange: field intensity in both cases is maintained the same. As a matter of fact, when a field is induced in a dielectric, it does an additional work associated with polarization. Therefore, under the energy of the field in the dielectric we must understand the sum of the electric energy proper and an additional work which is accomplished during polarization of the dielectric.

In order to prove this, let us substitute into (4.10) the quantity $\varepsilon_0\mathbf{E} + \mathbf{P}$ for $\mathbf{D}$, which gives

$$w = \frac{\varepsilon_0 E^2}{2} + \frac{\mathbf{E}\cdot\mathbf{P}}{2}. \tag{4.11}$$

The first term on the right-hand side coincides with the energy density of the field **E** in a vacuum. It remains for us to verify that the "additional" energy $\mathbf{E}\cdot\mathbf{P}/2$ is associated with polarization.

Let us calculate the work done by the electric field for polarization of a unit volume of the dielectric, i.e. for the

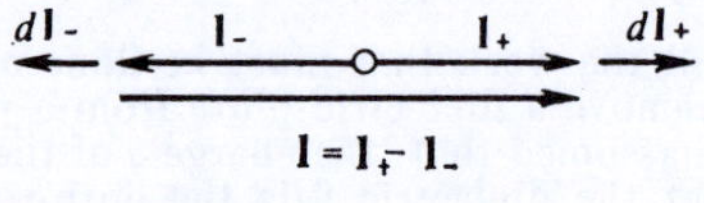

Fig. 4.3

displacements of charges ρ'_+ and ρ'_- respectively along and against the field upon an increase in the field intensity from **E** to **E** + d**E**. Neglecting the second-order terms, we write

$$\delta A = \rho'_+\mathbf{E}\cdot d\mathbf{l}_+ + \rho'_-\mathbf{E}\cdot d\mathbf{l}_-$$

where $d\mathbf{l}_+$ and $d\mathbf{l}_-$ are additional displacements due to an increase in the field by d**E** (Fig. 4.3). Considering that $\rho'_- = -\rho'_+$, we obtain

$$\delta A = \rho'_+ (d\mathbf{l}_+ - d\mathbf{l}_-)\cdot\mathbf{E} = \rho'_+\, d\mathbf{l}\cdot\mathbf{E},$$

where $d\mathbf{l} = d\mathbf{l}_+ - d\mathbf{l}_-$ is the additional displacement of the positive charges relative to the negative charges. According to (3.4), $\rho'_+ d\mathbf{l} = d\mathbf{P}$ and we get

$$\delta A = \mathbf{E}\cdot d\mathbf{P}. \tag{4.12}$$

Since $\mathbf{P} = \varkappa\varepsilon_0\mathbf{E}$, we have

$$\delta A = \mathbf{E}\cdot\varkappa\varepsilon_0\, d\mathbf{E} = d\left(\frac{\varkappa\varepsilon_0 E^2}{2}\right) = d\left(\frac{\mathbf{E}\cdot\mathbf{P}}{2}\right).$$

Hence, the total work spent for the polarization of a unit volume of a dielectric is

$$A = \mathbf{E}\cdot\mathbf{P}/2, \tag{4.13}$$

which coincides with the second term in formula (4.11).

Thus, the volume energy density $w = \mathbf{E}\cdot\mathbf{D}/2$ includes the energy $\varepsilon_0 E^2/2$ of the field proper and the energy $\mathbf{E}\cdot\mathbf{P}/2$ associated with the polarization of the substance.

4.4. A System of Two Charged Bodies

Suppose that we have a system of two charged bodies in a vacuum. Let one body create in the surrounding space the field $\mathbf{E}_1$, while the other body, the field $\mathbf{E}_2$. The resultant field is $\mathbf{E} = \mathbf{E}_1 + \mathbf{E}_2$, and the square of this quantity is

$$E^2 = E_1^2 + E_2^2 + 2\mathbf{E}_1 \cdot \mathbf{E}_2.$$

Therefore, according to (4.9), the total energy W of the given system is equal to the sum of three integrals:

$$W = \int \frac{\varepsilon_0 E_1^2}{2}\, dV + \int \frac{\varepsilon_0 E_2^2}{2}\, dV + \int \varepsilon_0 \mathbf{E}_1 \cdot \mathbf{E}_2\, dV, \qquad (4.14)$$

which coincides with formula (4.5) and reveals the field meaning of the terms appearing in this sum. The first two integrals in (4.14) are the intrinsic energies of the first and second charged bodies (W_1 and W_2), while the last integral is the energy of their interaction (W_{12}).

The following important circumstances should be mentioned in connection with formula (4.14).

1. The intrinsic energy of each charged body is an essentially positive quantity. The total energy (4.9) is also always positive. This can be readily seen from the fact that the integrand contains essentially positive quantities. However, the energy of interaction can be either positive or negative.

2. The intrinsic energy of bodies remains constant upon all possible displacements that do not change the configuration of charges on each body, and consequently this energy can be assumed to be an additive constant in the expression for the total energy W. In such cases, the changes in W are completely determined only by the changes in the interaction energy W_{12}. In particular, this is just the mode of behaviour of the energy of a system consisting of two point charges upon a change in the distance between them.

3. Unlike vector $\mathbf{E}$, the energy of the electric field is not an additive quantity, i.e. the energy of a field $\mathbf{E}$ which is the sum of fields $\mathbf{E}_1$ and $\mathbf{E}_2$ is generally not equal to the sum of the energies of these fields in view of the presence of the interaction energy W_{12}. In particular, if E increases n times everywhere, the energy of the field increases n^2 times.

4.5. Forces Acting in a Dielectric

Electrostriction. Experiments show that a dielectric in an electric field experiences the action of forces (sometimes these forces are called *ponderomotive*). These forces appear when the dielectric is neutral as a whole. Ponderomotive forces appear in the long run due to the action of a nonuniform electric field on dipole molecules of the polarized dielectric (it is known that a dipole in a nonuniform electric field is acted upon by a force directed towards the increasing field). In this case, the forces are caused by the nonuniformity of not only the macroscopic field but the microscopic field as well, which is created mainly by the nearest molecules of the polarized dielectric.

Under the action of these electric forces, the polarized dielectric is deformed. This phenomenon is called *electrostriction*. As a result of electrostriction, mechanical stresses appear in the dielectric.

Owing to electrostriction, not only the electric force (which depends on the charges) acts on a conductor in a polarized dielectric, but also an additional mechanical force caused by the dielectric. In the general case, the effect of a dielectric on the resultant force acting on a conductor cannot be taken into account by any simple relations, and the problem of calculating the forces with simultaneous analysis of the mechanism of their appearance is, as a rule, rather complicated. However, in many cases these forces can be calculated in a sufficiently simple way without a detailed analysis of their origin by using the law of conservation of energy.

Energy Method for Calculating Forces. This method is the most general. It allows us to take into account automatically all force interactions (both electric and mechanical) ignoring their origin, and hence leads to a correct result.

Let us consider the essence of the energy method for calculating forces. The simplest case corresponds to a situation when charged conductors are disconnected from the power supply. In this case, the charges on the conductors remain unchanged, and we may state that the work A of all internal forces of the system upon slow displacements of the conductors and dielectrics is done completely at the expense of a decrease in the electric energy W of the system (or its field). Here we assume that these displacements do not cause the transformation of electric energy into other kinds of energy. To be more precise, it is assumed that such transformations are negligibly small. Thus, for infinitesimal displacements we can write

$$\delta A = -dW|_q, \tag{4.15}$$

where the symbol q emphasizes that the decrease in the energy of the system must be calculated when charges on the conductors are constant.

Equation (4.15) is the initial equation for determining the forces acting on conductors and dielectrics in the electric field. This can be done as follows. Suppose that we are interested in the force acting on a given body (a conductor or a dielectric). Let us displace this body by an infinitesimal distance dx in the direction X we are interested in. Then the work of the required force $\mathbf{F}$ over the distance dx is $\delta A = F_x dx$, where F_x is the projection of the force $\mathbf{F}$ onto the positive direction of the X-axis. Substituting this expression for δA into (4.15) and dividing both parts of (4.15) by dx, we obtain

$$\boxed{F_x = -\left.\frac{\partial W}{\partial x}\right|_q}. \tag{4.16}$$

We must pay attention to the following circumstance. It is well known that the force depends only on the position of bodies and on the distribution of charges at a given instant. It cannot depend on *how* the energy process will develope if the system starts to move under the action of forces. And this means that in order to calculate F_x by formula (4.16), we do not have to select conditions under which all the charges of the conductor are necessarily constant (q = const). We must simply find the increment dW under the condition that q = const, which is a purely mathematical operation.

It should be noted that if a displacement is performed at constant potential on the conductors, the corresponding calculation leads to another expression for the force: $F_x = + \partial W/\partial x\ |_\varphi$. However (and it is important!) the result of the calculation of F_x with the help of this formula or (4.16) is the *same*, as should be expected. Therefore, henceforth we shall confine ourselves to the application of only formula (4.16) and will use it for any conditions, including those where $q \neq$ const upon small displacements. We must not be confused: the derivative $\partial W/\partial x$ will be calculated at $q =$ const in such cases as well.

Example. Find the force acting on one of the plates of a parallel-plate capacitor in a liquid dielectric, if the distance between the plates is h, the capacitance of the capacitor under given conditions is C and the voltage U is maintained across its plates.

In this case, if we mentally move the plates apart, the voltage U remains constant, while the charge q of the capacitor changes (this follows from the relation $C = q/U$). In spite of this, we shall calculate the force under the assumption that $q =$ const, i.e. with the help of formula (4.16). Here the most convenient expression for the energy of the capacitor is

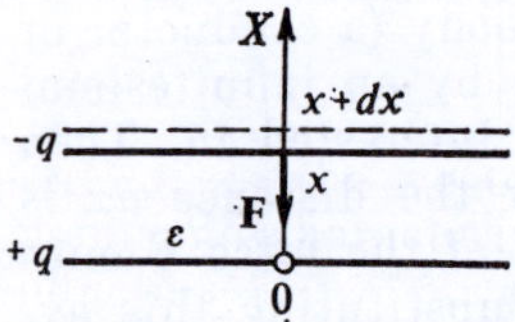

Fig. 4.4

$$W = \frac{q^2}{2C} = \frac{q^2}{2\varepsilon\varepsilon_0 S}\, x,$$

where ε is the permittivity of the dielectric, S is the area of each plate, and x is the distance between them ($x = h$).

Next, let us choose the positive direction of the X-axis as is shown in Fig. 4.4. According to (4.16), the force acting on the upper plate of the capacitor is

$$F_x = -\left.\frac{\partial W}{\partial x}\right|_q = -\frac{q^2}{2\varepsilon\varepsilon_0 S}. \qquad (1)$$

The minus sign in this formula indicates that vector $\mathbf{F}$ is directed towards the negative values on the X-axis, i.e. the force is attractive by nature. Considering that $q = \sigma S = DS = \varepsilon\varepsilon_0 ES$ and $E = U/h$, we transform (1) to

$$F_x = -CU^2/2h.$$

Forces in a Liquid Dielectric. Formula (1) of the last example shows that the force of interaction between the plates of a parallel-plate capacitor in a liquid dielectric is smaller than the corresponding force in a vacuum by a factor of ε (in vacuum $\varepsilon = 1$). Experiment shows that this result can be generalized: if the entire space occupied by a field is filled by a liquid or gaseous dielectric, the forces of interaction between charged conductors (at constant charges on them) decrease by a factor of ε:

$$F = F_0/\varepsilon. \tag{4.17}$$

Hence it follows that two point charges q_1 and q_2 separated by a distance r and placed into an infinite liquid or gaseous dielectric interact with the force

$$F = \frac{1}{4\pi\varepsilon_0} \frac{|q_1 q_2|}{\varepsilon r^2}, \tag{4.18}$$

which is smaller than the force in a vacuum by a factor of ε. This formula expresses the Coulomb law for point charges in an infinite dielectric.

We must pay a special attention to the fact that point charges in this law are extraneous charges concentrated on *macroscopic* bodies whose dimensions are small in comparison with the distance between them. Thus, the law (4.18) has, unlike the Coulomb law in vacuum, a very narrow field of application: the dielectric must be homogeneous, infinite, liquid or gaseous, while the interacting bodies must be pointlike in the macroscopic sense.

It is interesting to note that the electric field intensity **E**, as well as the force **F** acting on a point charge q in a homogeneous liquid or gaseous dielectric filling the entire space of the field, are by a factor of ε smaller than the values of $\mathbf{E}_0$ and $\mathbf{F}_0$ in the absence of dielectric. This means that the force **F** acting on the point charge q in this case is determined by the same formula as in vacuum:

$$\mathbf{F} = q\mathbf{E}, \tag{4.19}$$

where **E** is the field intensity in the *dielectric* at the point where the extraneous charge q is placed. Only in this case formula (4.19) makes it possible to determine the field **E** in the dielectric from the known force **F**. It should be noted that *another* field differing from the field in the dielectric will be acting on the extraneous charge itself (which is located on some small body). Nevertheless, formula (4.19) gives the correct result, strange as it may seem.

Surface Density of a Force. We shall be speaking of the force acting on a unit surface area of a charged conductor in a liquid or gaseous dielectric. For this purpose, let us consider a parallel-plate capacitor in a liquid dielectric.

Suppose that the capacitor is charged and then disconnected from the power supply to maintain the charge and the field **E** of the capacitor constant when the plates are moved apart.

Let us consider once again Fig. 4.4. The energy of the capacitor is the energy of the field within it. In accordance with (4.9), this energy is $W = (1/2)\,EDSx$, where S is the surface area of each plate and x is the distance between them (Sx is the volume occupied by the field). By formula (4.16), the force acting on the upper plate is

$$F_x = -\partial W/\partial x|_q = -\frac{1}{2}EDS,$$

whence the surface density of the force is

$$\boxed{F_u = \frac{\mathbf{E}\cdot\mathbf{D}}{2}.} \tag{4.20}$$

We have obtained an interesting and important result of a general nature (for a liquid or gaseous dielectric). It turns out that the surface density of the force acting on a conductor is equal to the volume density of the electric energy near the surface. This force is directed always outward along the normal to the surface of the conductor (tending to stretch it) regardless of the sign of the surface charge.

Problems

● **4.1. Energy of interaction.** A point charge q is at a distance l from an infinite conducting plane. Find the energy W of interaction between this charge and the charges induced on the plane.

Solution. Let us mentally "freeze" the charge distributed over the plane and displace under these conditions the point charge q to infinity. In this case the charge q will move in the potential field which is equivalent to the field of a fixed fictitious point charge $-q$, located at a fixed distance l on the other side of the plane. We can write straightaway

$$W = -\frac{1}{4\pi\varepsilon_0}\,\frac{q^2}{2l}.$$

● **4.2. Intrinsic, mutual, and total energies.** A system consists of two concentric metallic shells of radii R_1 and R_2 with charges q_1 and q_2 respectively. Find intrinsic energies W_1 and W_2 of each shell, the energy W_{12} of interaction between the shells, and the total electric energy of the system W, if $R_2 > R_1$.

Solution. In accordance with (4.6), the intrinsic energy of each shell is equal to $q\varphi/2$, where φ is the potential of a shell created only by the charge q on it, i.e. $\varphi = q/4\pi\varepsilon_0 R$, where R is the shell radius. Thus, the intrinsic energy of each shell is

$$W_{12} = \frac{1}{4\pi\varepsilon_0} \frac{q_{1,2}^2}{2R_{1,2}}.$$

The energy of interaction between the charged shells is equal to the charge q of one shell multiplied by the potential φ created by the charge of the other shell in the point of location of the charge q: $W_{12} = q\varphi$.

In our case ($R_2 > R$), we have

$$W_{12} = q_1 \frac{1}{4\pi\varepsilon_0} \frac{q_2}{R_2} = \frac{1}{4\pi\varepsilon_0} \frac{q_1 q_2}{R_2}.$$

The total electric energy of the system is

$$W = W_1 + W_2 + W_{12} = \frac{1}{4\pi\varepsilon_0} \left(\frac{q_1^2}{2R_1} + \frac{q_2^2}{2R_2} + \frac{q_1 q_2}{R_2} \right).$$

● **4.3.** Two small metallic balls of radii R_1 and R_2 are in vacuum at a distance considerably exceeding their dimensions and have a certain total charge. Find the ratio q_1/q_2 between the charges of the balls at which the energy of the system is minimal. What is the potential difference between the balls in this case?

Solution. The electric energy of this system is

$$W = W_1 + W_2 + W_{12} = \frac{1}{4\pi\varepsilon_0} \left(\frac{q_1^2}{2R_1} + \frac{q_2^2}{2R_2} + \frac{q_1 q_2}{l} \right),$$

where W_1 and W_2 are the intrinsic electric energies of the balls ($q\varphi/2$), W_{12} is the energy of their interaction ($q_1\varphi_2$ or $q_2\varphi_1$), and l is the distance between the balls. Since $q_2 = q - q_1$, where q is the total charge of the system, we have

$$W = \frac{1}{4\pi\varepsilon_0} \left[\frac{q_1^2}{2R_1} + \frac{(q-q_1)^2}{2R_2} + \frac{q_1(q-q_1)}{l} \right].$$

The energy W is minimal when $\partial W/\partial q_1 = 0$. Hence

$$q_1 \approx q \frac{R_1}{R_1 + R_2} \quad \text{and} \quad q_2 \approx q \frac{R_2}{R_1 + R_2},$$

where we took into account that R_1 and R_2 are considerably smaller than l and

$$q_1/q_2 = R_1/R_2.$$

The potential of each ball (they can be considered isolated) is $\varphi \propto q/R$. Hence it follows from the above relation that $\varphi_1 = \varphi_2$, i.e. the potential difference is equal to zero for such a distribution.

● **4.4. Energy localization in the field.** A charge q is uniformly distributed inside a sphere of radius R. Assuming that the dielectric constant is equal to unity everywhere, find the intrinsic electric energy of the sphere and the ratio of the energy W_1 localized inside the sphere to the energy W_2 in the surrounding space.

Solution. Let us first find the fields inside and outside the sphere with the help of the Gauss theorem:

$$E_1=\frac{q}{4\pi\varepsilon_0 R^3}r \ (r\leqslant R), \quad E_2=\frac{q}{4\pi\varepsilon_0}\frac{1}{r^2} \ (r\geqslant R).$$

We can now calculate the intrinsic electric energy of the sphere:

$$W=W_1+W_2=\int_0^R \frac{\varepsilon_0 E_1^2}{2}4\pi r^2\,dr+\int_R^\infty \frac{\varepsilon_0 E_2^2}{2}4\pi r^2\,dr=\frac{q^2}{8\pi\varepsilon_0 R}\left(\frac{1}{5}+1\right).$$

Hence, it follows that

$$W=\frac{1}{4\pi\varepsilon_0}\frac{3q^2}{5R}, \quad \frac{W_1}{W_2}=\frac{1}{5}.$$

It is interesting to note that the ratio W_1/W_2 does not depend on the radius of the sphere.

● **4.5.** A spherical shell is uniformly charged by a charge q. A point charge q_0 is at its centre. Find the work of electric forces upon the expansion of the shell from radius R_1 to R_2.

Solution. The work of electric forces is equal to the decrease in the electric energy of the system:

$$A=W_1-W_2.$$

In order to find the difference W_1-W_2, we note that upon the expansion of the shell (Fig. 4.5), the electric field, and hence the energy localized in it, changed only in the hatched spherical layer. Consequently,

$$W_1-W_2=\int_{R_1}^{R_2}\frac{\varepsilon_0}{2}(E_1^2-E_2^2)\,4\pi r^2\,dr,$$

where E_1 and E_2 are the field intensities (in the hatched region at a distance r from the centre of the system) before and after the expansion of the shell. By using the Gauss theorem, we find

$$E_1=\frac{1}{4\pi\varepsilon_0}\frac{q+q_0}{r^2}, \quad E_2=\frac{1}{4\pi\varepsilon_0}\frac{q_0}{r^2}.$$

As a result of integration, we obtain

$$A=\frac{q\,(q_0+q/2)}{4\pi\varepsilon_0}\left(\frac{1}{R_1}-\frac{1}{R_2}\right).$$

Remark. If we try to calculate the work in terms of the potential as $A = q\,(\varphi_1 - \varphi_2)$, where φ is the potential created by the charge q_0 at the point of location of the charge q, the answer would be different

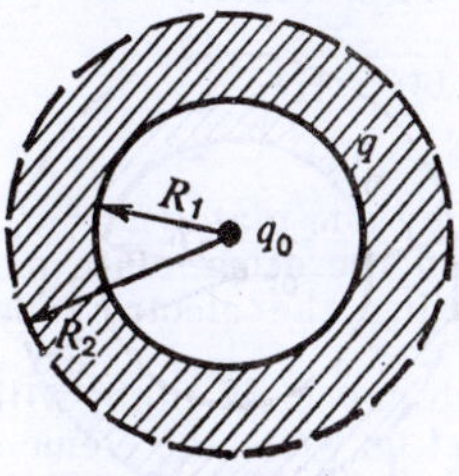

Fig. 4.5

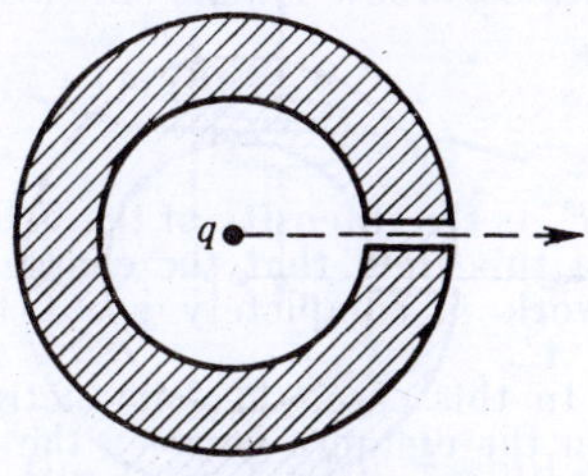

Fig. 4.6

and incorrect. This is due to the fact that this approach does not take into account the additional work performed by electric forces upon the change in the configuration of the charge q located on the *expanding* shell.

● **4.6.** A point charge q is at the centre of a spherical uncharged conducting layer whose inner and outer radii are a and b respectively. Find the work done by electric forces in this system upon the removal of the charge q from its original position through a small hole (Fig. 4.6) to a very large distance from the spherical layer.

Solution. We shall proceed from the fact that the work of electric forces is equal to the decrease in the electric energy of the system. As is well known, the latter is localized in the field itself. Thus, the problem is reduced to determination of the change in the field as a result of this process.

It can be easily seen that the field around the charge q will change only within the spherical layer with the inner radius a and the outer radius b. Indeed, in the initial position of the charge there was no field in this region, while in the final position there is a certain field (since the conducting spherical layer is far from the charge q). Consequently, the required work is

$$A=0-W_l=-\int_a^b \frac{\varepsilon_0 E^2}{2}\,dV.$$

Considering that $E = q/4\pi\varepsilon_0 r^2$ and $dV = 4\pi r^2\,dr$, and integrating, we obtain

$$A=\frac{q^2}{8\pi\varepsilon_0}\,\frac{a-b}{ab}<0.$$

● **4.7. The work done upon moving capacitor plates apart.** A parallel-plate air capacitor has the plates of area S each. Find the work A' against the electric forces, done to increase the distance between the plates from x_1 to x_2, if (1) the charge q of the capacitor and (2) its voltage U are maintained constant. Find the increments of the electric energy of the capacitor in the two cases.

Solution. (1) The required work is

$$A' = qE_1 (x_2 - x_1) = \frac{q^2}{2\varepsilon_0 S} (x_2 - x_1),$$

where E_1 is the intensity of the field created by one plate ($E = \sigma/2\varepsilon_0$). It is in this field that the charge located on the other plate moves. This work is completely spent for increasing the electric energy: $\Delta W = A'$.

(2) In this case, the force acting on each capacitor plate will depend on the distance between the plates. Let us write the elementary work of the force acting on a plate during its displacement over a distance dx relative to the other plate:

$$\delta A' = qE_1\, dx = \frac{\varepsilon_0 S U^2}{2} \frac{dx}{x^2},$$

where we took into account that $q = CU$, $E_1 = U/2x$, and $C = \varepsilon_0 S/x$. After integration, we obtain

$$A' = \frac{\varepsilon_0 S U^2}{2} \left(\frac{1}{x_1} - \frac{1}{x_2} \right) > 0.$$

The increment of the electric energy of the capacitor is

$$\Delta W = \frac{(C_2 - C_1) U^2}{2} = \frac{\varepsilon_0 S U^2}{2} \left(\frac{1}{x_2} - \frac{1}{x_1} \right) < 0.$$

It should be noted that $\Delta W = -A'$.

Thus, by moving the plates apart, we perform a positive work (against the electric forces). The energy of the capacitor decreases in this case. In order to understand this, we must consider a source maintaining the potential difference of the capacitor at a constant value. This source also accomplishes the work A_s. According to the law of conservation of energy, $A_s + A' = \Delta W$, whence $A_s = \Delta W - A' = -2A' < 0$.

● **4.8. Forces acting between conductors in a dielectric.** A parallel-plate capacitor is immersed, in the horizontal position, into a liquid dielectric with the dielectric constant ε, filling the gap of width h between the plates. Then the capacitor is connected to a source of permanent voltage U. Find the force f' acting on a unit surface of the plate from the dielectric.

Solution. The resultant force f acting per unit area of each plate can be represented as

$$f = f_0 - f', \tag{1}$$

where f_0 is the electric force acting per unit area of a plate from the other plate (it is just the force per unit area when the dielectric is absent). In our case, we have

$$f = f_0/\varepsilon, \quad f_0 = \sigma E = \sigma^2/2\varepsilon_0, \tag{2}$$

where E is the field intensity in the region occupied by one plate, created by the charges of the *other* plate. Considering that $\sigma = D = \varepsilon\varepsilon_0 U/h$ and substituting (2) into (1), we obtain

$$f' = f_0 (1 - 1/\varepsilon) = \varepsilon (\varepsilon - 1) \varepsilon_0 U^2/2h^2.$$

For example, for $U = 500$ V, $h = 1.0$ mm and $\varepsilon = 81$ (water), we get $f' = 7$ kPa (0.07 atm).

● **4.9. The force acting on a dielectric.** A cylindrical layer of a homogeneous dielectric with the dielectric constant ε is introduced into a cylindrical capacitor so that the layer fills the gap of width d between the plates. The mean radius of the plates is R such that $R \gg d$. The capacitor is connected to a source of a permanent voltage U. Find the force pulling the dielectric inside the capacitor.

Solution. Using the formula $W = q^2/2C$ for the energy of a capacitor, we find that, in accordance with (4.16), the required force is

$$F_x = -\left.\frac{\partial W}{\partial x}\right|_q = \frac{q^2}{2}\frac{\partial C/\partial x}{C^2} = \frac{U^2}{2}\frac{\partial C}{\partial x}. \tag{1}$$

Since $d \ll R$ the capacitance of the given capacitor can be calculated by the formula for a parallel-plate capacitor. Therefore, if the dielectric is introduced to a depth x and the capacitor length is l, we have

$$C = \frac{\varepsilon\varepsilon_0 x\cdot 2\pi R}{d} + \frac{\varepsilon_0 (l-x)\cdot 2\pi R}{d} = \frac{\varepsilon_0\cdot 2\pi R}{d}(\varepsilon x + l - x). \tag{2}$$

Substituting (2) into (1), we obtain

$$F_x = \varepsilon_0 (\varepsilon - 1) \pi R U^2/d.$$

● **4.10.** A capacitor consists of two fixed plates in the form of a semicircle of radius R and a movable plate of thickness h made of a dielectric with the dielectric constant ε, placed between them. The latter plate can freely rotate about the axis O (Fig. 4.7) and practically fills the entire gap between the fixed plates. A constant voltage U is maintained between the plates. Find the moment M about the axis O of forces acting on the movable plate when it is placed as shown in the figure.

Solution. The work performed by the moment of forces M upon the rotation of the plate through an angle element $d\alpha$ is equal to the decrease in the electric energy of the system at q = const [see (4.16)]:

$$M_z\, d\alpha = -dW|_q,$$

where $W = q^2/2C$. Hence

$$M_z = -\left.\frac{\partial W}{\partial \alpha}\right|_q = \frac{q^2}{2}\frac{\partial C/\partial \alpha}{C^2}. \qquad (1)$$

In the case under consideration, $C = C_1 + C_\varepsilon$, where C_1 and C_ε are the capacitances of the parts of the capacitor with and without the dielectric. The area of a sector with an angle α is determined as $S = \alpha R^2/2$, and hence

$$C = \varepsilon_0 \alpha R^2/2h + \varepsilon\varepsilon_0 (\pi - \alpha)\, R^2/2h.$$

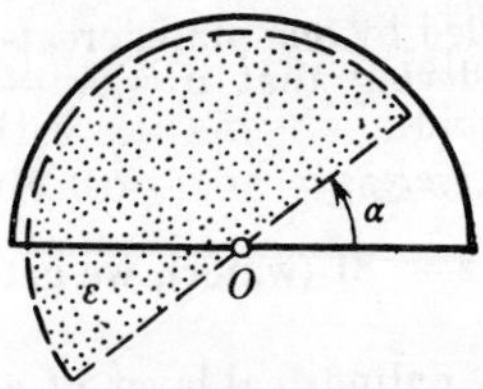

Fig. 4.7

Differentiating with respect to α, we find $\partial C/\partial \alpha = (\varepsilon_0 R^2/2h)(1-\varepsilon)$. Substituting this expression into formula (1) and considering that $C = q/U$, we obtain

$$M_z = \frac{U^2}{2}\frac{\varepsilon_0 R^2}{2h}(1-\varepsilon) = -(\varepsilon - 1)\frac{\varepsilon_0 R^2 U^2}{4h} < 0.$$

The negative sign of M_z indicates that the moment of the force is acting clockwise (oppositely to the positive direction of the angle α; see Fig. 4.7). This moment tends to pull the dielectric inside the capacitor.

It should be noted that M_z is independent of the angle α. However, in equilibrium, when $\alpha = 0$, the moment $M_z = 0$. This discrepancy is due to the fact that for small values of α we cannot ignore edge effects as was done in the solution of this problem.

5. Direct Current

5.1. Current Density. Continuity Equation

Electric Current. In this chapter, we shall confine ourselves to an analysis of conduction current in a conducting medium, especially in metals. It is well known that electric current is the transfer of charge through a certain surface S (say, the cross section of a conductor).

In a conducting medium, current can be carried by electrons (in metals), ions (in electrolytes), or some other particles. In the absence of electric field, current carriers perform chaotic motion and on average the same number of carriers of either sign passes through each side of any imaginary surface S. Thus, the current passing through S in this case is zero. However, when an electric field is applied, an ordered motion with a certain average velocity **u** is imposed on the chaotic motion of the carriers, and a current flows through the surface S. Thus, an electric current is essentially an ordered transfer of electric charges.

The quantitative measure of electric current is *intensity* I, defined as the charge transferred across the surface S in a unit time:

$$I = dQ/dt.$$

The unit of current is the *ampere* (A).

Current Density. Electric current may be distributed nonuniformly over the surface through which it passes. Hence, in order to characterize the current in greater detail, current density vector **j** is introduced. The magnitude of this vector is equal to the ratio of the current dI through a surface element perpendicular to the direction of motion of charge carriers to the area $dS_\perp$ of this element: $j = dI/dS_\perp$. For the direction of vector **j** we take the direction of velocity vector **u** of the ordered motion of *positive* carriers (or the direction opposite to that of the velocity vector of the ordered motion of negative carriers). If the current consists of positive and negative charges, its density is defined by the formula

$$\mathbf{j} = \rho_+\mathbf{u}_+ + \rho_-\mathbf{u}_-, \tag{5.1}$$

where ρ_+ and ρ_- are the volume densities of positive and negative charge carriers respectively, and $\mathbf{u}_+$ and $\mathbf{u}_-$ are the velocities of their ordered motion. In conductors, where the charge is carried only by electrons ($\rho_- < 0$ and $\mathbf{u}_+ = 0$), the current density is given by

$$\mathbf{j} = \rho_-\mathbf{u}_-. \tag{5.2}$$

The field of vector **j** can be graphically represented with the help of lines of current (lines of vector **j**), which are drawn in the same way as for vector **E**.

Knowing the current density vector at each point of the surface S under consideration, we can also find the current passing through this surface as the flux of vector **j**:

$$I = \int \mathbf{j}\cdot d\mathbf{S}. \tag{5.3}$$

Current I is a scalar algebraic quantity. It can be seen from formula (5.3) that besides other factors, the sign of the current is determined by the choice of the direction of the normal at each point on the surface S, i.e. by the choice of the direction of vectors $d\mathbf{S}$. If the directions of all the vectors $d\mathbf{S}$ are reversed, the current I changes sign.

Continuity Equation. Let us imagine a closed surface S in a conducting medium through which a current is passing. It is customary to choose the outward direction for vectors normal to closed surfaces, and hence, for vectors $d\mathbf{S}$. Consequently, the integral $\oint \mathbf{j}\, d\mathbf{S}$ gives the charge leaving the volume V (enveloped by the surface S) per unit time. According to the law of conservation of charge, this integral is equal to the decrease in the charge inside volume V per unit time:

$$\boxed{\oint \mathbf{j}\, d\mathbf{S} = -\frac{dq}{dt}.} \tag{5.4}$$

This relation is called the *continuity equation*. It is an expression for the law of conservation of charge.

In the case of a steady-state (direct) current, the charge distribution in space must remain unchanged. In other words, on the right-hand side of Eq. (5.4) $dq/dt = 0$. Consequently, for direct current we have.

$$\oint \mathbf{j}\, d\mathbf{S} = 0. \tag{5.5}$$

This means that in this case, the lines of vector $\mathbf{j}$ do not start or terminate anywhere. The field of vector $\mathbf{j}$ is said to have no sources in the case of direct current.

Differential form of the continuity equation. Let us write Eqs. (5.4) and (5.5) in the differential form. For this purpose, we express the charge q as $\int \rho\, dV$ and the right-hand side of Eq. (5.4) as $-\frac{d}{dt}\int \rho\ dV = -\int \frac{\partial \rho}{\partial t}\, dV$. Here, we have taken the partial time derivative of ρ, since ρ can depend on time as well as on coordinates. Thus,

$$\oint \mathbf{j}\, d\mathbf{S} = -\int \frac{\partial \rho}{\partial t}\, dV.$$

Proceeding in the same way as for the flux of vector **E** in Sec. 1.4, we find that the divergence of vector **j** at a certain point is equal to the decrease in the charge density per unit time at the same point:

$$\boxed{\nabla \cdot \mathbf{j} = -\partial\rho/\partial t.} \tag{5.6}$$

This leads to the *steady-state condition* (when $\partial\rho/\partial t = 0$):

$$\boxed{\nabla \cdot \mathbf{j} = 0.} \tag{5.7}$$

This means that for direct current the field of vector **j** does not have sources.

5.2. Ohm's Law for a Homogeneous Conductor

Ohm's law, which was discovered experimentally, states: *the current passing through a homogeneous conductor is proportional to the potential difference across its terminals* (or *to the voltage U*):

$$\boxed{I = U/R,} \tag{5.8}$$

where R is the *electric resistance* of the conductor.

The unit of resistance is the *ohm* (Ω).

Resistance R depends on the shape and size of the conductor, on its material and temperature, and above all on the configuration (distribution) of the current through the conductor. The meaning of resistance is quite clear when we are dealing with a wire. In a more general case of the volume distribution of current, it is meaningless to speak of resistance until we know either the position of the leads attached to the conductor or the current configuration.

In the simplest case of a homogeneous cylindrical conductor, resistance is given by

$$R = \rho \frac{l}{S}, \tag{5.9}$$

where l is the length of the conductor, S is its cross-sectional area, and ρ is the *resistivity*, which depends on the material of the conductor and its temperature. Resistivity is measured in ohm·metres (Ω·m).

Resistivity of the best conductors like copper and aluminium is of the order of 10^{-8} Ω·m at room temperature.

Differential Form of Ohm's Law. Let us establish a relation between the current density **j** and the field **E** at a point of a conducting medium. We shall confine ourselves to an isotropic conductor in which the directions of vectors **j** and **E** coincide.

Let us mentally isolate around a certain point in a conducting medium a cylindrical volume element with generatrices parallel to vector **j**, and hence to vector **E**. If the cylinder has a cross-sectional area dS and length dl, we can write on the basis of (5.8) and (5.9) the following expression for such cylindrical element:

$$j\,dS = \frac{E\,dl}{\rho\,dl/dS}.$$

After cancelling out common terms we obtain the following equation in vector form:

$$\boxed{\mathbf{j} = \frac{1}{\rho}\mathbf{E} = \sigma\mathbf{E},} \qquad (5.10)$$

where $\sigma = 1/\rho$ is the *conductivity* of the medium. The unit reciprocal to an ohm is called a *mho* (Siemens). Consequently, the unit for the measurement of σ is mho per metre.

Equation (5.10) is the differential form of Ohm's law. It does not contain any differentials (derivatives), but is called the differential form since it establishes a relation between quantities corresponding to the same point in a conductor. In other words, Eq. (5.10) is an expression for Ohm's law at a point.

Methods for Calculating Resistance (R). There are several methods for measuring resistance and all of them are ultimately based on the application of relations (5.8)-(5.10). The expedience of using a particular method is determined by the formulation of the problem and by its symmetry. The practical applications of these methods are described in Problems 5.1-5.3 and 5.6.

Charge in a Current-carrying Conductor. If the current is *direct* (constant) the excess charge inside a *homogeneous* conductor is equal to zero everywhere. Indeed, Eq. (5.5) is valid for direct current. Taking into account Ohm's law in the form (5.10) we can rewrite (5.5) as follows:

$$\oint \sigma \mathbf{E}\, d\mathbf{S} = 0,$$

where the integration is performed over any closed surface S *inside* the conductor. For a homogeneous conductor, the quantity σ can be taken out of the integral:

$$\sigma \oint \mathbf{E}\, d\mathbf{S} = 0.$$

According to the Gauss theorem, the remaining integral is proportional to the algebraic sum of the charges inside the closed surface S, i.e. proportional to the excess charge in this surface. However, it can be directly seen from the last equation that this integral is equal to zero (since $\sigma \neq 0$). This means that the excess charge is also equal to zero. Since the surface S is chosen arbitrarily, the excess charge under these conditions is always equal to zero inside a conductor.

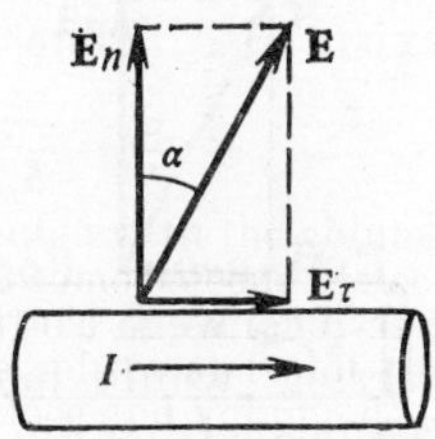

Fig. 5.1

The excess charge can appear only at the surface of a homogeneous conductor at places where it comes in contact with other conductors, as well as in the regions where the conductor has inhomogeneities.

Electric Field of a Current-carrying Conductor. Thus, when current flows, an excess charge appears on the surface of a conductor (inhomogeneity region). In accordance with (2.2), this means that the normal component of vector $\mathbf{E}$ appears at the outer surface of the conductor. Further, from the continuity of the tangential component of vector $\mathbf{E}$, we conclude that a tangential component of this vector also exists near the surface of the conductor. Thus, near the surface of the conductor, vector $\mathbf{E}$ forms (in the presence of the current) a nonzero angle α with the vertical (Fig. 5.1). When the current is absent $\alpha = 0$.

If the current flows in a steady state, the distribution of electric charges in the conducting medium (generally speaking, inhomogeneous) does not change in time, although the charges are in motion: at each point, new charges continuously replace the departing ones. These moving charges create a Coulomb field which is identical to the one created by stationary charges of the same configuration. This means that the electric field of stationary currents is a potential field.

At the same time, electric field in the case of stationary currents differs significantly from electrostatic, or Coulomb, field of fixed charges. If charges are in equilibrium, the electrostatic field inside a conductor is always equal to zero. The electric field of stationary currents is also an electrostatic (Coulomb) field, although the charges inducing this field are in motion. Hence, the field **E** of steady-state currents exists also inside a current-carrying conductor.

5.3. Generalized Ohm's Law

Extraneous Forces. If all the forces acting on charge carriers were reduced to electrostatic forces, the positive carriers would move under the action of these forces from regions with a higher potential to those with a lower potential, while the negative carriers would move in the opposite direction. This would lead to the equalization of potentials, and the potentials of all mutually connected conductors would become the same, thus terminating the current. In other words, under the action of Coulomb forces alone, a *stationary* field must be a *static* field.

In order to prevent this, a direct current circuit must contain, besides the subcircuits in which positive carriers move in the direction of decreasing potential φ, subcircuits in which positive carriers move in the direction of increasing φ or against the electric field forces. In these subcircuits, the transfer of carriers is possible only through forces that are not electrostatic. We shall call such forces *extraneous*.

Thus, a direct current can be sustained only with the help of extraneous forces which are applied either to certain parts of the circuit, or to the entire circuit. The physical nature of extraneous forces may be quite diverse. Such forces may be created by physical or chemical inhomogeneities in a conductor, for example, forces resulting from a contact of conductors of different types (galvanic cells, accumulators) or conductors at different temperatures (thermocouples).

Generalized Ohm's Law. In order to describe extraneous forces quantitatively, the concepts of extraneous force field and its strength $\mathbf{E}^*$ are introduced. The latter is numerically equal to the extraneous force acting on a unit positive charge.

Let us now turn to current density. If an electric field $\mathbf{E}$ generates in a conductor a current of density $\mathbf{j} = \sigma\mathbf{E}$, it is obvious that under the combined action of the field $\mathbf{E}$ and the extraneous force field $\mathbf{E}^*$, the current density will be given by

$$\boxed{\mathbf{j} = \sigma(\mathbf{E} + \mathbf{E}^*).} \tag{5.11}$$

This equation generalizes the law (5.10) to the case of inhomogeneous regions of a conducting medium, and is called the *generalized Ohm's law* in the differential form.

Ohm's Law for a Nonuniform Subcircuit. Nonuniform subcircuits are characterized by the presence of extraneous forces in them.

Let us consider a special, but practically quite important case when a current flows along a *thin wire*. In this case, the direction of the current will coincide with the axis of the wire and the current density j can be assumed to be the same at all points on the cross section of the wire. Suppose that the cross-sectional area of the wire is equal to S (S need not be the same over the entire length of the wire).

Dividing both sides of Eq. (5.11) by σ, forming a scalar product of the resulting expression with the element $d\mathbf{l}$ of the wire along its axis from cross section *1* to cross section *2* (this direction is taken as the positive direction) and integrating over the length of the wire between the two cross sections, we obtain

$$\int_1^2 \frac{\mathbf{j}\, d\mathbf{l}}{\sigma} = \int_1^2 \mathbf{E}\, d\mathbf{l} + \int_1^2 \mathbf{E}^*\, d\mathbf{l}. \tag{5.12}$$

Let us transform the integrand in the first integral: we replace σ by $1/\rho$ and $\mathbf{j}\, d\mathbf{l}$ by $j_l dl$, where j_l is the projection of vector $\mathbf{j}$ onto the direction of vector $d\mathbf{l}$. Further, we note that j_l is an algebraic quantity and depends on the orientation of vector **j** relative to $d\mathbf{l}$: if $\mathbf{j} \uparrow\uparrow d\mathbf{l}$ then $j_l > 0$; if $\mathbf{j} \uparrow\downarrow d\mathbf{l}$, then $j_l < 0$. Finally, we replace j_l by I/S, where I, the current, is also an algebraic quantity (like j_l). Since I is the same in all sections of the circuit when we are dealing with direct current, this quantity can be taken out of the integral. Cosequently, we get

$$\int_1^2 \frac{\mathbf{j}\,d\mathbf{l}}{\sigma} = I \int_1^2 \rho \frac{dl}{S}. \tag{5.13}$$

The expression $\rho\, dl/S$ defines the resistance of the subcircuit of length dl, while the integral of this expression is the total resistance R of the circuit between cross sections *1* and *2*.

Let us now consider the right-hand side of (5.12). Here, the first integral is the potential difference $\varphi_1 - \varphi_2$, while the second integral is the *electromotive force* (e.m.f.) $\mathscr{E}$, acting in the given subcircuit:

$$\mathscr{E}_{12} = \int_1^2 \mathbf{E}^*\, d\mathbf{l}. \tag{5.14}$$

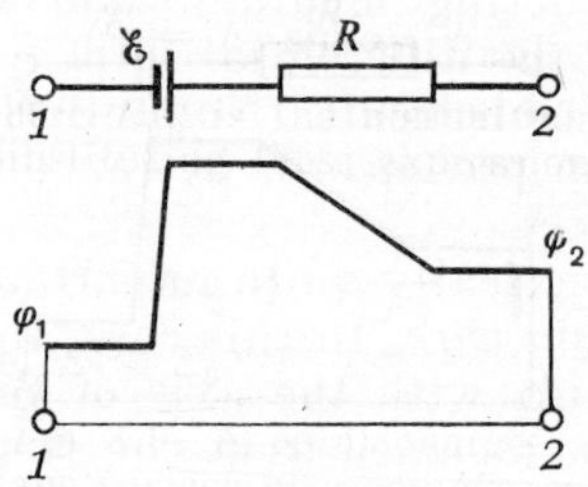

Fig. 5.2

Like the current I, the electromotive force is an algebraic quantity: if e.m.f. facilitates the motion of positive carriers in a certain direction, then $\mathscr{E}_{12} > 0$; if, the e.m.f. hinders this motion $\mathscr{E}_{12} < 0$.

As a result of all the transformations described above, (5.12) assumes the following form:

$$\boxed{RI = \varphi_1 - \varphi_2 + \mathscr{E}_{12}.} \tag{5.15}$$

This equation is the *integral form of Ohm's law* for a *nonuniform* subcircuit [cf. Eq. (5.11) which describes the same law in the differential form].

Example. Consider the subcircuit shown in Fig. 5.2. The resistance is nonzero only in the segment R. The lower part of the figure shows the variation of potential φ over this region. Let us analyse the course of events in this circuit.

Since the potential decreases over the segment R from left to right, $I > 0$. This means that the current flows in the positive direction (from *1* to *2*). In the present case, $\varphi_1 < \varphi_2$, but the current flows from point *1* to point *2*, i.e. towards increasing potential. This is possible only because an e.m.f. $\mathscr{E}$ is acting in this circuit from *1* to *2*, i.e. in the positive direction.

Let us consider Eq. (5.15). It follows from this equation that points *1* and *2* are identical for a closed circuit, i.e. $\varphi_1 = \varphi_2$. In this case, the equation acquires a simpler form:

$$RI = \mathscr{E}, \quad (5.16)$$

where R is the total resistance of the closed circuit, and $\mathscr{E}$ is the algebraic sum of all the e.m.f.'s in the circuit.

Next, we consider the subcircuit containing the source of the e.m.f. between terminals *1* and *2*. Then, R in (5.15) will be the internal resistance of the e.m.f. source in the subcircuit under consideration, while $\varphi_1 - \varphi_2$ the potential difference across its terminals. If the source is disconnected, $I = 0$ and $\mathscr{E} = \varphi_2 - \varphi_1$. In other words, the e.m.f. of the source can be defined as the potential difference across its terminals in the open circuit.

The potential difference across the terminals of an e.m.f. source connected to an external resistance is always less than its e.m.f. and depends on the load.

Example. The external resistance of a circuit is η times higher than the internal resistance of the source. Find the ratio of the potential difference across the terminals of the source to its e.m.f.

Let R_i be the internal resistance of the source and R_a the external resistance of the circuit. According to (5.15), $\varphi_2 - \varphi_1 = \mathscr{E} - R_i I$, while according to (5.16), $(R_i + R_a) I = \mathscr{E}$. From these two equations, we get

$$\frac{\varphi_2 - \varphi_1}{\mathscr{E}} = 1 - \frac{R_i I}{\mathscr{E}} = 1 - \frac{R_i}{R_i + R_a} = \frac{R_a}{R_i + R_a} = \frac{\eta}{1+\eta}.$$

It follows from here that the higher the value of η, the closer the potential difference across the current terminals to its e.m.f., and vice versa.

Concluding this section, let us consider a picture illustrating the flow of direct current in a closed circuit. Figure 5.3 shows the distribution of potential along a closed circuit containing the e.m.f. source on segment AB. For the sake of clarity of representation, potential φ is plotted along the generatrices of a cylindrical surface resting on the current contour. Points A and B correspond to the positive and negative terminals of the source. It can be seen from the figure that the flow of current can be described as follows: the positive charge carriers "slide" along the inclined "chute" from point φ_A to point φ_B along the external subcircuit, while inside the source, carriers "rise" from point φ_B to φ_A with the help of extraneous force shown by an arrow.

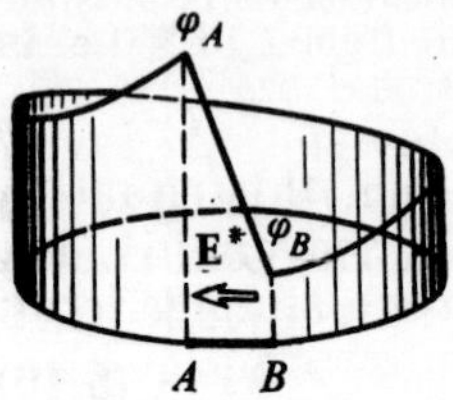

Fig. 5.3

5.4. Branched Circuits. Kirchhoff's Laws

Calculations of branched circuits, for example, determination of current in individual branches, can be considerably simplified by using the following two *Kirchhoff's laws.*

Kirchhoff's First Law pertains to the junctions, i.e. branch points in a circuit, and states that *the algebraic sum of the currents meeting at a junction is equal to zero*:

$$\sum I_k = 0. \tag{5.17}$$

Here, currents converging at a junction and diverging from it are supposed to have opposite signs; for example, we can assume the former to be positive and the latter to be negative (or vice versa, this is immaterial). When applied to Fig. 5.4, Eq. (5.17) assumes the form $I_1 - I_2 + I_3 = 0$.

Equation (5.17) is a consequence of the steady-state condition (5.7); otherwise, the charge at a junction would change and the currents would not be in a steady state.

Kirchhoff's Second Law. This law is applicable to any closed contour in a branched circuit and states that *the algebraic sum of the products of current and resistance in each part of a network is equal to the algebraic sum of e.m.f.'s in the circuit*:

$$\sum I_k R_k = \sum \mathcal{E}_k. \tag{5.18}$$

In order to prove this law, it is sufficient to consider the case when the circuit consists of three subcircuits (Fig. 5.5). Let us assume the direction of circumvention to be clockwise,

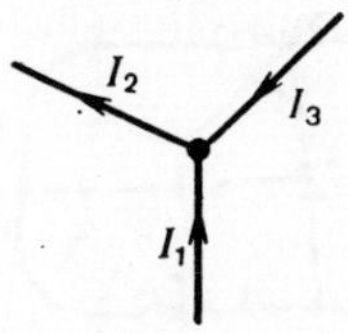

Fig. 5.4

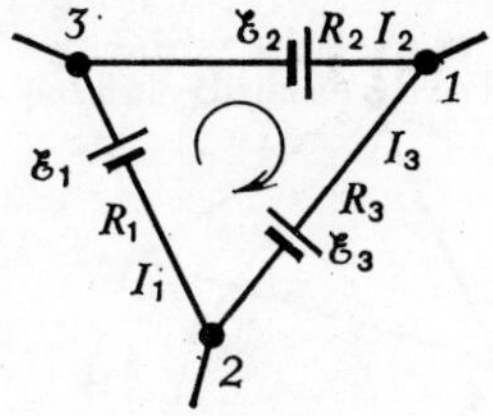

Fig. 5.5

as shown in the figure. Applying Ohm's law (5.15) to each of the three subcircuits, we obtain

$$I_1R_1 = \varphi_2 - \varphi_3 + \mathscr{E}_1,$$
$$I_2R_2 = \varphi_3 - \varphi_1 + \mathscr{E}_2,$$
$$I_3R_3 = \varphi_1 - \varphi_2 + \mathscr{E}_3.$$

Adding these equations and cancelling all potentials, we arrive at formula (5.18), i.e. Kirchhoff's second law.

Thus, Eq. (5.18) is a consequence of Ohm's law for non-uniform subcircuits.

Setting up of a System of Equations. In each specific case, Kirchhoff's laws lead to a complete system of algebraic equations which can be used, say, for finding all the unknown currents in the circuit.

The number of equations of the form (5.17) and (5.18) must be equal to the number of unknown quantities. Care should be taken to ensure that none of the equations is a corollary of any other equation in the system:

(1) if a branched circuit has N junctions, independent equations of type (5.17) can be set up only for $N - 1$ junctions, and the equation for the last junction will be a corollary of the preceding equations;

(2) if a branched circuit contains several closed loops, the independent equations of type (5.18) can be set up only for loops which cannot be obtained by superimposing the loops considered before. For example, such equations will be independent for loops *124* and *234* of the

circuit shown in Fig. 5.6. Equation for the loop *1234* will follow from the two preceding ones. It is possible to set up autonomous equations for two other loops, say, *124* and *1234*. Then the equation for contour *234* will be a

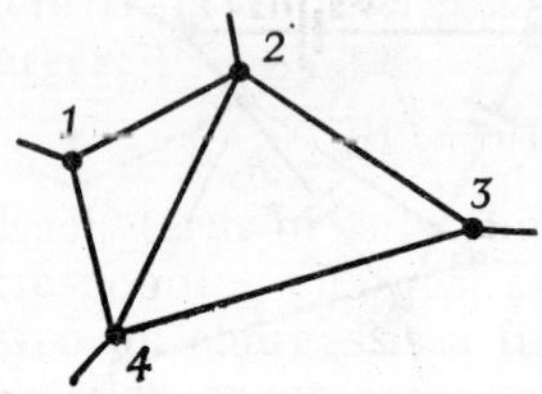

Fig. 5.6

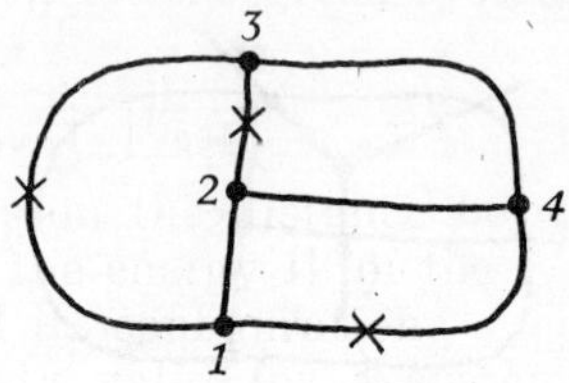

Fig. 5.7

consequence of these two equations. The number of autonomous equations of type (5.18) will be equal to the smallest number of discontinuities which must be created in a circuit in order to break all the loops. This number is equal to the number of subcircuits bounded by conductors if the circuit can be drawn in a plane without intersections.

For example, it is necessary to set up three equations of type (5.17) and three of type (5.18) for a circuit (Fig. 5.7) containing four junctions. This is so because the number of discontinuities (marked by crosses in the circuit) breaking all the loops is three (the number of subcircuits is also equal to three in this case). If we assume the currents to be unknown the number of discontinuities will be six in accordance with the number of subcircuits between junctions, which corresponds to the number of independent equations.

The following procedure should be adopted while setting up equations of type (5.17) and (5.18).

1. The directions of currents are marked hypothetically by arrows, without caring for the direction of these arrows. If a certain current turns out to be positive as a result of calculations, this means that the direction chosen for it is correct. If, however, the current is found to be negative, its actual direction will be opposite to the one pointed by the arrow.

2. Having selected an arbitrary closed loop, we circumvent its sections in one direction, say, clockwise. If the assumed direction for any current coincides with the

direction of circumvention, the corresponding term IR in (5.18) should be taken with the positive sign; in the opposite case, the minus sign should be used. The same procedure is applicable to $\mathscr{E}$: if an e.m.f. increases potential in the direction of circumvention, it should be assigned the positive sign; the negative sign should be used in the opposite case.

Example. Find the magnitude and direction of the current passing through resistor R in the circuit shown in Fig. 5.8. All resistances and e.m.f.'s are assumed to be known.

There are three subcircuits, and hence, three unknown currents I, I_1 and I_2 in this circuit. We mark (arbitrarily) the supposed directions of these currents by arrows (at the right junction).

The circuit contains $N = 2$ junctions. This means that there is only one independent equation of type (5.17):

$$I + I_1 + I_2 = 0.$$

Let us now set up equations of type (5.18). According to the number of subcircuits, there should be two of them. Let us consider the loop containing R and R_1 and the loop with R and R_2. Taking the clockwise direction for circumventing each of these loops, we can write

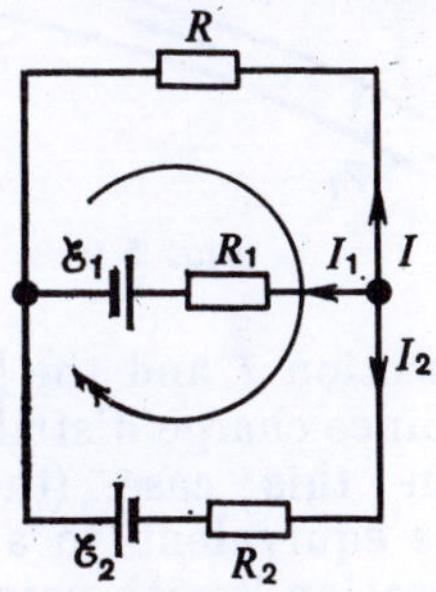

Fig. 5.8

$$-IR + I_1R_1 = -\mathscr{E}_1, \quad -IR + I_2R_2 = \mathscr{E}_2.$$

We can verify that the corresponding equation for the loop with R_1 and R_2 can be obtained from these two equations. Solving the system of three equations, we obtain

$$I = \frac{-R_1\mathscr{E}_2 + R_2\mathscr{E}_1}{R_1R_2 + RR_1 + RR_2}.$$

If we find that as a result of substitution of numerical values into this equation $I > 0$, the current actually flows as shown in Fig. 5.8. If $I < 0$, the current flows in the opposite direction.

5.5. Joule's Law

The passage of current through a conductor having a resistance is invariably accompanied by liberation of heat (heating of conductors). Our task is to find the quantity of heat liberated per unit time in a certain part of the circuit. We shall consider two cases which are possible, viz. uniform and nonuniform subcircuits. The problem is considered on the basis of the law of conservation of energy and Ohm's law.

Uniform Subcircuit. Suppose that we are interested in the region between cross sections *1* and *2* of a conductor (Fig. 5.9). Let us find the work done by the field during a time interval dt in the region *12*.

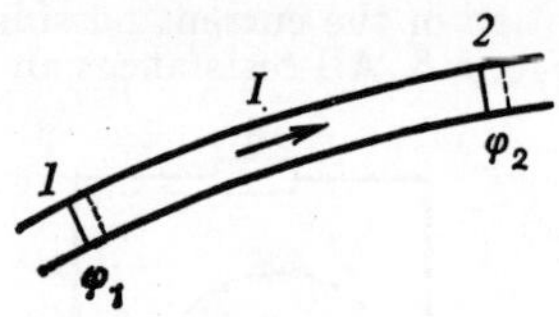

Fig. 5.9

If the current through the conductor is equal to I, a charge $dq = I\,dt$ will pass through each cross section of the conductor during the time dt. In particular, this charge dq will enter cross section *1* and the same charge will leave cross section *2*. Since charge distribution in the conductor remains unchanged in this case (the current is direct), the whole process is equivalent to a transfer of charge dq from section *1* to section *2* with potentials φ_1 and φ_2 respectively.

Hence the work done by the field in such a charge transfer is

$$\delta A = dq\,(\varphi_1 - \varphi_2) = I\,(\varphi_1 - \varphi_2)\,dt.$$

According to the law of conservation of energy, an equivalent amount of energy must be liberated in another form. If the conductor is stationary and no chemical transformations take place in it, this energy must be liberated in the form of internal (thermal) energy. Consequently, the conductor gets heated. The mechanism of this transformation is quite simple: as a result of the work done by the field, charge carriers (for example, electrons in metals) acquire an additional kinetic energy which is then spent on exciting lattice vibrations due to collisions of the carriers with the lattice sites' atoms.

Thus, in accordance with the law of conservation of energy, the elementary work $\delta A = \dot{Q}\,dt$, where $\dot{Q}$ is the heat liberated per unit time (thermal power). A comparison of this equation with the preceding one gives

$$\dot{Q} = I\,(\varphi_1 - \varphi_2).$$

Since in accordance with Ohm's law $\varphi_1 - \varphi_2 = RI$, we get

$$\boxed{\dot{Q} = RI^2.} \tag{5.19}$$

This is the expression for the well-known *Joule's law*.

We shall now derive an expression for the differential form of this law, characterizing the liberation of heat at different parts of a conducting medium. For this purpose, we isolate a volume element of this medium in the form of a cylinder with its generatrices parallel to vector **j**, viz. the current density at the given point. Let dS and dl be the cross-sectional area and length of the small cylinder respectively. The amount of heat liberated in this volume during the time dt will be given, in accordance with Joule's law, by

$$\delta Q = RI^2\,dt = \frac{\rho\,dl}{dS}(j\,dS)^2\,dt = \rho j^2\,dV\,dt,$$

where $dV = dS\,dl$ is the volume of the small cylinder. Dividing the last equation by $dV\,dt$, we obtain a formula for the amount of heat liberated per unit time in a unit volume of the conducting medium, or the *thermal power density of the current*:

$$\boxed{\dot{Q}_{\mathrm{d}} = \rho j^2.} \tag{5.20}$$

This formula expresses *Joule's law in the differential form*: the *thermal power density of current at any point is proportional to the square of the current density and to the resistivity of the medium at that point.*

Equation (5.20) is the most general form of Joule's law, applicable to all conductors irrespective of their shape or homogeneity, as well as the nature of forces which induce the electric current. If the charge carriers are subjected to electric forces only, we can write, on the basis of Ohm's law (5.10),

$$\dot{Q}_{\mathrm{d}} = \mathbf{j}\cdot\mathbf{E} = \sigma E^2. \tag{5.21}$$

This equation is of a less general nature than (5.20).

Nonuniform Subcircuit. If a subcircuit contains a source of e.m.f., the charge carriers will be subjected not only to electric forces, but to extraneous forces as well. In accordance with the law of conservation of energy, the amount of heat liberated in this case will be equal to the algebraic sum of the works done by the electric and extraneous forces. The same applies to the corresponding powers: the thermal power must be equal to the algebraic sum of the powers due to electric and extraneous forces. This can be easily verified by multiplying (5.15) by I:

$$RI^2 = (\varphi_1 - \varphi_2)\, I + \mathscr{E}I. \qquad (5.22)$$

The left-hand side of this expression is the thermal power $\dot{Q}$ liberated in the subcircuit. In the presence of extraneous forces, the value of $\dot{Q}$ is determined by the same formula (5.19) which is applied to a uniform subcircuit. The last term on the right-hand side is the power generated by extraneous forces in the subcircuit under consideration. It should also be noted that the last term ($\mathscr{E}I$) is an algebraic quantity and, unlike RI^2, it reverses its sign when the direction of the current I is reversed.

Thus, (5.22) indicates that the heat liberated in the region between points *1* and *2* is equal to the algebraic sum of the electric and extraneous powers. The sum of these powers, i.e. the right-hand side of this equation, is called the *electric power* developed in the subcircuit. It can then be stated that for a stationary subcircuit, the thermal power evolved is equal to the electric power developed in the subcircuit by the current.

Applying (5.22) to the entire unbranched circuit ($\varphi_1 = \varphi_2$), we get

$$\dot{Q} = \mathscr{E}I. \qquad (5.23)$$

In other words, the total Joule heat liberated per unit time in the entire circuit is equal to the power developed by extraneous forces only. This means that heat is generated only by extraneous forces. The role of the electric field is reduced to redistribution of this heat over different parts of the circuit.

We shall now derive Eq. (5.22) in the differential form. For this purpose, we multiply both sides of Eq. (5.11) by

j, and consider that $\sigma = 1/\rho$ and $\rho j^2 = \dot{Q}_d$ (see (5.20)). The thermal power density of current in an inhomogeneous medium can then be written in the form

$$\boxed{\dot{Q}_d = \rho j^2 = \mathbf{j}\cdot(\mathbf{E} + \mathbf{E}^*).} \tag{5.24}$$

5.6. Transient Processes in a Capacitor Circuit

Transient Processes. These are the processes involving a transition from one stationary regime of the circuit to another. An example of such a process is the charging and discharging of a capacitor, which will be considered in detail in this section.

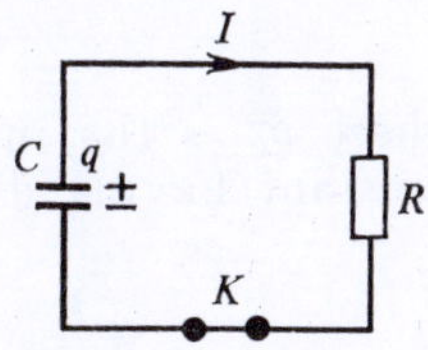

Fig. 5.10

So far, we have considered only direct currents. It turns out, however, that most of the laws obtained in this case are also applicable to alternating currents. This applies to all cases in which the variation of current is not too rapid. The instantaneous value of current will then practically be the same at all cross sections of the circuit. Such currents and fields associated with them are called *quasistationary* (a more exact criterion for quasistationary currents and fields is given in Sec. 11.1).

Quasistationary currents can be described by the laws obtained for constant currents by applying these laws to instantaneous values of quantities.

Let us now consider the discharging and charging of a capacitor, assuming that the currents in these processes are quasistationary.

Discharging of a Capacitor. If the plates of a charged capacitor with capacitance C are connected through a resistor R, a current will flow through the resistor. Let I, q and U be the instantaneous values of the current, charge of the positive plate, and the potential difference between the plates (voltage).

Assuming that the current I is positive when it flows from the positive plate to the negative plate (Fig. 5.10), we can write $I = -dq/dt$. According to Ohm's law, we obtain the following expression for the external subcircuit having resistance R:

$$RI = U.$$

Considering that $I = -dq/dt$ and $U = q/C$, we can transform the above equation as follows:

$$\frac{dq}{dt} + \frac{q}{RC} = 0. \tag{5.25}$$

After separating the variables in this differential equation and integrating, we get

$$q = q_0 e^{-t/\tau}, \tag{5.26}$$

where q_0 is the initial charge on the capacitor and τ is a constant having dimensions of time:

$$\tau = RC. \tag{5.27}$$

This constant is called *relaxation time*. It can be seen from (5.26) that τ is the time during which the capacitor charge decreases to 1/e of its initial value.

Differentiating (5.26) with respect to time, we obtain the law of variation of current:

$$I = -\frac{dq}{dt} = I_0 e^{-t/\tau}, \tag{5.28}$$

where $I_0 = q_0/\tau$ is the current at the instant $t = 0$.

Figure 5.11 shows the dependence of the capacitor charge q on time t. The I vs. t dependence also has the same form.

Charging of a Capacitor. Consider a circuit in which a capacitor C, a resistor R and a source of e.m.f. $\mathscr{E}$ are connected in series (Fig. 5.12). To begin with, the capacitor is not charged (key K is open). At the instant $t = 0$ the key is closed, and a current starts to flow through the circuit, thus charging the capacitor. The increasing charge on the capacitor plates will obstruct the passage of the current and gradually decrease it.

The current in the circuit will now be assumed positive if it flows toward the positive plate of the capacitor: $I = dq/dt$. Applying Ohm's law for a nonuniform subcircuit to the loop $1\mathscr{E}R2$, we get

$$RI = \varphi_1 - \varphi_2 + \mathscr{E},$$

where R is the total resistance of the circuit, including the internal resistance of the e.m.f. source. Considering

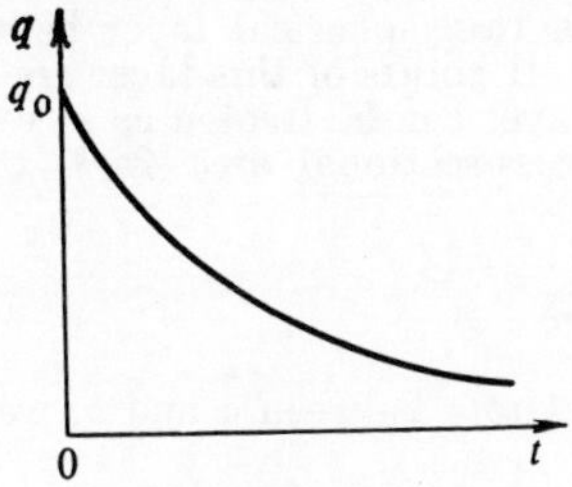

Fig. 5.11

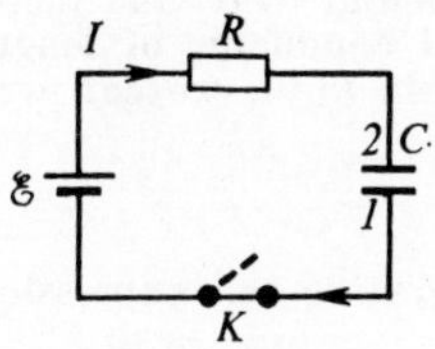

Fig. 5.12

that $I = dq/dt$ and $\varphi_2 - \varphi_1 = U = q/C$, we can rewrite the last equation in the form

$$\frac{dq}{dt} = \frac{\mathscr{E} - q/C}{R}.$$

Separating the variables, we get

$$\frac{R\,dq}{\mathscr{E} - q/C} = dt.$$

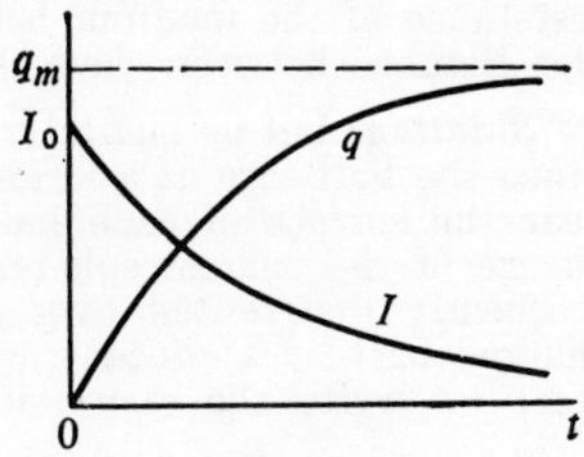

Fig. 5.13

Integrating this equation under the initial condition $q = 0$ at $t = 0$, we obtain

$$RC \ln\left(1 - \frac{q}{\mathscr{E}C}\right) = -t,$$

whence

$$q = q_m (1 - e^{-t/\tau}). \tag{5.29}$$

Here $q_m = \mathscr{E}C$ is the limiting value of the capacitor charge (for $t \to \infty$), and $\tau = RC$.

The current varies with time according to the following law:

$$I = \frac{dq}{dt} = I_0 e^{-t/\tau}, \tag{5.30}$$

where $I_0 = \mathscr{E}/R$.

The dependences of q and I on t are shown in Fig. 5.13.

Problems

● **5.1. Resistance of a conducting medium.** A metallic sphere of radius a is surrounded by a concentric thin metallic shell of radius b. The space between these two electrodes is filled with a homogeneous poorly conducting medium of resistivity ρ. Find the resistance of the gap between the two electrodes.

Solution. Let us mentally isolate a thin spherical layer between radii r and $r + dr$. Lines of current at all points of this layer are perpendicular to it, and therefore such a layer can be treated as a cylindrical conductor of length dr and a cross-sectional area $4\pi r^2$. Using formula (5.9), we can write

$$dR = \rho \frac{dr}{4\pi r^2}.$$

Integrating this expression with respect to r between a and b, we get

$$R = \frac{\rho}{4\pi}\left(\frac{1}{a} - \frac{1}{b}\right).$$

● **5.2.** Two identical metallic balls of radius a are placed in a homogeneous poorly conducting medium with resistivity ρ. Find the resistance of the medium between the balls under the condition that the distance between them is much larger than their size.

Solution. Let us mentally impart charges $+q$ and $-q$ to the balls. Since the balls are at a large distance from one another, electric field near the surface of each ball is practically determined only by the charge of the nearest sphere, and its charge can be considered to be uniformly distributed over the surface. Surrounding the positively charged ball by a concentric sphere adjoining directly the ball's surface, we write the expression for the current through this sphere;

$$I = 4\pi a^2 j,$$

where j is the current density. Using Ohm's law ($j = E/\rho$) and the formula $E = q/4\pi\varepsilon_0 a^2$, we obtain

$$I = q/\varepsilon_0 \rho.$$

Let us now find the potential difference between the balls:

$$U = \varphi_+ - \varphi_- \approx 2q/4\pi\varepsilon_0 a.$$

The sought resistance is given by

$$R = U/I = \rho/2\pi a.$$

This result is valid regardless of the magnitude of the dielectric constant of the medium.

● **5.3.** Two conductors of arbitrary shape are placed into an infinite homogeneous slightly conducting medium with the resistivity ρ and the dielectric constant ε. Find the value of the product RC for this system, where R is the resistance of the medium between the conductors and C is the mutual capacitance of the conductors in the presence of the medium.

Solution. Let us mentally supply the charges $+q$ and $-q$ to the conductors. Since the medium between them is poorly conducting, the surfaces of the conductors are equipotential, and the field configuration is the same as in the absence of the medium.

Let us surround, for example, the positively charged conductor by a closed surface S directly adjoining the conductor and calculate R and C separately:

$$R=\frac{U}{I}=\frac{U}{\oint j_n\,dS}=\frac{U}{\sigma\oint E_n\,dS},$$

$$C=\frac{q}{U}=\frac{\oint D_n\,dS}{U}=\frac{\varepsilon\varepsilon_0\oint E_n\,dS}{U},$$

where the integrals are taken over the given surface S. While calculating R, we used Ohm's law in the form $\mathbf{j}=\sigma\mathbf{E}$, and C was calculated with the help of the Gauss theorem.

Multiplying these expressions, we obtain

$$RC=\varepsilon_0\varepsilon/\sigma=\varepsilon_0\varepsilon\rho.$$

● **5.4. Conditions on the boundary of a conductor.** A conductor with resistivity ρ borders on a dielectric whose dielectric constant is ε. The electric induction at a certain point A at the surface of the conductor is equal to D, vector $\mathbf{D}$ being directed away from the conductor and forming angle α with the normal to the surface. Find the surface charge density on the conductor and the current density near the point A.

Solution. The surface charge density on the conductor is given by

$$\sigma=D_n=D\cos\alpha.$$

The current density can be found with the help of Ohm's law: $\mathbf{j}=\mathbf{E}/\rho$. It follows from the continuity equation (5.5) that the normal components of vector $\mathbf{j}$ are equal, and since in the dielectric $j_n=0$ (there is no current), in the conductor we also have $j_n=0$. Hence, vector $\mathbf{j}$ in the conductor is tangent to its surface. The same applies to vector $\mathbf{E}$ inside the conductor.

On the other hand, it follows from the theorem on circulation of vector $\mathbf{E}$ that its tangential components on different sides of the interface are equal, and hence $E=E_\tau=D\sin\alpha/\varepsilon\varepsilon_0$, where E_τ is the tangential component of vector $\mathbf{E}$ in the dielectric. Taking these arguments into account, we obtain

$$j=\frac{E}{\rho}=\frac{D\sin\alpha}{\varepsilon_0\varepsilon\rho}.$$

● 5.5. The gap between the plates of a parallel-plate capacitor is filled consecutively by two dielectric layers *1* and *2* with thicknesses l_1 and l_2, dielectric constants ε_1 and ε_2, and resistivities ρ_1 and ρ_2. The capacitor is under a permanent voltage U, the electric field being directed from layer *1* to *2*. Find the surface density of extraneous charges at the interface between the dielectric layers.

Solution. The required surface charge density is given by

$$\sigma = D_{2n} - D_{1n} = \varepsilon_0\varepsilon_2 E_2 - \varepsilon_0\varepsilon_1 E_1. \quad (1)$$

In order to determine E_1 and E_2, we shall make use of two conditions: since $j_1 = j_2$, it follows that $E_1/\rho_1 = E_2/\rho_2$, and besides $E_1 l_1 + E_2 l_2 = U$. Solving the two equations, we find E_1 and E_2. Substituting these values into (1), we obtain

$$\sigma = \frac{\varepsilon_2\rho_2 - \varepsilon_1\rho_1}{\rho_1 l_1 + \rho_2 l_2}\,\varepsilon_0 U.$$

Hence it follows that $\sigma = 0$ for $\varepsilon_1\rho_1 = \varepsilon_2\rho_2$.

● **5.6. Nonhomogeneous conductor.** A long conductor of a circular cross section of area S is made of a material whose resistivity depends only on the distance r from the conductor axis as $\rho = \alpha/r^2$, where α is a constant. The conductor carries current I. Find (1) field intensity E in the conductor and (2) the resistance of the unit length of the conductor.

Solution. (1) In accordance with Ohm's law, field intensity E is related to current density j, while j is related to current I. Hence we can write

$$I = \int j 2\pi r\, dr = \int (E/\rho)\, 2\pi r\, dr.$$

The field intensity E is the same at all poi ts of the cross section of the given conductor, i.e. is independent of r. We can easily verify this by selecting a rectangular contour in the conductor so that one side of the contour coincides with, for example, the conductor's axis, and then applying to this contour the theorem on circulation of vector **E**.

Thus, E can be taken out of the integral, and as a result of integration we obtain

$$E = 2\pi\alpha I/S^2.$$

(2) The resistance of a unit length of the conductor can be found with the help of the formula $R = U/I$. Dividing both sides of this equation by length l of the section of the conductor, having resistance R and voltage U, we find

$$R_u = E/I = 2\pi\alpha/S^2.$$

● **5.7. Ohm's law for nonuniform subcircuit.** In the circuit shown in Fig. 5.14 the e.m.f.'s $\mathscr{E}$ and $\mathscr{E}_0$ of sources, the resistances R and R_0, and the capacitance C of a capacitor are known. The internal resistances of sources are negligibly small. Find the charge on plate *1* of the capacitor.

Solution. In accordance with Ohm's law for a closed circuit containing resistors R and R_0, we can write

$$(R + R_0)I = \mathscr{E} - \mathscr{E}_0,$$

where the positive direction is chosen clockwise. On the other hand, for the nonuniform subcircuit aRb of the circuit we have

$$RI = \varphi_a - \varphi_b + \mathscr{E},$$

while for the subcircuit aCb we have

$$\mathscr{E} + \varphi_2 - \varphi_1 = \varphi_b - \varphi_a.$$

The joint solution of these equations gives

$$\varphi_1 - \varphi_2 = \frac{R}{R+R_0}(\mathscr{E} - \mathscr{E}_0).$$

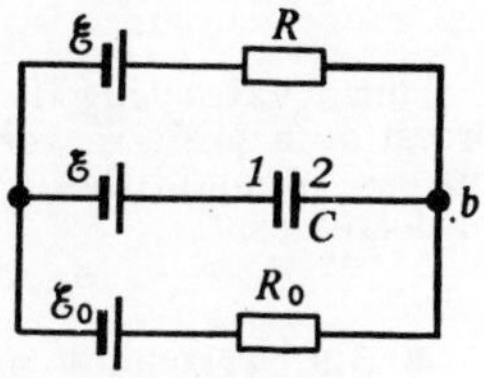

Fig. 5.14

The charge on plate *1* is determined by the formula $q_1 = C(\varphi_1 - \varphi_2)$. Therefore, the final result is

$$q_1 = \frac{RC}{R+R_0}(\mathscr{E} - \mathscr{E}_0).$$

It can be seen that $q_1 > 0$ for $\mathscr{E} > \mathscr{E}_0$ and vice versa.

● **5.8. The work of an e.m.f. source.** A glass plate completely fills the gap between the plates of a parallel-plate capacitor whose capacitance is equal to C_0 when the plate is absent. The capacitor is connected to a source of permanent voltage U. Find the mechanical work which must be done against electric forces for extracting the plate out of the capacitor.

Solution. According to the law of conservation of energy, we can write

$$A_m + A_s = \Delta W, \tag{1}$$

where A_m is the mechanical work accomplished by extraneous forces against electric forces, A_s is the work of the voltage source in this process, and ΔW is the corresponding increment in the energy of the capacitor (we assume that contributions of other forms of energy to the change in the energy of the system is negligibly small).

Let us find ΔW and A_s. It follows from the formula $W = CU^2/2 = qU/2$ for the energy of a capacitor that for $U = \text{const}$

$$\Delta W = \Delta C\, U^2/2 = \Delta q\, U/2. \tag{2}$$

Since the capacitance of the capacitor decreases upon the removal of the plate ($\Delta C < 0$), the charge of the capacitor also decreases ($\Delta q < 0$). This means that the charge has passed through the source against the direction of the action of extraneous forces, and the source has done negative work

$$A_s = \Delta q \cdot U. \tag{3}$$

Comparing formulas (3) and (2), we obtain

$$A_s = 2\Delta W.$$

Substitution of this expression into (1) gives

$$A_m = -\Delta W \quad \text{or } A_m = \frac{1}{2}(\varepsilon - 1)\, C_0 U^2.$$

Thus, extracting the plate out of the capacitor, we (extraneous forces) do a positive work (against electric forces). The e.m.f. source in this case accomplishes a negative work, and the energy of the capacitor decreases:

$$A_m > 0, \quad A_s < 0, \quad \Delta W < 0.$$

● **5.9. Transient processes.** A circuit consists of a permanent source of e.m.f. $\mathscr{E}$, and a resistor R and capacitor C connected in series. The internal resistance of the source is negligibly small. At the moment $t = 0$, the capacitance of the capacitor was abruptly (jumpwise) decreased by a factor of η. Find the current in the circuit as a function of time.

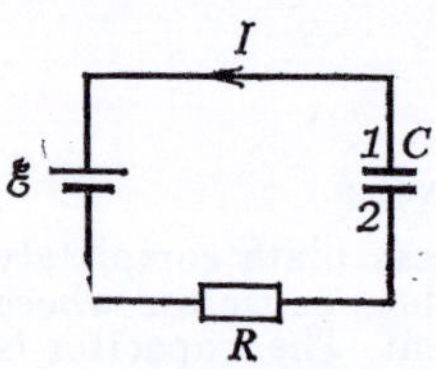

Fig. 5.15

Solution. We write Ohm's law for the inhomogeneous part $1\mathscr{E}R2$ of the circuit (Fig. 5.15):

$$RI = \varphi_1 - \varphi_2 - \mathscr{E} = U - \mathscr{E}.$$

Considering that $U = q/C'$, where $C' = C/\eta$, we obtain

$$RI = \eta q/C - \mathscr{E}. \tag{1}$$

We differentiate this equation with respect to time, considering that in our case (q decreases) $dq/dt = -I$:

$$R\frac{dI}{dt} = -\frac{\eta}{C} I, \qquad \frac{dI}{I} = -\frac{\eta}{RC}\, dt.$$

Integration of this equation gives

$$\ln\frac{I}{I_0} = -\frac{\eta t}{RC}, \qquad I = I_0 e^{-\eta t/RC},$$

where I_0 is determined by condition (1). Indeed, we can write

$$RI_0 = \eta q_0/C - \mathscr{E},$$

where $q_0 = \mathscr{E}C$ is the charge of the capacitor before its capacitance has changed. Therefore,

$$I_0 = (\eta - 1)\,\mathscr{E}/R.$$

● **5.10.** A charge q_0 was supplied to a capacitor with capacitance C, which at the moment $t = 0$ was shunted by resistor R. Find the amount of heat liberated in the resistor as a function of time t.

Solution. The required amount of heat is given by

$$Q = \int_0^t RI^2\,dt, \tag{1}$$

from which it follows that first of all we must find the time dependence $I(t)$. For this purpose, we shall use Ohm's law for the part *1R2* of the circuit (Fig. 5.16):

$$RI = \varphi_1 - \varphi_2 = U,$$

or

$$RI = q/C. \tag{2}$$

Let us differentiate (2) with respect to time:

$$R\frac{dI}{dt} = \frac{1}{C}I, \qquad \frac{dI}{I} = \frac{dt}{RC}.$$

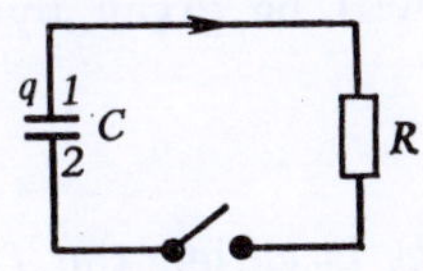

Fig. 5.16

Integrating the last equation, we obtain

$$\ln\frac{I}{I_0} = \frac{t}{RC}, \qquad I = I_0 e^{-t/RC}, \tag{3}$$

where I_0 is determined by condition (2) for $q = q_0$, i.e. $I_0 = q_0/RC$.

Substituting (3) into (1) and integrating over time, we get

$$Q = \frac{q_0^2}{2C}(1 - e^{-2t/RC}).$$

6. Magnetic Field in a Vacuum

6.1. Lorentz Force. Field B

Lorentz Force. Experiments show that the force **F** acting on a point charge q generally depends not only on the position of this charge but also on its velocity **v**. Accordingly, the force **F** is decomposed into two components, viz. the electric component $\mathbf{F}_e$ (which does not depend on the motion of the charge) and the magnetic component $\mathbf{F}_m$ (which depends on the charge velocity). The direction and magnitude of the magnetic force at any point of space depend on the velocity **v** of the charge, this force being always perpendicular to vector **v**. Besides, at any point, the magnetic force is perpendicular to the direction specified at this point. Finally, the magnitude of this force is proportional to the velocity component which is perpendicular to this direction.

All these properties of magnetic force can be described by introducing the concept of *magnetic field*. Characterizing this field by vector **B** which determines the specific direction at each point of space, we can write the expression for magnetic force in the form

$$\mathbf{F}_m = q\,[\mathbf{v} \times \mathbf{B}]. \tag{6.1}$$

Then the total electromagnetic force acting on charge q will be given by

$$\boxed{\mathbf{F} = q\mathbf{E} + q\,[\mathbf{v} \times \mathbf{B}].} \tag{6.2}$$

It is called the *Lorentz force*. Expression (6.2) is *universal*: it is valid for constant as well as for varying electric and magnetic fields for any velocity **v** of the charge.

The action of the Lorentz force on a charge can in principle be used for determining the magnitudes and directions of vectors **E** and **B**. Hence, the expression for the Lorentz force can be considered as the definition of electric and magnetic fields (as was done in the case of electric field).*

It should be emphasized that *magnetic field does not act on an electric charge at rest*. In this respect, magnetic field essentially differs from electric field. Magnetic field acts only on moving charges.

Vector **B** characterizes the force acting due to magnetic field on a moving charge, and hence in this respect it is an analog of vector **E** characterizing the force acting due to electric field.

A distinctive feature of magnetic force is that it is always perpendicular to the velocity vector of the charge. Therefore, *no work is done over the charge*. This means that the energy of a charged particle moving in a permanent magnetic field always remains unchanged irrespective of the motion of the particle.

In the nonrelativistic approximation, the Lorentz force (6.2), like any other force, does not depend on the choice

---

* Several methods of measuring field **B** have been worked out. In the final analysis, they all are based on the phenomena that can be described by Eq. (6.2).

of the (inertial) reference system. On the other hand, the magnetic component of the Lorentz force varies upon a transition from one frame of reference to another (because of **v**). Therefore, the electric component $q\mathbf{E}$ must also change. Hence it follows that the decomposition of the total force **F** (the Lorentz force) into the electric and magnetic components depends on the choice of the reference system. Without specifying the reference system, such a decomposition is meaningless.

Magnetic Field of a Uniformly Moving Charge. Experiments show that magnetic field is generated by moving charges (currents). As a result of the generalization of experimental results, the elementary law defining the field **B** of a point charge q moving at a constant nonrelativistic velocity **v** was obtained. This law is written in the form*

$$\boxed{\mathbf{B}=\frac{\mu_0}{4\pi}\frac{q\,[\mathbf{v}\times\mathbf{r}]}{r^3},} \tag{6.3}$$

where μ_0 is the *magnetic constant*, the coefficient

$$\mu_0/4\pi = 10^{-7}\ \mathrm{H/m},$$

and **r** is the radius vector from the point charge q to the point of observation. The tail of the radius vector **r** is fixed in a given frame of reference, while its tip moves with velocity **v** (Fig. 6.1). Consequently, vector **B** in the given reference frame depends not only on the position of the point of observation but on time as well.

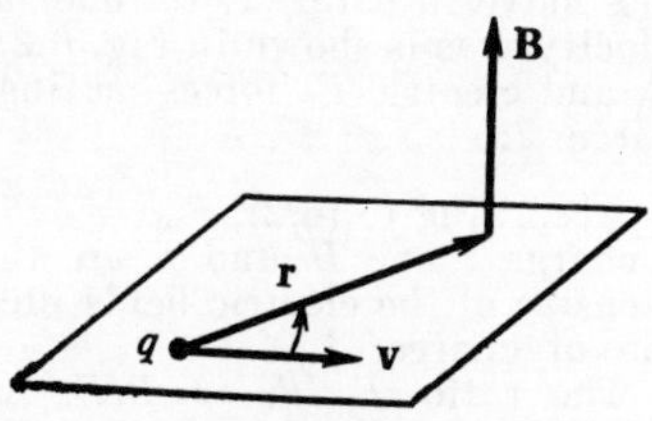

Fig. 6.1

In accordance with formula (6.3), vector **B** is directed normally to the plane containing vectors **v** and **r**, the rotation around vector **v** in the direction of vector **B** forming

* Formula (6.3) is also valid in the case when the charge moves with an acceleration, but only at sufficiently small distances r from the charge (so small that the velocity **v** of the charge does not noticeably change during the time r/c).

the right-handed system with vector **v** (Fig. 6.1). It should be noted that vector **B** is axial (pseudovector).

The quantity **B** is called *magnetic induction.*

Magnetic induction is measured in *teslas* (T).

The electric field of a point charge q moving at a nonrelativistic velocity is described by the same law (1.2). Hence expression (6.3) can be written in the form

$$\mathbf{B} = \varepsilon_0\mu_0\,[\mathbf{v}\times\mathbf{E}] = [\mathbf{v}\times\mathbf{E}]/c^2, \qquad (6.4)$$

where c is the *electrodynamic constant* ($c = 1/\sqrt{\varepsilon_0\mu_0}$), equal

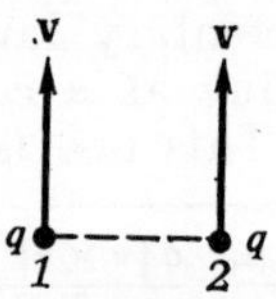

Fig. 6.2

to the velocity of light in vacuum (this coincidence is not accidental).

Example. Comparison of forces of magnetic and electric interaction between moving charges. Let two point charges q of a sufficiently large mass move in parallel to one another with the same nonrelativistic velocity **v** as is shown in Fig. 6.2. Find the ratio between the magnetic F_m and electric F_e forces acting, for example, from charge *1* on charge *2*.

According to (6.2), $F_m = qvB$ and $F_e = qE$ where v is the velocity of charge *2* and B and E are the induction of the magnetic and the intensity of the electric fields created by charge *1* at the point of location of charge *2*.

The ratio $F_m/F_e = vB/E$. According to (6.4), in our case $B = vE/c^2$, and hence

$$F_m/F_e = (v/c)^2. \qquad (6.5)$$

Even at sufficiently high velocities, e.g. $v = 300$ km/s, this ratio is equal to 10^{-6}, i.e. the magnetic component of the force is a millionth fraction of the electric component and constitutes a negligible correction to the electric force.

This example may give rise to the following question: Are such forces worth investigating? It turns out that they are, and there are two sound reasons behind this.

First, we have to deal with beams of particles moving at velocities close to the velocity of light, for which this "correction" and the electric force become comparable (it should be noted that relation (6.5) is also valid for relativistic velocities).

Second, during the motion, say, of electrons, along wires, their directional velocity amounts to several tens of a millimeter per second at normal densities, while the ratio $(v/c)^2 \approx 10^{-24}$. It is indeed a negligible correction to the electric force! But as a matter of fact, in this case the magnetic force is practically the *only* force since electric forces disappeared as a result of an almost ideal balance (much more perfect than 10^{-24}) of negative and positive charges in the wires. The participation of a vast number of charges in creating current compensates for the smallness of this term.

In other words, the excess charges on the wires are negligibly small in comparison with the total charge of carriers. For this reason, magnetic forces in this case considerably exceed the electric forces acting on excess charges of the wires.

6.2. The Biot-Savart Law

The Principle of Superposition. Experiments show that magnetic fields, as well as electric fields, obey the principle of superposition: the magnetic field created by several moving charges or currents is equal to the vector sum of the magnetic fields created by each charge or current separately:

$$\boxed{\mathbf{B} = \sum \mathbf{B}_i.} \tag{6.6}$$

The Biot-Savart Law. Let us consider the problem of determining the magnetic field created by a *direct* electric current. We shall solve this problem on the basis of law (6.3) determining the induction **B** of the field of a uniformly moving point charge. We substitute into (6.3) the charge $\rho\, dV$ for q (where dV is the volume element and ρ is the volume charge density) and take into account that in accordance with (5.2) $\rho\mathbf{v} = \mathbf{j}$. Then formula (6.3) becomes

$$d\mathbf{B} = \frac{\mu_0}{4\pi} \frac{[\mathbf{j} \times \mathbf{r}]\, dV}{r^3}. \tag{6.7}$$

If the current I flows along a *thin* wire with the cross-sectional area ΔS, we have

$$j\, dV = j\, \Delta S\, dl = I\, dl,$$

where dl is an element of the length of the wire. Introducing vector $d\mathbf{l}$ in the direction of the current I, we can write this expression as follows:

$$\mathbf{j}\, dV = I\, d\mathbf{l}. \tag{6.8}$$

Vectors $\mathbf{j}\, dV$ and $I\, d\mathbf{l}$ are called *volume* and *linear current elements* respectively. Replacing in formula (6.7) the volume current element by the linear one, we obtain

$$d\mathbf{B} = \frac{\mu_0}{4\pi} \frac{I\, [d\mathbf{l} \times \mathbf{r}]}{r^3}. \tag{6.9}$$

Formulas (6.7) and (6.9) express the *Biot-Savart law*.

In accordance with the principle of superposition the total field $\mathbf{B}$ is found as a result of integration of Eq. (6.7) or (6.9) over all current elements:

$$\mathbf{B} = \frac{\mu_0}{4\pi} \int \frac{[\mathbf{j} \times \mathbf{r}]\, dV}{r^3}, \quad \mathbf{B} = \frac{\mu_0}{4\pi} \oint \frac{I\, [d\mathbf{l} \times \mathbf{r}]}{r^3}. \tag{6.10}$$

Generally, calculation of the magnetic induction of a current of an arbitrary configuration by these formulas is complicated. However, the calculations can be considerably simplified if current distribution has a certain symmetry. Let us consider several simple examples of determining the magnetic induction of current.

Example 1. Magnetic field of the line current, i.e. the current flowing along a thin straight wire of infinite length (Fig. 6.3).

In accordance with (6.9), vectors $d\mathbf{B}$ from all current elements have at a point A the same direction, viz. are directed behind the plane of the figure. Therefore, the summation of vectors $d\mathbf{B}$ can be replaced by the summation of their magnitudes dB, where

$$dB = \frac{\mu_0}{4\pi} \frac{I\, dl \cos\alpha}{r^2}.$$

It is clear from the figure that $dl\cos\alpha = r\,d\alpha$ and $r = b/\cos\alpha$. Hence

$$dB = \frac{\mu_0}{4\pi}\,\frac{I\cos\alpha\,d\alpha}{b}.$$

Integrating this expression over all current elements, which is equivalent to the integration over α between $-\pi/2$ and $\pi/2$, we find

$$B = \frac{\mu_0}{4\pi}\,\frac{2I}{b}. \tag{6.11}$$

Example 2. Magnetic field at the axis of circular current. Figure 6.4 shows vector $d\mathbf{B}$ from the current element $I\,dl$ located to the

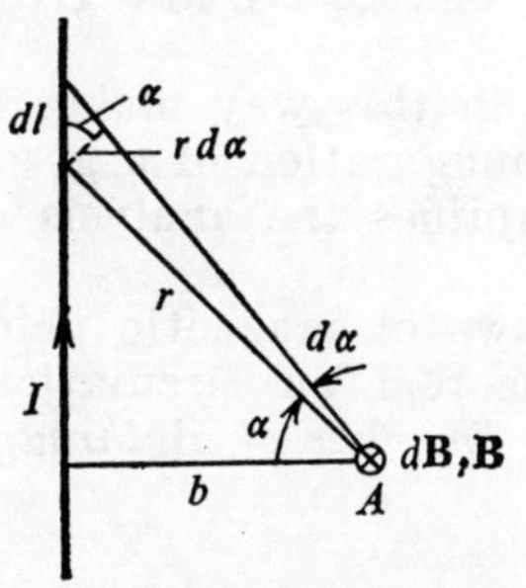

Fig. 6.3

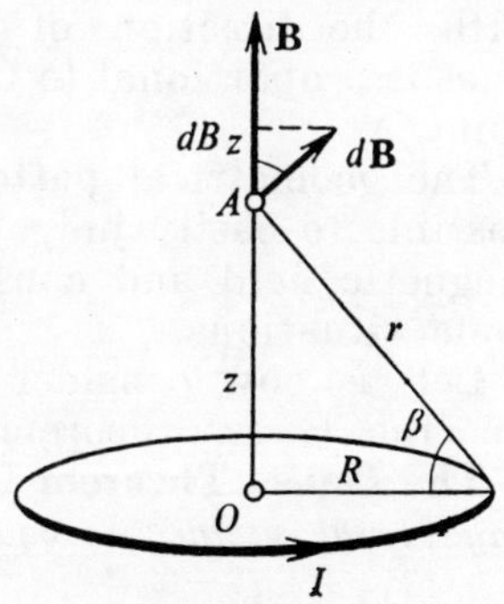

Fig. 6.4

right. All current elements will form the cone of vectors $d\mathbf{B}$, and it can be easily seen that the resultant vector $\mathbf{B}$ at point A will be directed upwards along the Z-axis. This means that in order to find the magnitude of vector $\mathbf{B}$, it is sufficient to sum up the components of vectors $d\mathbf{B}$ along the Z-axis. Each such projection has the form

$$dB_z = dB\cos\beta = \frac{\mu_0}{4\pi}\,\frac{I\,dl}{r^2}\cos\beta,$$

where we took into account that the angle between the element dl and the radius vector $\mathbf{r}$ is equal to $\pi/2$, and hence the sine is equal to unity. Integrating this expression over dl (which gives $2\pi R$) and taking into account that $\cos\beta = R/r$ and $r = (z^2 + R^2)^{1/2}$, we get

$$B = \frac{\mu_0}{4\pi}\,\frac{2\pi R^2 I}{(z^2+R^2)^{3/2}}. \tag{6.12}$$

Hence it follows that the magnitude of vector $\mathbf{B}$ at the centre of the current ring ($z = 0$) and at a distance $z \gg R$ is given by

$$B_{z=0} = \frac{\mu_0}{4\pi}\,\frac{2\pi I}{R}, \quad B_{z\gg R} \approx \frac{\mu_0}{4\pi}\,\frac{2\pi R^2 I}{z^3}. \tag{6.13}$$

6.3. Basic Laws of Magnetic Field

Like electric field, magnetic field has two very important properties. These properties, which are also related to the flux and circulation of a vector field, express the basic laws of magnetic field.

Before analysing these laws, we should consider the *graphic representation* of field **B**. Just as any vector field, field **B** can be visually represented with the help of the lines of vector **B**. They are drawn in a conventional way, viz. so that the tangent to these lines at any point coincides with the direction of vector **B**, and the density of the lines is proportional to the magnitude of vector **B** at a given point.

The geometrical pattern obtained in this way makes it possible to easily judge about the configuration of a given magnetic field and considerably simplifies the analysis of some situations.

Let us now consider the basic laws of magnetic field, i.e. the Gauss theorem and the theorem on circulation.

The Gauss Theorem for Field B. *The flux of* **B** *through any closed surface is equal to zero*:

$$\boxed{\oint \mathbf{B}\, d\mathbf{S} = 0.} \tag{6.14}$$

This theorem is essentially a generalization of experience. It expresses in the form of a postulate the experimental result that the lines of vector **B** have neither beginning nor end. Therefore, the number of lines of vector **B**, emerging from any volume bounded by a closed surface S, is always equal to the number of lines entering this volume.

Hence follows an *important corollary* which will be repeatedly used below: *the flux of* **B** *through a closed surface* S *bounded by a certain contour does not depend on the shape of the surface* S. This can be easily grasped with the help of the concept of the lines of vector **B**. Since these lines are not discontinued anywhere, their number through the surface S bounded by a given contour (i.e. the flux of **B**) indeed must be independent of the shape of the surface S.

Law (6.14) also expresses the fact that there are no magnetic charges in nature, on which the lines of vector **B** begin or terminate. In other words, magnetic field has no sources, in contrast to electric field.

Theorem on Circulation of Vector B (for the magnetic field of a direct current in a vacuum). *Circulation of vector* **B** *around an arbitrary contour* Γ *is equal to the product of* μ_0 *by the algebraic sum of the currents enveloped by the contour* Γ:

$$\boxed{\oint \mathbf{B}\, d\mathbf{l} = \mu_0 I,} \tag{6.15}$$

where $I = \sum I_k$, and I_k are algebraic quantities. The current is assumed positive if its direction is connected with the direction of the circumvention of the contour through the right-hand screw rule. The current having the opposite direction is considered to be negative. This rule is illustrated in Fig. 6.5: here currents I_1 and I_3 are positive since their directions are connected with the direction of contour circumvention through the right-hand screw rule, while current I_2 is negative.

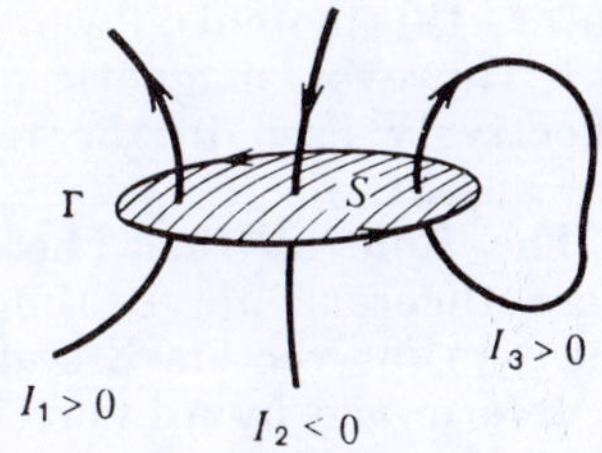

Fig. 6.5

The theorem on circulation (6.15) can be proved on the basis of the Biot-Savart law. In the general case of arbitrary currents, the proof is rather cumbersome and will not be considered here. We shall treat statement (6.15) as a postulate verified experimentally.

Let us make one more remark. If current I in (6.15) is distributed over the volume where contour Γ is located, it can be represented in the form

$$I = \int \mathbf{j}\, d\mathbf{S}. \tag{6.16}$$

In this expression, the integral is taken over an arbitrary surface S stretched on contour Γ. Current density $\mathbf{j}$ in the integrand corresponds to the point where the area element $d\mathbf{S}$ is located, vector $d\mathbf{S}$ forming the right-handed system with the direction of the contour circumvention.

Thus, in the general case Eq. (6.15) can be written as

follows:

$$\oint \mathbf{B}\, d\mathbf{l} = \mu_0 \int \mathbf{j}\, d\mathbf{S} = \mu_0 \int j_n\, dS. \tag{6.17}$$

The fact that circulation of vector **B** generally differs from zero indicates that in contrast to electrostatic field, field **B** is not a potential field. Such fields are called *vortex*, or *solenoidal* fields.

Since circulation of vector **B** is proportional to the current I enveloped by the contour, in general we cannot ascribe to magnetic field a scalar potential which would be related to vector **B** through an expression similar to $\mathbf{E} = -\nabla\varphi$. This potential would not be single-valued: upon each circumvention of the current and return to the initial point, this potential would acquire an increment equal to $\mu_0 I$. However, magnetic potential φ_m can be introduced and effectively used in the region of space where the currents are absent.

The Role of the Theorem on Circulation of Vector B. This theorem plays almost the same role as the Gauss theorem for vectors **E** and **D**. It is well known that field **B** is determined by all currents, while circulation of vector **B**, only by the currents enveloped by the given contour. In spite of this, in certain cases (in the presence of a special symmetry) the theorem on circulation proves to be quite effective since it allows us to determine **B** in a very simple way.

This can be done in the cases when the calculation of circulation of vector **B** can be reduced, by an appropriate choice of the contour, to the product of B (or B_l) by the length of the contour or its part. Otherwise, field **B** must be calculated by some other methods, for example, with the help of the Biot-Savart law or by solving the corresponding differential equations, and the calculation becomes much more difficult.

6.4. Applications of the Theorem on Circulation of Vector B

Let us consider several practically important examples illustrating the effectiveness of the application of the theorem on circulation of vector **B** and then see whether this method is universal.

Example 1. Magnetic field of a straight current. Direct current I flows along an infinitely long straight wire with a circular cross section of radius a. Find the magnetic field induction **B** outside and inside the wire.

It follows from the symmetry of the problem that the lines of vector **B** in this case have the form of circles with the centre at the

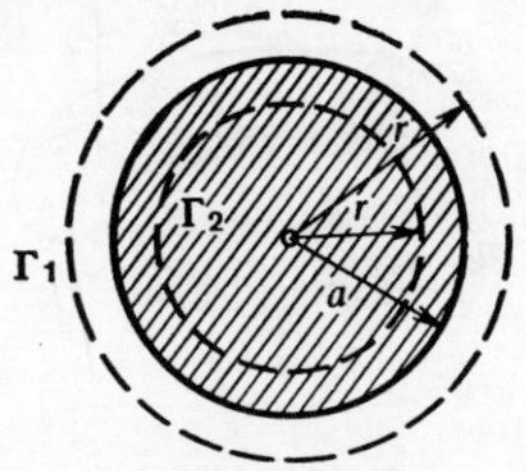

Fig. 6.6

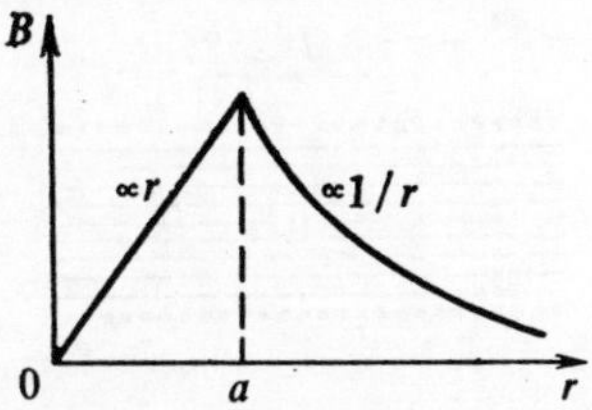

Fig. 6.7

wire axis. The magnitude of vector **B** must be the same at all points at a distance r from the axis of the wire. Therefore, in accordance with the theorem on circulation of vector **B** for a circular contour Γ_1 (Fig. 6.6), we have $B\cdot 2\pi r = \mu_0 I$, whence it follows that outside the wire

$$B = (\mu_0/2\pi)\, I/r \quad (r \geqslant a). \tag{6.18}$$

It should be noted that a direct solution of this problem (with the help of the Biot-Savart law) turns out to be more complicated.

From the same symmetry considerations it follows that inside the wire the lines of vector **B** are also circles. According to the theorem on circulation of vector **B** for a circular contour Γ_2 (see Fig. 6.6), $B\cdot 2\pi r = \mu_0 I_r$, where $I_r = I\,(r/a)^2$ is the current enveloped by the given contour. Whence we find that inside the wire

$$B = (\mu_0/2\pi)\, Ir/a^2 \quad (r \leqslant a) \tag{6.19}$$

The dependence $B\,(r)$ is shown in Fig. 6.7.

If the wire has the shape of a tube, the induction B outside it is determined by formula (6.18), while inside the tube magnetic field is absent. This can also be easily shown with the help of the theorem on circulation of vector **B**.

Example 2. Magnetic field of a solenoid. Let current I flow along a conductor helically wound on the surface of a cylinder. Such cylinder streamlined by the current is called a *solenoid*. Suppose that a unit length of the solenoid contains n turns of the conductor. If the pitch of the helix is sufficiently small, each turn of the solenoid can be approximately replaced by a closed loop. We shall also assume that the conductor cross-sectional area is so small that the current in the solenoid can be considered flowing over its surface.

Experiments and calculations show that the longer the solenoid, the less the magnetic induction outside it. For an infinitely long solenoid, magnetic field outside it is absent at all.

It is clear from symmetry considerations that the lines of vector **B** inside the solenoid are directed along its axis, vector **B** forming the right-handed system with the direction of the current in the solenoid.

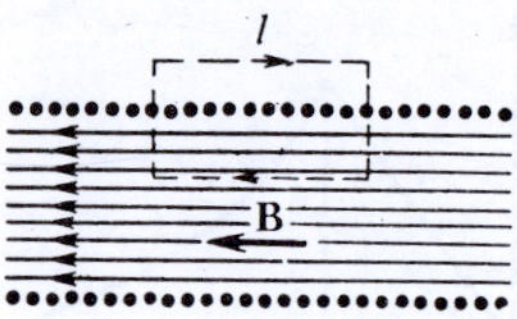

Fig. 6.8

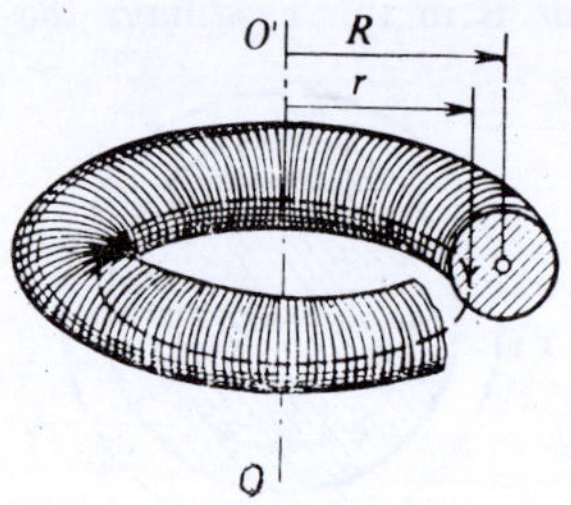

Fig. 6.9

Even the above considerations about the magnetic field configuration of the solenoid indicate that we must choose a rectangular contour as shown in Fig. 6.8. Circulation of vector **B** around this contour is equal to Bl, and the contour envelops the current nlI. According to the theorem on circulation, $Bl = \mu_0 nlI$, whence it follows that inside a long solenoid

$$B = \mu_0 nI, \tag{6.20}$$

i.e. the field inside a long solenoid is uniform (with the exception of the regions adjoining the solenoid's endfaces, which is often ignored in calculations). The product nI is called the *number of ampere-turns*. For $n = 2000$ turns/m and $I = 2$ A, the magnetic field inside the solenoid is equal to 5 mT.

Example 3. Magnetic field of a toroid. A toroid is a wire wound around a torus (Fig. 6.9).

It can be easily seen from symmetry considerations that the lines of vector **B** must be circles whose centres lie on the axis OO' of the toroid. Hence it is clear that for the contour we must take one of such circles.

If a contour lies inside the toroid, it envelops the current NI, where N is the number of turns in the toroidal coil and I is the current in the wire. Let the contour radius be r; then, according to the theorem on circulation $B \cdot 2\pi r = \mu_0 NI$, whence it follows that inside the toroid

$$B = (\mu_0/2\pi)\, NI/r. \tag{6.21}$$

A comparison of (6.21) and (6.18) shows that the magnetic field inside the toroid coincides with the magnetic field NI of the straight current flowing along the OO' axis. Tending N and radius R of the toroid to infinity (at a constant toroid cross section), we obtain in the limit the expression (6.20) for the magnetic field of an infinitely long solenoid.

If the chosen circular contour passes outside the toroid, it does not envelop any currents, and hence $B \cdot 2\pi r = 0$ for such a contour. This means that outside the toroid magnetic field is absent.

In the above considerations it was assumed that the lines of current lie in meridional planes, i.e. the planes passing through the axis OO' of the toroid. In real toroids, the lines of current (turns) do not lie strictly in these planes, and hence there is a current component around the OO' axis. This component creates an additional field similar to the field of a circular current.

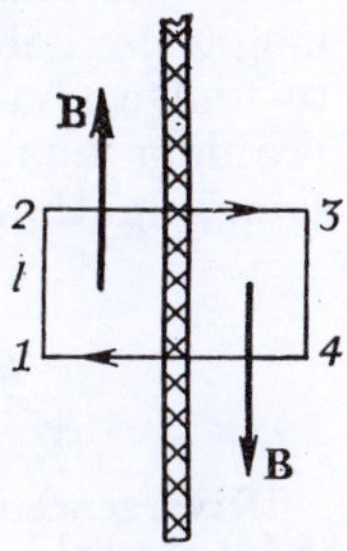

Fig. 6.10

Example 4. Magnetic field of a current-carrying plane. Let us consider an infinite conducting plane with undirectional current uniformly distributed over its surface. Figure 6.10 shows the cross section of such a plane when the current flows behind the plane of the figure (this is marked by crosses). Let us introduce the concept of *linear current density* **i** directed along the lines of current. The magnitude of this vector is the current per unit length which plays the role of the "cross-sectional area".

Mentally dividing the current-carrying plane into thin current filaments, we can easily see that the resultant field **B** will be directed parallel to the plane, downwards to the right of the plane and upwards to the left of it (Fig. 6.10). These directions can be easily established with the help of the right-hand screw rule.

In order to determine induction **B** of the field, we shall use the theorem on circulation of vector **B**. Knowing the arrangement of the lines of vector **B**, we shall choose the contour in the form of rectangle *1234* (Fig. 6.10). Then, in accordance with the theorem on circulation, $2Bl = \mu_0 il$, where l is the length of the contour side parallel to the conducting plane. This gives

$$B = \frac{1}{2}\mu_0 i. \tag{6.22}$$

This formula shows that the magnetic field is uniform on both sides of the plane. This result is also valid for a bounded current-carrying plane, but only for the points lying near the plane and far from its ends.

General Considerations. The results obtained in the above examples could be found directly with the help of the Biot-Savart law. However, the theorem on circulation makes it possible to obtain these results much more simply and quickly.

However, the simplicity with which the field was calculated in these examples must not produce an erroneous impression about the potentialities of the method based on the application of the theorem on circulation. Just as in the case of the Gauss theorem for electric field, the number of problems that can be easily solved by using the theorem on circulation of vector **B** is quite limited. It is sufficient to say that even for such a symmetric current configuration as the current ring, the theorem on circulation becomes helpless. Despite an apparently high symmetry, the magnetic field configuration does not allow us, however, to find a simple contour required for calculations, and the problem has to be solved by other, much more cumbersome, methods.

6.5. Differential Forms of Basic Laws of Magnetic Field

Divergence of field B. The differential form of the Gauss theorem (6.14) for field **B** can be written as

$$\boxed{\nabla \cdot \mathbf{B} = 0,} \tag{6.23}$$

i.e. the *divergence of field* **B** *is equal to zero everywhere*. As was mentioned above, this means that magnetic field has no sources (magnetic charges). Magnetic field is generated by electric currents and not by magnetic charges which do not exist in nature.

Law (6.23) is of *fundamental* nature: it is valid not only for constant but for varying fields as well.

Curl of field B. The important property of magnetic field expressed by the theorem on circulation of vector **B** motivates the representation of this theorem in the differential form which broadens its potentialities as a tool for investigations and calculations.

For this purpose, we consider the ratio of circulation of vector **B** to the area S bounded by a contour. It turns out that this ratio tends to a certain limit as $S \to 0$. This limit depends on the contour orientation at a given point of space. The contour orientation is specified by vector **n** (the normal to the plane of the contour around which the circulation is calculated) the direction of **n** being connected with the direction of contour circumvention through the right-hand screw rule.

The limit obtained as a result of this operation is a scalar quantity which behaves as the projection of a certain vector onto the direction of the normal **n** to the plane of the contour around which the circulation is taken. This vector is called the *curl of field* **B** and is denoted as curl **B**. Thus, we can write

$$\lim_{S \to 0} \frac{\oint \mathbf{B}\, d\mathbf{l}}{S} = (\mathrm{curl}\, \mathbf{B})_n, \tag{6.24}$$

where on the right-hand side we have the projection of curl **B** onto normal **n**.

Hence, each point of vector field **B** can be characterized by curl **B** whose direction and magnitude are determined by the properties of the field itself at a given point. The direction of curl **B** is determined by the direction of normal **n** to the surface element S, for which the quantity (6.24), which defines the magnitude of curl **B**, attains its maximum value.

In mathematics, curl **B** is obtained in coordinate representation. However, for our purposes another fact is more important: it turns out that formally curl **B** can be considered as the cross product of the operator ∇ by vector **B**, i.e. $\nabla \times \mathbf{B}$. We shall be using the latter, more convenient notation since it allows us to represent the cross product $\nabla \times \mathbf{B}$ with the help of the determinant:

$$\nabla \times \mathbf{B} = \begin{vmatrix} \mathbf{e}_x & \mathbf{e}_y & \mathbf{e}_z \\ \partial/\partial x & \partial/\partial y & \partial/\partial z \\ B_x & B_y & B_z \end{vmatrix}, \tag{6.25}$$

where $\mathbf{e}_x$, $\mathbf{e}_y$, $\mathbf{e}_z$ are the unit vectors of Cartesian coordinates. This expression is valid for the curl of not only field **B**, but of any other vector field also, in particular, of field **E**.

Let us now consider the theorem on circulation of vector **B**. According to (6.24), Eq. (6.17) can be represented in the form

$$\lim_{S\to 0} \frac{\oint \mathbf{B}\, d\mathbf{l}}{S} = \mu_0 j_n,$$

or $(\nabla \times \mathbf{B})_n = \mu_0 j_n$. Hence

$$\boxed{\nabla \times \mathbf{B} = \mu_0 \mathbf{j}.} \tag{6.26}$$

This is precisely the differential form of the theorem on circulation of vector **B**. Obviously, the direction of curl **B** coincides with that of vector **j**, viz. the current density at this point, while the magnitude of $\nabla \times \mathbf{B}$ is equal to $\mu_0 j$.

In an electrostatic field, circulation of vector **E** is equal to zero, and hence

$$\boxed{\nabla \times \mathbf{E} = 0.} \tag{6.27}$$

A vector field whose curl is equal to zero everywhere is a potential field. Otherwise it is called the *solenoidal field*. Consequently, *electrostatic field is a potential field, while magnetic field is a solenoidal field.*

6.6. Ampère's Force

Ampère's Law. Each charge carrier experiences the action of a magnetic force. The action of this force is transmitted to a conductor through which charges are moving.

As a result, magnetic field acts with a certain force on a current-carrying conductor. Let us find this force.

Suppose that the volume density of charges which are the carriers of current (e.g. electrons in a metal) is equal to ρ. We mentally isolate a volume element dV in the conductor. It contains the charge (current carrier) equal to $\rho\, dV$. Then the force acting on the volume element dV of the conductor can be written by formula (6.1) as follows:

$$d\mathbf{F} = \rho\,[\mathbf{u} \times \mathbf{B}]\, dV.$$

Since $\mathbf{j} = \rho\mathbf{u}$, we have

$$\boxed{d\mathbf{F} = [\mathbf{j} \times \mathbf{B}]\, dV.} \tag{6.28}$$

If the current flows along a thin conductor, then, according to (6.8), $\mathbf{j}\, dV = I\, d\mathbf{l}$, and

$$\boxed{d\mathbf{F} = I\,[d\mathbf{l} \times \mathbf{B}],} \tag{6.29}$$

where $d\mathbf{l}$ is the vector coinciding in direction with the current and characterizing an element of length of the thin conductor.

Formulas (6.28) and (6.29) express *Ampère's law*. Integrating these expressions over the current elements (volume or linear), we can find the magnetic force acting on a certain volume of a conductor or on its linear part.

The forces acting on currents in a magnetic field are called *Ampère's forces*.

Example. The force of interaction between parallel currents. Find the Ampère's force with which two infinitely long wires with currents I_1 and I_2 interact in a vacuum, if the distance between them is b. Calculate the force per unit length of the system.

Each current element of I_2 is in the magnetic field of the current I_1, i.e. according to (6.19), the field $B_1 = (\mu_1/4\pi)\, 2I_1/b$. The angle between the element of current I_2 and vector $\mathbf{B}_1$ is $\pi/2$. Hence, it follows from formula (6.29) that the force acting per unit length of the conductor with current I_2 is $F_u = I_2 B_1$, or

$$F_u = \frac{\mu_0}{4\pi}\,\frac{2I_1 I_2}{b} \tag{6.30}$$

Naturally, the same expression can be obtained for the force acting per unit length of the conductor with current I_1.

Finally, it can be easily seen that currents of the same direction are attracted, while the opposite currents repell each other. Here we speak only about a magnetic force. It should be recalled, however, that besides magnetic forces there are electric forces, i.e. the forces due to excess charges on the surface of conductors. Consequently, if we speak about the total force of interaction between the wires, it can either be attractive or repulsive, depending on the ratio between the magnetic and electric components of the total force (see Problem 6.7).

The Force Acting on a Current Loop. The resultant Ampère's force acting on a current loop in a magnetic field is defined, in accordance with (6.29), by

$$\mathbf{F} = I \oint [d\mathbf{l} \times \mathbf{B}], \tag{6.31}$$

where the integration is performed along the given loop with current I.

If the magnetic field is uniform, vector $\mathbf{B}$ can be taken out of the integral, and the problem is reduced to the calculation of the integral $\oint d\mathbf{l}$. This integral is a closed chain of elementary vectors $d\mathbf{l}$ and hence is equal to zero. Consequently, $\mathbf{F} = 0$, i.e. the resultant Ampère's force in a uniform magnetic field is equal to zero.

If, however, the magnetic field is nonuniform, the resultant force (6.31) generally differs from zero and in each concrete case is determined with the help of (6.31). The most interesting case for further analysis is that of a plane current loop of sufficiently small size. Such a current loop is called *elementary*.

The behaviour of an elementary current loop can be conveniently described with the help of *magnetic moment* $\mathbf{p}_m$. By definition, magnetic moment $\mathbf{p}_m$ is given by

$$\mathbf{p}_m = IS\mathbf{n}, \tag{6.32}$$

where I is the current, S is the area bounded by the loop and $\mathbf{n}$ is the normal to the loop whose direction is connected with the direction of the current in the loop through the right-hand screw rule (Fig. 6.11). In terms of magnetic field, the elementary loop is completely characterized by its magnetic moment $\mathbf{p}_m$.

An involved calculation with the help of formula (6.31), taking into account the small size of the loop, yields the following expression for the force acting on an elementary current loop in a nonuniform magnetic field:

$$\boxed{\mathbf{F} = p_m \frac{\partial \mathbf{B}}{\partial n},} \tag{6.33}$$

where p_m is the *magnitude* of the magnetic moment of the loop, and $\partial\mathbf{B}/\partial n$ is the derivative of vector $\mathbf{B}$ along the

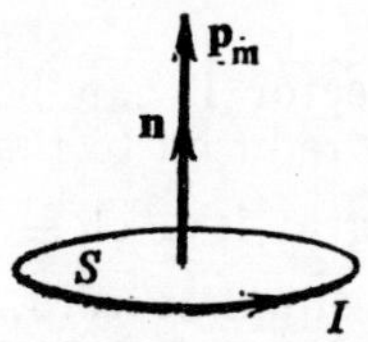

Fig. 6.11

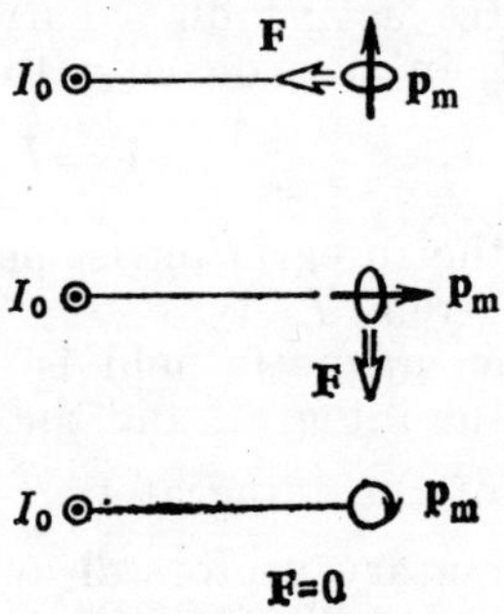

Fig. 6.12

direction of normal $\mathbf{n}$ or along the direction of vector $\mathbf{p}_m$. This formula is similar to expression (1.39) for the force acting on an electric dipole in an electric field.

It follows from formula (6.33) that, like in the case of electric dipole,

(1) in a uniform magnetic field $\mathbf{F} = 0$ since $\partial\mathbf{B}/\partial n = 0$;

(2) the direction of vector $\mathbf{F}$ generally does not coincide with vector $\mathbf{B}$ or with vector $\mathbf{p}_m$; vector $\mathbf{F}$ coincides in direction only with the elementary *increment* of vector $\mathbf{B}$ taken in the direction of vector $\mathbf{p}_m$ at the locus of the loop. This is illustrated by Fig. 6.12, where three arrangements of the loop in the magnetic field of straight current I_0 are shown. The vector of the resultant force $\mathbf{F}$ acting on the loop in each case is also shown in the figure (it is useful to verify independently that it is really so).

If we are interested in the projection of force $\mathbf{F}$ onto a certain direction X, it is sufficient to write expression

(6.33) in terms of the projections onto this direction. This gives

$$F_x = p_m \frac{\partial B_x}{\partial n} \tag{6.34}$$

where $\partial B_x/\partial n$ is the derivative of the corresponding projection of vector **B** again with respect to normal **n** (or with respect to $\mathbf{p}_m$) to the loop.

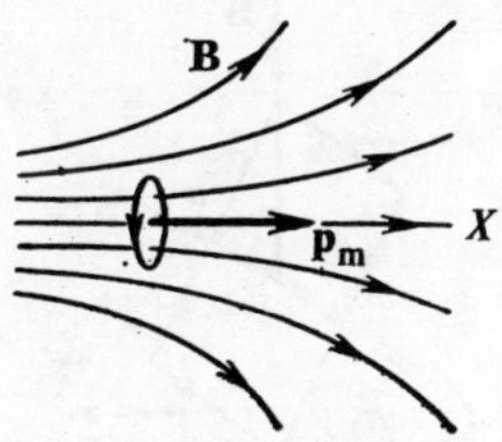

Fig. 6.13

Example. An elementary current loop having a magnetic moment $\mathbf{p}_m$ is arranged perpendicularly to the symmetry axis of a nonuniform magnetic field, vector $\mathbf{p}_m$ being directed along vector **B**. Let us choose the positive direction of the X-axis as is shown in Fig. 6.13. Since the increment of projection B_x will be negative in the direction of vector $\mathbf{p}_m$, $F_x < 0$. Hence vector **F** is directed to the left, i.e. to the side where **B** is greater. If we rotate the current loop (and vector $\mathbf{p}_m$) through 90° so that the loop centre coincides with the symmetry axis of field **B**, in this position $F_x = 0$, and vector **F** is directed perpendicularly to the X-axis and to the same side as vector $\mathbf{p}_m$.

6.7. Torque Acting on a Current Loop

Let us consider a plane loop with a current I in a *uniform* magnetic field **B**. It was shown above (see p. 157) that the resultant force (6.31) acting on the current loop in a uniform magnetic field is equal to zero. It is known from mechanics that if the resultant of the forces acting on any system is equal to zero, the resultant moment of these forces does not depend on the position of point O relative to which the torques are determined. Therefore, we can speak about the resultant moment of Ampère's forces in our case.

By definition, the resultant moment of Ampère's forces is given by

$$\mathbf{M} = \oint [\mathbf{r} \times d\mathbf{F}], \tag{6.35}$$

where $d\mathbf{F}$ is defined by formula (6.29). If we perform calculation by formula (6.35) (it is cumbersome and hardly interesting, hence we omit it here), it turns out that the moment of forces for an arbitrary shape of the loop can be represented in the form

$$\boxed{\mathbf{M} = [\mathbf{p}_m \times \mathbf{B}],} \tag{6.36}$$

where $\mathbf{p}_m$ is the magnetic moment of the current loop ($\mathbf{p}_m = IS\mathbf{n}$ for a plane loop).*

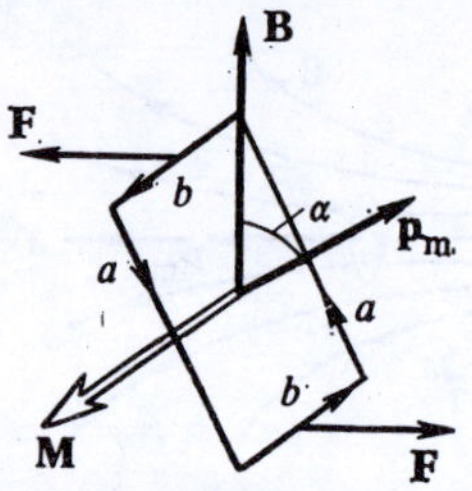

Fig. 6.14

It is clear from (6.36) that the moment $\mathbf{M}$ of Ampère's forces acting on a current loop in a uniform magnetic field is perpendicular both to vector $\mathbf{p}_m$ and to vector $\mathbf{B}$. The magnitude of vector $\mathbf{M}$ is $M = p_m B \sin \alpha$, where α is the angle between vectors $\mathbf{p}_m$ and $\mathbf{B}$. When $\mathbf{p}_m \uparrow\uparrow \mathbf{B}$, $\mathbf{M} = 0$ and it is not difficult to see that the position of the loop is stable. When $\mathbf{p}_m \downarrow\uparrow \mathbf{B}$, $\mathbf{M}$ also equals zero but such a position of the loop is unstable: the slightest deviation from it leads to the appearance of the moment of force that tends to deviate the loop from the initial position still further.

Example. Let us verify the validity of formula (6.36) by using a simple example of a rectangular current loop (Fig. 6.14).

It can be seen from the figure that the forces acting on sides a are perpendicular to them and to vector $\mathbf{B}$. Hence these forces are directed along the horizontal (they are not shown in the figure) and only strive to stretch (or compress) the loop. Sides b are perpendicular to $\mathbf{B}$, and hence each of them is acted upon by the force

$$F = IbB.$$

These forces tend to rotate the loop so that vector $\mathbf{p}_m$ becomes directed like vector $\mathbf{B}$. Hence, the loop is acted upon by a couple of forces whose torque is equal to the product of the arm $a \sin \alpha$ of the couple by the force F, i.e.

$$M = IbBa \sin \alpha.$$

---

* If the loop is not plane, its magnetic moment is $\mathbf{p}_m = I \int d\mathbf{S}$, where the integral is taken over the surface S stretched over the current loop. This integral does not depend on the choice of the surface S and depends only on the loop over which it is stretched.

Considering that ab is the area bounded by the loop and $Iba = p_m$, we obtain

$$M = p_m B \sin \alpha,$$

which in the vector form is written as (6.36).

Concluding this section, we note that expression (6.36) is valid for nonuniform magnetic fields as well. The only requirement is that the size of the current loop must be sufficiently small. Then the effect of nonuniformity on the rotational moment (torque) **M** can be ignored. This precisely applies to the elementary current loop.

An elementary current loop behaves in a nonuniform magnetic field in the same way as an electric dipole in an external nonuniform electric field: it will be rotated towards the position of stable equilibrium (for which $\mathbf{p}_m \uparrow\uparrow \mathbf{B}$) and, besides, under the action of the resultant force **F** it will be pulled into the region where induction **B** is larger.

6.8. Work Done upon Displacement of Current Loop

When a current loop is in an external magnetic field (we shall assume that this field is constant), certain elements of the loop are acted upon by Ampère's forces which hence will do work during the displacement of the loop. We shall show here that the work done by Ampère's forces upon an elementary displacement of the loop with current I is given by

$$\delta A = I d\Phi, \qquad (6.37)$$

where $d\Phi$ is the increment of the magnetic flux through the loop upon the given displacement.

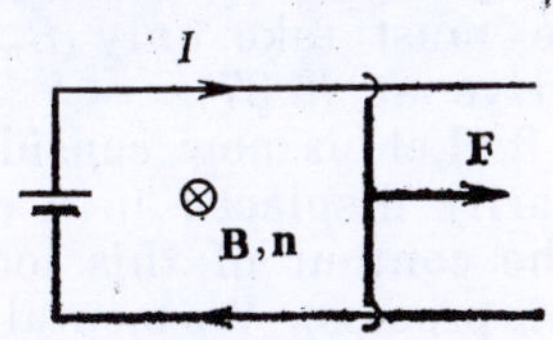

Fig. 6.15

We shall prove this formula in three stages.

1. Let us first consider a particular case: a circuit (Fig. 6.15) with a movable jumper of length l is in a uniform magnetic field perpendicular to the circuit plane and directed behind the plane of the figure. In accordance with (6.29), the jumper is acted upon by Ampère's force $F = IlB$. Displacing the jumper to the right by dx, this force accomplishes positive

work

$$\delta A = F\,dx = IBl\,dx = IB\,dS, \qquad (6.38)$$

where dS is the increment of the area bounded by the contour.

In order to determine the sign of magnetic flux Φ, we shall *agree* to take the normal $\mathbf{n}$ to the surface bounded by the contour always in such a way that it forms with the direction of the current in the circuit a right-handed system (Fig. 6.15). In this case, the current I will always be a positive quantity. On the other hand, the flux Φ may be either positive or negative. In our case, however, both Φ and $d\Phi = B\,dS$ are positive quantities (if the field $\mathbf{B}$ were directed towards us or the jumper were displaced to the left, $d\Phi < 0$). In any of these cases expression (6.38) can be represented in the form (6.37).

2. The obtained result is valid for an arbitrary direction of the field $\mathbf{B}$ as well. To prove this, let us decompose vector $\mathbf{B}$ into three components: $\mathbf{B} = \mathbf{B}_n + \mathbf{B}_l + \mathbf{B}_x$. The component $\mathbf{B}_l$ directed along the jumper is parallel to the current in it and does not produce any force acting on the jumper. The component $\mathbf{B}_x$ (along the displacement) is responsible for the force perpendicular to the displacement, which does not accomplish any work. The only remaining component is $\mathbf{B}_n$ which is normal to the plane in which the jumper moves. Hence, in formula (6.38) instead of B we must take only B_n. But $B_n\,dS = d\Phi$, and we again arrive at (6.37).

3. Let us now consider any current loop which is arbitrarily displaced in a constant nonuniform magnetic field (the contour of this loop may be arbitrarily deformed in this process). We mentally divide the given loop into infinitely small current elements and consider their infinitesimal displacements. Under these conditions, the magnetic field in which each element of current is displaced can be assumed uniform. For such a displacement, we can apply to each element of current the expression $dA = I\,d'\Phi$ for the elementary work, where $d'\Phi$ expresses the contribution of a given element of current to the increment of the flux through the contour. Summing up elementary works for all the elements of the loop, we again obtain expression

(6.37), where $d\Phi$ is the increment of the magnetic flux through the entire circuit.

In order to find the work done by Ampère's forces upon the total displacement of the current loop from the initial position *1* to the final position *2*, it is sufficient to integrate expression (6.37):

$$A = \int_1^2 I\, d\Phi. \tag{6.39}$$

If the current I is maintained constant during this displacement we obtain

$$A = I(\Phi_2 - \Phi_1), \tag{6.40}$$

where Φ_1 and Φ_2 are magnetic fluxes through the contour in the initial and final positions. Thus, the work of Ampère's forces in this case is equal to the product of the current by the increment of the magnetic flux through the circuit. Expression (6.40) gives not only the magnitude but also the sign of the accomplished work.

Example. A plane loop with current I is rotated in magnetic field $\mathbf{B}$ from the position in which $\mathbf{n} \downarrow\uparrow \mathbf{B}$ to the position in which $\mathbf{n} \uparrow\uparrow \mathbf{B}$, $\mathbf{n}$ being the normal to the loop (it should be recalled that the direction of the normal is connected with the direction of the current through the right-hand screw rule). The area bounded by the loop is S. Find the work of Ampère's forces upon such a displacement, assuming that the current I is maintained constant.

In accordance with (6.40), we have

$$A = I\,[BS - (-BS)] = 2IBS.$$

In the given case, the work $A > 0$, while upon the reverse rotation $A < 0$.

It should be noted that work (6.40) is accomplished not at the expense of the energy of the external magnetic field (which does not change) but at the expense of the e.m.f. source maintaining the current in the loop. (This question will be considered in greater detail in Chap. 9.)

Problems

● 6.1. **Direct calculation of induction B.** Current I flows along a thin conductor bent as is shown in Fig. 6.16. Find the magnetic induction B at point O. The required data are presented in the figure.

Solution. The required quantity $B = B_{-} + B_{\smile}$, where B_{-} is the magnetic field created by the rectilinear part of the loop and $B_{\smile}$, by its curvilinear part. According to the Biot-Savart law (see Example 1 on p. 146), we have

$$B_{-}=2\int_0^{\alpha_0}\frac{\mu_0 I\cos\alpha\,d\alpha}{4\pi a\cos\alpha_0}=\frac{\mu_0 I}{2\pi a}\,\mathrm{tg}\,\alpha_0,$$

$$B_{\smile}=\frac{\mu_0}{4\pi}\,\frac{I\,(2\pi-2\alpha_0)\,a}{a^2}=\frac{\mu_0 I}{2\pi a}\,(\pi-\alpha_0).$$

As a result, we obtain

$$B=(\pi-\alpha_0+\mathrm{tg}\,\alpha_0)\,\mu_0 I/2\pi a.$$

It is interesting to show that for $\alpha_0\to 0$, we arrive at the familiar expression (6.13).

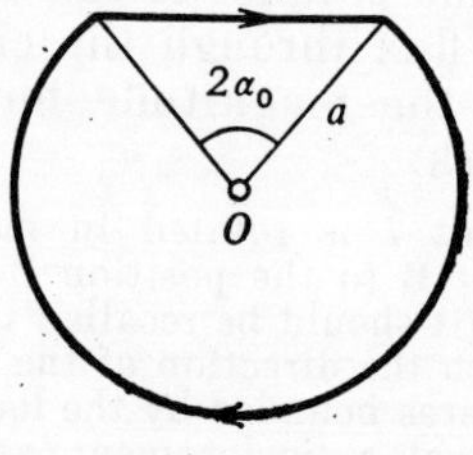

Fig. 6.16

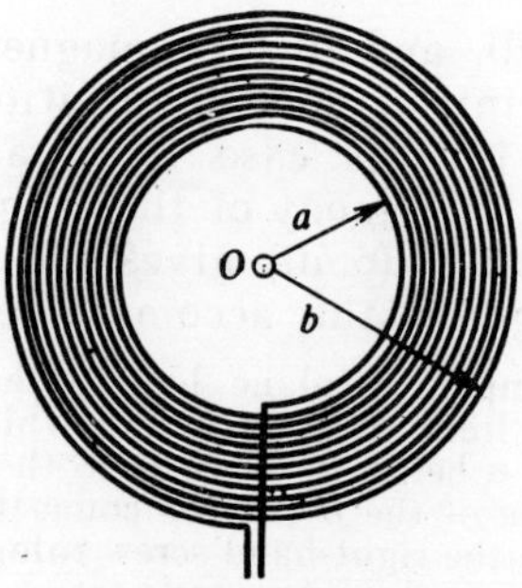

Fig. 6.17

● **6.2.** A thin isolated wire forms a plane spiral consisting of a large number N of closely packed turns through which a direct current I is flowing. The radii of the internal and external loops are a and b (Fig. 6.17). Find (1) the magnetic induction B at the centre of the spiral (point O), and (2) the magnetic moment of the spiral for the given current.

Solution. (1) According to (6.13), the contribution of one turn of radius r to the magnetic induction is equal to

$$B_1=\mu_0 I/2r, \tag{1}$$

and of all the turns,

$$B=\int B_1\,dN, \tag{2}$$

where dN is the number of turns in the interval $(r,\ r+dr)$,

$$dN=\frac{N}{b-a}\,dr.$$

Substituting (1) and (3) into (2) and integrating the result over r between a and b, we obtain

$$B=\frac{\mu_0 IN}{2(b-a)}\ln\frac{b}{a}.$$

(2) The magnetic moment of a turn of radius r is $p_{m1}=I\pi r^2$, and of all the turns, $p_m=\int p_{m1}\,dN$, here dN is defined by formula (3). The integration gives

$$p_m=\frac{\pi IN}{b-a}\,\frac{b^3-a^3}{3}=\frac{\pi IN}{3}(a^2+ab+b^2).$$

● **6.3.** Current I flows in a long straight conductor having the shape of a groove with a cross section in the form of a thin half-ring

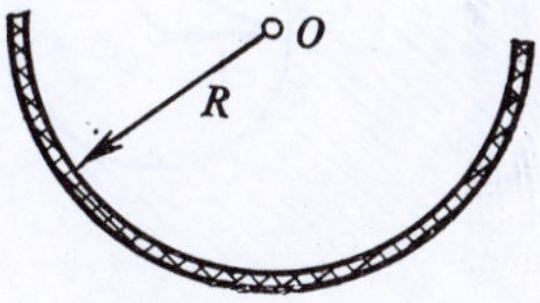

Fig. 6.18

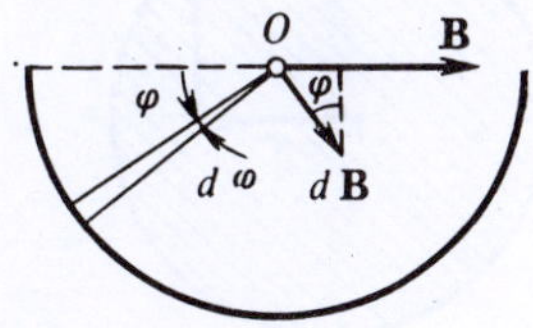

Fig. 6.19

of radius R (Fig. 6.18). The current is directed from the reader behind the plane of the drawing. Find the magnetic induction B on the axis O.

Solution. First of all, let us determine the direction of vector **B** at point O. For this purpose, we mentally divide the entire conductor into elementary filaments with currents dI. Then it is clear that the sum of any two symmetric filaments gives vector $d\mathbf{B}$ directed to the right (Fig. 6.19). Consequently, vector **B** will also be directed to the right.

Therefore, for calculating the field **B** at point O it is sufficient to find the sum of the projections of elementary vectors $d\mathbf{B}$ from each current filament onto the direction of vector **B**:

$$B=\int dB\sin\varphi \qquad (1)$$

In accordance with (6.11), we have

$$dB=\mu_0\,dI/2\pi R, \qquad (2)$$

where $dI=(I/\pi)\,d\varphi$ (see Fig. 6.19). Substituting (2) into (1), we get

$$B=\frac{\mu_0 I}{2\pi^2 R}\int_0^{\pi}\sin\varphi\,d\varphi=\frac{\mu_0 I}{\pi^2 R}.$$

● **6.4. Theorem on circulation of B and principle of superposition.** Inside a homogeneous long straight wire of circular cross section, there is a circular cylindrical cavity whose axis is parallel to the conductor axis and displaced relative to it by a distance **l**. A direct current of density **j** flows along the wire. Find magnetic induction **B** inside the cavity.

Solution. In accordance with the principle of superposition, the required quantity can be represented as follows:

$$\mathbf{B} = \mathbf{B}_0 - \mathbf{B}', \tag{1}$$

where $\mathbf{B}_0$ is the magnetic induction of the conductor without cavity,

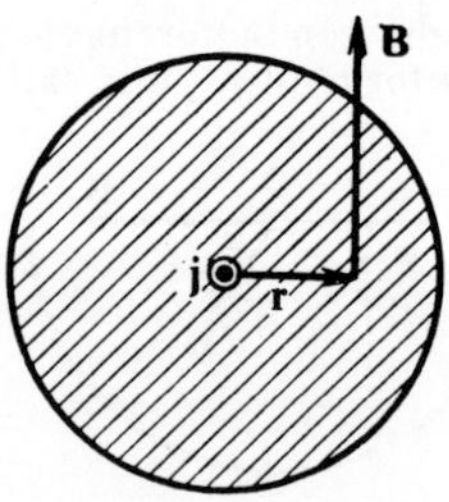

Fig. 6.20

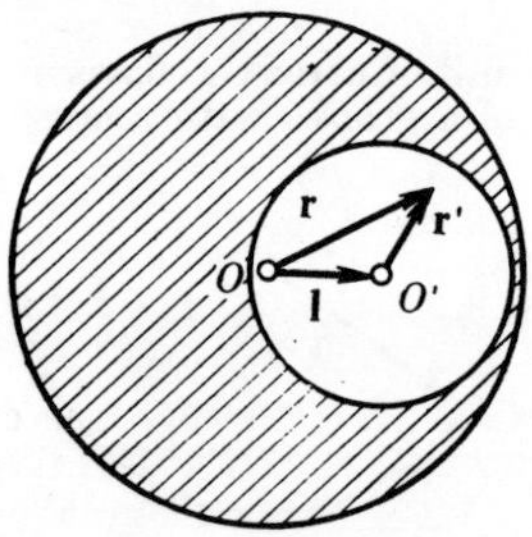

Fig. 6.21

while $\mathbf{B}'$ is the magnetic induction of the field at the same point due to the current flowing through the part of the conductor which has been removed in order to create the cavity.

Thus, the problem implies first of all the calculation of magnetic induction **B** inside the solid conductor at a distance r from its axis. Using the theorem on circulation, we can write $2\pi r B = \mu_0 \pi r^2 j$, whence $B = (1/2)\,\mu_0 r j$. This expression can be represented with the help of Fig. 6.20 in the vector form:

$$\mathbf{B} = \frac{1}{2}\mu_0\,[\mathbf{j}\times\mathbf{r}].$$

Using now this formula for $\mathbf{B}_0$ and $\mathbf{B}'$, we find their difference (1):

$$\mathbf{B} = \frac{\mu_0}{2}[\mathbf{j}\times\mathbf{r}] - \frac{\mu_0}{2}[\mathbf{j}\times\mathbf{r}'] = \frac{\mu_0}{2}[\mathbf{j}\times(\mathbf{r}-\mathbf{r}')].$$

Figure 6.21 shows that $\mathbf{r} = \mathbf{l} + \mathbf{r}'$, whence $\mathbf{r} - \mathbf{r}' = \mathbf{l}$, and

$$\mathbf{B} = \frac{1}{2}\mu_0\,[\mathbf{j}\times\mathbf{l}].$$

Thus, in our case the magnetic field **B** in the cavity is uniform, and if the current is flowing towards us (Fig. 6.21), the field **B** lies in the plane of the figure and is directed upwards.

● **6.5. Principle of superposition.** Current I flows through a long solenoid, whose cross-sectional area is S and the number of turns n. Find the magnetic flux through the endface of this solenoid.

Solution. Let the flux of $\mathbf{B}$ through the solenoid endface be Φ. If we place another similar solenoid in contact with the endface of our

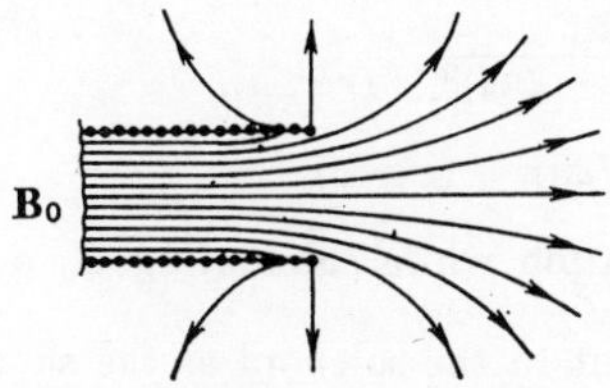

Fig. 6.22

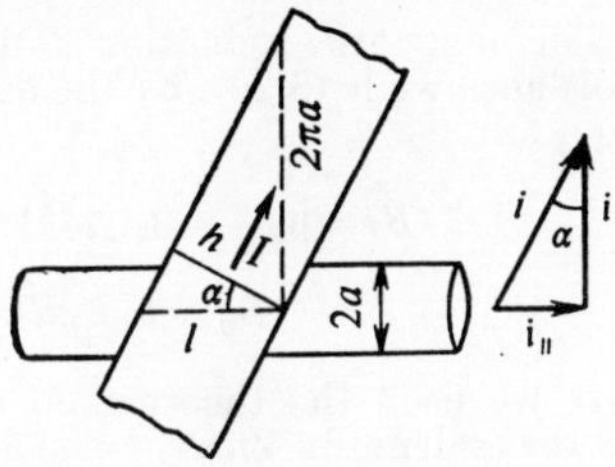

(a) (b)

Fig. 6.23

solenoid, the flux through the contacting endfaces will be $\Phi + \Phi = \Phi_0$, where Φ_0 is the flux through the solenoid's cross section far from its endface. Then we have

$$\Phi = \Phi_0/2 = \mu_0 nIS/2.$$

By the way, we must pay attention to the following properties of the field $\mathbf{B}$ at the endface of a long solenoid.

1. The lines of $\mathbf{B}$ are arranged as shown in Fig. 6.22. This can be easily shown with the help of superposition principle: if we place on the right one more solenoid, field $\mathbf{B}$ outside the thus formed solenoid must vanish, which is possible only with the field configuration shown in the figure.

2. From the principle of superposition, it follows that the normal component B_n will be the same over the area of the endface, since when we form a composite solenoid, $B_n + B_n = B_0$, where B_0 is the field inside the solenoid away from its endfaces. At the centre of the endface $B = B_n$, and we obtain $B = B_0/2$.

● **6.6. The field of a solenoid.** The winding of a long solenoid of radius a is a thin conducting band of width h, wound in one layer practically without gaps. Direct current I flows along the band. Find magnetic field $\mathbf{B}$ inside and outside the solenoid as a function of distance r from its axis.

Solution. The vector of linear current density $\mathbf{i}$ can be represented as the sum of two components:

$$\mathbf{i} = \mathbf{i}_\perp + \mathbf{i}_\parallel .$$

The meaning of vectors $\mathbf{i}_\perp$ and $\mathbf{i}_\parallel$ is clear from Fig. 6.23*b*. In our case, the magnitudes of these vectors can be found with the help of

Fig. 6.23*a* by the formulas

$$i_\perp = i\cos\alpha = i\sqrt{1-\sin^2\alpha} = (I/h)\sqrt{1-(h/2\pi a)^2}.$$

$$i_{||} = i\sin\alpha = I/2\pi a.$$

The magnetic induction B inside the solenoid is determined, in accordance with (6.20), by the quantity $i_\perp$, while outside the solenoid, by $i_{||}$:

$$B_i = \mu_0 i_\perp = (\mu_0 I/h)\sqrt{1-(h/2\pi a)^2} \quad (r < a),$$

$$B_a = \mu_0 i_{||} a/r = \mu_0 I/2\pi r \quad (r > a),$$

where we used the theorem on circulation while calculating B_a outside the solenoid: $2\pi r B_a = \mu_0 2\pi a i_{||}$.

Thus, having represented the current in the solenoid as the superposition of "transverse" and "longitudinal" components, we arrive at the conclusion that only the longitudinal component of the field **B** exists inside such a solenoid and only the transverse component outside it (as in the case of a straight current).

Besides, if we decrease the band width maintaining the current density unchanged, $I \to 0$ as $h \to 0$, but $I/h = \text{const}$. In this case, only the field inside the solenoid remains, i.e. the solenoid becomes "ideal"

● **6.7. Interaction of parallel currents.** Two long wires with negligible resistance are shunted at their ends by resistor R and at the other ends are connected to a source of constant voltage. The radius of the cross section of each wire is smaller than the distance between their axes by a factor of $\eta = 20$. Find the value of the resistance R at which the resultant force of interaction between the wires vanishes.

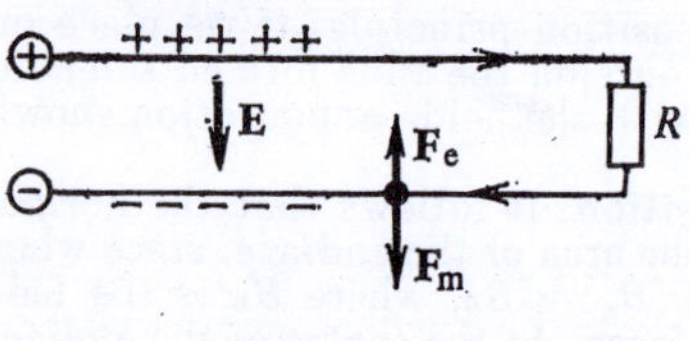

Fig.

Solution. There are excess surface charges on each wire (irrespective of whether or not the current is flowing through them) (Fig. 6.24). Hence, in addition to the magnetic force F_m, we must take into account the electric force F_e. Suppose that an excess charge λ corresponds to a unit length of the wire. Then the electric force exerted per unit length of the wire by the other wire can be found with the help of the Gauss theorem:

$$F_e = \lambda E = \lambda \frac{1}{4\pi\varepsilon_0}\frac{2\lambda}{l} = \frac{2\lambda^2}{4\pi\varepsilon_0 l},$$

where l is the distance between the axes of the wires. The magnetic force acting per unit length of the wire can be found with the help of the theorem on cirulation of vector $\mathbf{B}$:

$$F_m = (\mu_0/4\pi)\ 2I^2/l,$$

where I is the current in the wire.

It should be noted that the two forces, electric and magnetic, are directed oppositely. The electric force is responsible for the attraction between the wires, while the magnetic force causes their repulsion. Let us find the ratio of these forces:

$$F_m/F_e = \varepsilon_0\mu_0 I^2/\lambda^2. \tag{1}$$

There is a certain relation between the quantities I and λ (see Problem 2.8):

$$\lambda = C_1 U = \frac{\pi\varepsilon_0}{\ln\eta} U, \tag{2}$$

where $U = RI$. Hence it follows from relation (2) that

$$I/\lambda = \ln\eta/\pi\varepsilon_0 R. \tag{3}$$

Substituting (3) into (1), we obtain

$$\frac{F_m}{F_e} = \frac{\mu_0}{\varepsilon_0}\frac{\ln^2\eta}{\pi^2 R^2}. \tag{4}$$

The resultant force of interaction vanishes when this ratio is equal to unity. This is possible when $R = R_0$, where

$$R_0 = \sqrt{\frac{\mu_0}{\varepsilon_0}}\frac{\ln\eta}{\pi} = 360\ \Omega.$$

If $R < R_0$ then $F_m > F_e$, and the wires repel each other. If, on the contrary, $R > R_0$, $F_m < F_e$, and the wires attract each other. This can be observed experimentally.

Thus, the statement that current-carrying wires attract each other is true only in the case when the electric component of the interaction can be neglected, i.e. for a sufficiently small resistance R in the circuit shown in Fig. 6.24.

Besides, by measuring the force of interaction between the wires (which is always resultant), we generally cannot determine current I. This should be borne in mind to avoid confusion.

● **6.8. The moment of Ampère's forces.** A loop with current I is in the field of a long straight wire with current I_0 (Fig. 6.25). The plane of the loop is perpendicular to the straight wire. Find the moment of Ampère's forces acting on this loop. The required dimensions of the system are given in the figure.

Solution. Ampère's forces acting on curvilinear parts of the loop are equal to zero. On the other hand, the forces acting on the rectilinear parts form a couple of forces. We must calculate the torque of this couple.

Let us isolate two small elements of the loop (Fig. 6.26). It can be seen from the figure that the torque of the couple of forces corresponding to these elements is

$$dM = 2x \tan \varphi \, dF, \quad (1)$$

where the elementary Ampère's force is given by

$$dF = I \, dlB. \quad (2)$$

The dependence of the magnetic induction B on the distance r

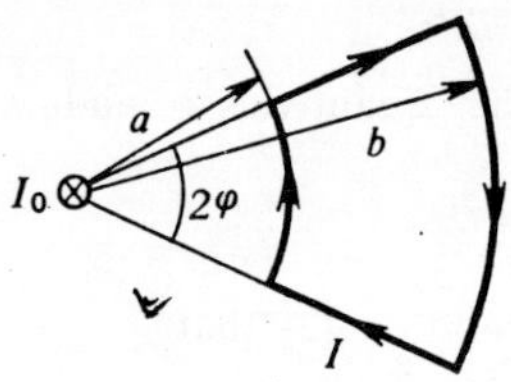

Fig. 6.25

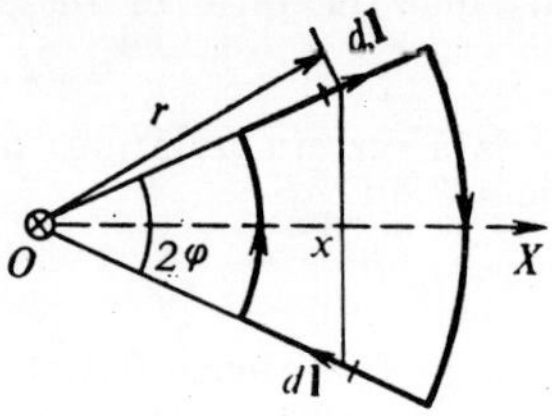

Fig. 6.26

from the straight wire can be found with the help of the theorem on circulation:

$$B = \mu_0 I / 2\pi r. \quad (3)$$

Let us now substitute (3) into (2), then (2) into (1) and, conside-ing that $dl = dr$ and $x = r \cos \varphi$, integrate the obtained expression over r between a and b. This gives

$$M = (\mu_0/\pi) \, I I_0 \, (b - a) \sin \varphi,$$

vector **M** being directed to the left (Fig. 6.26).

● **6.9.** A small coil with current, having the magnetic moment $\mathbf{p}_m$, is placed on the axis of a circular loop of radius R, along which current I is flowing. Find the force **F** acting on the coil if its distance from the centre of the loop is l and vector $\mathbf{p}_m$ is oriented as is shown in Fig. 6.27.

Solution. According to (6.33), the required force is defined as

$$\mathbf{F} = p_m \, \partial \mathbf{B} / \partial n, \quad (1)$$

where **B** is the magnetic induction of the field created by the loop at the locus of the coil. Let us select Z-axis in the direction of the vector $\mathbf{p}_m$. Then the projection of (1) onto this axis will be

$$F_z = p_m \, \partial B_z / \partial z = p_m \, \partial B / \partial z,$$

where we took into account that $B_z = B$ for the given direction of current in the loop. The magnetic induction B is defined by formula (6.12), whence

$$\frac{\partial B}{\partial z} = -\frac{3}{2} \frac{\mu_0 R^2 I l}{(l^2 + R^2)^{5/2}}.$$

Since $\partial B/\partial z < 0$, the projection of the force $F_z < 0$, i.e. vector **F** is directed towards the loop with current I. The obtained result can be represented in the vector form as follows:

$$\mathbf{F} = -\frac{3}{2}\frac{\mu_0 R^2 I l}{(l^2+R^2)^{5/2}}\mathbf{p}_m.$$

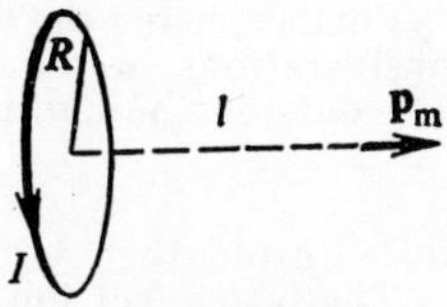

Fig. 6.27

It should be noted that if $\mathbf{p}_m$ (and hence Z-axis as well) were directed oppositely, then $B_z = -B$ and $\partial B_z/\partial z > 0$, and hence $F_z > 0$ and vector **F** would be directed to the right, i.e. again oppositely to the direction of $\mathbf{p}_m$. Hence, the obtained direction for **F** is valid for both orientations of $\mathbf{p}_m$.

● **6.10.** Current I flows in a long thin-walled circular cylinder of radius R. Find the pressure exerted on the cylinder walls.

Solution. Let us consider a surface current element **i** dS, where i is the linear current density and dS is the surface element. We shall

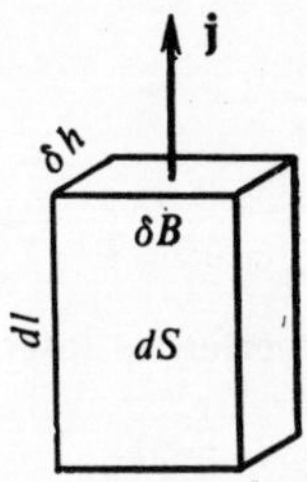

Fig. 6.28

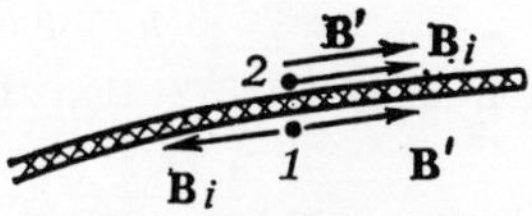

Fig. 6.29

find the relation between the surface and volume elements of the current:

$$j\,dV = j\delta h \cdot \delta b\; dl = i\,dS.$$

The meaning of the quantities appearing in this relation is clarified in Fig. 6.28. In the vector form, we can write

$$\mathbf{j}\,dV = \mathbf{i}\,dS. \tag{1}$$

The Ampère's force acting on the surface current element in this case is determined by the formula obtained from (6.28) with the help of substitution (1):

$$d\mathbf{F} = [\mathbf{i} \times \mathbf{B}']\,dS, \tag{2}$$

where $\mathbf{B}'$ is the magnetic induction of the field at the point of location of the given current element, but created by all other current elements excluding the given one. In order to find $\mathbf{B}'$, we proceed as for calculating electric force (Sec. 2.3). Let B_i be the magnetic induction of the field created by the surface element itself at a point very close to its surface (see Fig. 6.29, where it is assumed that the current is flowing from us). According to (6.22), $B_i = (1/2)\,\mu_0 i$.

Further, using the theorem on circulation of vector **B** and symmetry considerations, we can easily see that the magnetic induction of the field outside the cylinder near its surface is equal to

$$B = \mu_0 I/2\pi R, \tag{3}$$

while inside the cylinder the field is absent.

The latter fact indicates that the field **B**′ from the remaining current elements at points *1* and *2* very close to the cylinder surface (see Fig. 6.29) must be the same and satisfy the following conditions inside and outside the cylinder surface:

$$B' = B_i \quad \text{and} \quad B = B' + B_i = 2B'.$$

Hence it follows that

$$B' = B/2. \tag{4}$$

Substituting this result into (2), we obtain the following expression for the required pressure:

$$p = \frac{dF}{dS} = iB' = \frac{2B'}{\mu_0} B' = \frac{B^2}{2\mu_0}.$$

Taking into account (3), we finally get

$$p = \mu_0 I^2/8\pi^2 R^2.$$

It is clear from formula (2) that the cylinder experiences lateral compression.

7. Magnetic Field in a Substance

7.1. Magnetization of a Substance. Magnetization Vector J

Field in Magnetics. If a certain substance is introduced into a magnetic field formed by currents in conductors, the field will change. This is explained by the fact that each substance is a *magnetic*, i.e. it is magnetized (acquires a magnetic moment) under the action of magnetic field. A magnetized substance creates its own magnetic field **B**′, which forms, together with the primary field $\mathbf{B}_0$ created by conduction currents, the resultant field

$$\mathbf{B} = \mathbf{B}_0 + \mathbf{B}'. \tag{7.1}$$

Here **B**′ and **B** stand for the fields averaged over a physical infinitely small volume.

The field $\mathbf{B}'$, like the field $\mathbf{B}_0$ of conduction currents, has no sources (magnetic charges). Hence, for the resultant field $\mathbf{B}$ in the presence of a magnetic, the *Gauss theorem* is applicable:

$$\boxed{\oint \mathbf{B}\, d\mathbf{S} = 0.} \tag{7.2}$$

This means that the field lines of **B** remain continuous everywhere in the presence of a substance.

Mechanism of Magnetization. At present, it is established that molecules of many substances have intrinsic magnetic moments due to the motion of intrinsic charges in them. Each magnetic moment corresponds to a circular current creating a magnetic field in the surrounding space. In the absence of external magnetic field, the magnetic moments of molecules are oriented at random, and hence the resultant magnetic field due to these moments is equal to zero, as well as the total magnetic moment of the substance. This also applies to the substances that have no magnetic moments in the absence of an external field.

If a substance is placed into an external magnetic field, under the action of this field the magnetic moments of the molecules acquire a predominant orientation, and the substance is magnetized, viz. its resultant magnetic moment becomes other than zero. In this case, the magnetic moments of individual molecules no longer compensate each other, and as a result the field $\mathbf{B}'$ appears.

Magnetization of a substance whose molecules have no magnetic moments in the absence of external field proceeds differently. When such materials are introduced into an external field elementary circular currents are induced in the molecules, and the entire substance acquires a magnetic moment, which also results in the generation of the field $\mathbf{B}'$.

Most of materials are weakly magnetized when introduced into a magnetic field. Only ferromagnetic materials such as iron, nickel, cobalt and their many alloys have clearly pronounced magnetic properties.

Magnetization. The degree of magnetization of a magnetic is characterized by the magnetic moment of a unit volume. This quantity is called *magnetization* and denoted by **J**. By definition,

$$\mathbf{J} = \frac{1}{\Delta V} \sum \mathbf{p}_m, \tag{7.3}$$

where ΔV is a physical infinitely small volume surrounding a given point and $\mathbf{p}_m$ is the magnetic moment of an individual molecule. The summation is performed over all molecules in volume ΔV.

By analogy with what was done for polarization **P** [see (3.3)], magnetization can be represented in the form

$$\mathbf{J} = n \langle \mathbf{p}_m \rangle, \tag{7.4}$$

where n is molecular concentration and $\langle \mathbf{p}_m \rangle$ is the average magnetic moment of a molecule. This formula shows that vector **J** is collinear precisely to the average vector $\langle \mathbf{p}_m \rangle$. Hence for further analysis it is sufficient to know the behaviour of vector $\langle \mathbf{p}_m \rangle$ and assume that all the molecules within the volume ΔV have the same magnetic moment $\langle \mathbf{p}_m \rangle$. This will considerably simplify the understanding of questions associated with the phenomenon of magnetization. For example, an increase in magnetization **J** of a material implies the corresponding increase in vector $\langle \mathbf{p}_m \rangle$: if $\mathbf{J} = 0$, $\langle \mathbf{p}_m \rangle = 0$ as well.

If vector **J** is the same at each point of a substance, it is said that the substance is uniformly magnetized.

Magnetization Currents I'. As was mentioned above, magnetization of a substance is caused by the preferential orientation of magnetic moments of individual molecules in one direction. The same can be said about the elementary circular currents associated with each molecule and called *molecular currents*. It will be shown that such a behaviour of molecular currents leads to the appearance of macroscopic currents I' called *magnetization currents*. Ordinary currents flowing in the conductors are associated with the motion of charge carriers in a substance and are called *conduction currents* (I).

In order to understand how magnetization currents appear, let us first imagine a cylinder made of a *homogeneous* magnetic whose magnetization **J** is uniform and directed along the axis. Molecular currents in the magnetized magnetic are oriented as shown in Fig. 7.1. Molecular currents of adjacent molecules at the points of their contact have opposite directions and compensate each other. The only uncompensated molecular currents are those emerging on the lateral surface of the cylinder. These currents form macroscopic *surface* magnetization current I' circulating over the lateral

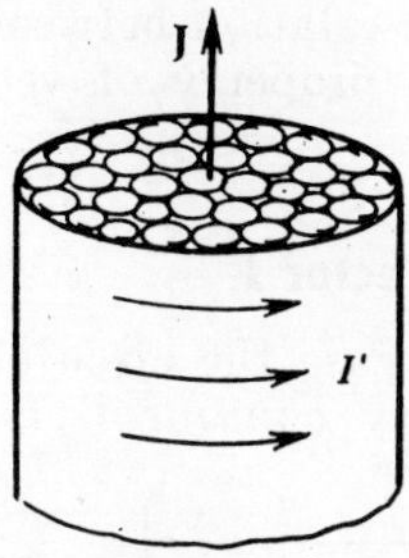

Fig. 7.1

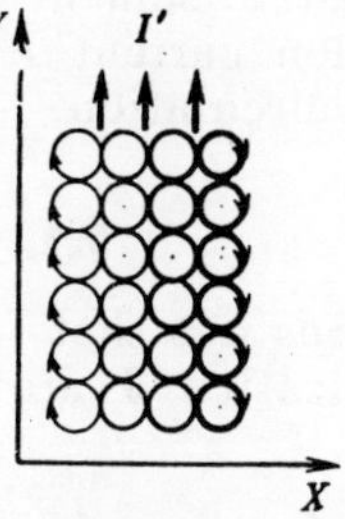

Fig. 7.2

surface of the cylinder. The current I' induces the same macroscopic magnetic field as that of all molecular currents taken together.

Let us now consider another case: a magnetized magnetic is *nonhomogeneous*. Let, for instance, the molecular currents be arranged as shown in Fig. 7.2, where the line thickness corresponds to the intensity of molecular currents. It follows from the figure that vector **J** is directed behind the plane of the figure and increases in magnitude with the coordinate x. It can be seen that the molecular currents are not compensated inside a nonhomogeneous magnetic, and as a result, the macroscopic *volume* magnetization current I' appears, which flows in the positive direction of the Y-axis. Accordingly, we can speak about the linear $\mathbf{i}'$ and surface $\mathbf{j}'$ current densities, i' being measured in A/m and j' in A/m^2.

Calculation of Field B in a Magnetic. It can be stated that the contribution to field **B** from a magnetized magnetic is equal to the contribution that would be created by the same distribution of currents I' in a vacuum. In other words, having established the distribution of magnetization currents I', we can find, with the help of the Biot-Savart law, the field $\mathbf{B}'$ corresponding to them, and then calculate the resultant field **B** by formula (7.1).

Unfortunately, the distribution of currents I' depends not only on the configuration and the properties of a magnetic but on the field $\mathbf{B}$ as well. For this reason, in the general case the problem of finding $\mathbf{B}$ in a magnetic cannot be solved directly. It remains for us to try and find another approach to the solution of this problem. The first step in this direction is the establishment of an important relation between magnetization current I' and a certain property of vector $\mathbf{J}$, viz. its circulation.

7.2. Circulation of Vector J

It turns out that for a stationary case the circulation of magnetization $\mathbf{J}$ around an arbitrary contour Γ is equal

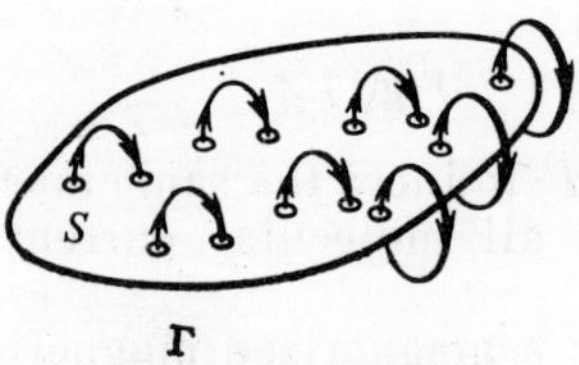

Fig. 7.3

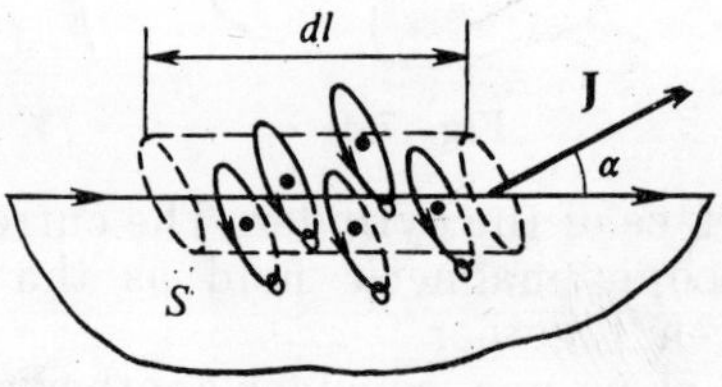

Fig. 7.4

to the algebraic sum of magnetization currents I' enveloped by the contour Γ:

$$\boxed{\oint \mathbf{J}\, d\mathbf{l} = I',} \tag{7.5}$$

where $I' = \int \mathbf{j}'\, d\mathbf{S}$ and integration is performed over an arbitrary surface stretched on the contour Γ.

In order to prove this theorem, we shall calculate the algebraic sum of molecular currents enveloped by contour Γ. Let us stretch an arbitrary surface S on the contour Γ (Fig. 7.3). It can be seen from the figure that some molecular currents intersect the surface S twice (once in one direction and for the second time in the opposite direction). Hence such currents make no contribution to the resultant magnetization current through the surface S.

However, the molecular currents that are wound around

the contour Γ intersect the surface S only once. Such molecular currents create a macroscopic magnetization current I' piercing the surface S.

Let each molecular current be equal to I_m and the area embraced by it be S_m. Then, as is shown in Fig. 7.4, the element dl of the contour Γ is wound by those molecular currents whose centres get into the oblique cylinder with the volume $dV = S_m \cos\alpha\, dl$, where α is the angle between the contour element $d\mathbf{l}$ and the direction of vector $\mathbf{J}$ at the point under consideration. All these molecular currents intersect the surface S once, and their contribution to the magnetization current is $dI' = I_m n\, dV$, where n is the molecular concentration. Substituting into this formula the expression for dV, we obtain

$$dI' = I_m S_m n \cos\alpha\, dl = J \cos\alpha\, dl = \mathbf{J}\, d\mathbf{l},$$

where we took into account that $I_m S_m = p_m$ is the magnetic moment of an individual molecular current, and $I_m S_m n$ is the magnetic moment of a unit volume of the material.

Integrating this expression over the entire contour Γ, we obtain (7.5). The theorem is proved.

It remains for us to note that if a magnetic is nonhomogeneous, magnetization current I' generally pierces the entire surface (see Fig. 7.3) and not only the region near its boundary, adjoining the contour Γ. This is why we can use the expression $I' = \int \mathbf{j}'\, d\mathbf{S}$, where integration is performed over the entire surface S, bounded by the contour Γ. In the above proof, we managed as if to "drive" the entire current I' to the boundary of the surface S. The only goal of this method is to simplify the calculation of the current I'.

Differential form of Eq.(7.5):

$$\boxed{\nabla \times \mathbf{J} = \mathbf{j}',} \tag{7.6}$$

i.e. the curl of magnetization **J** is equal to the magnetization current density at the same point of space.

A Remark about the Field of J. The properties of the field of vector **J**, expressed by Eqs. (7.5) and (7.6), do not imply at all that the field **J** itself is determined only by the currents I'. The field of **J** (which is bounded only by the spatial region filled by magnetic) depends on *all* currents, viz. both on the magnetization currents I' and the conduction currents I. However, in some cases with a definite symmetry the situation is such as if the field of vector **J** were determined only by currents I'.

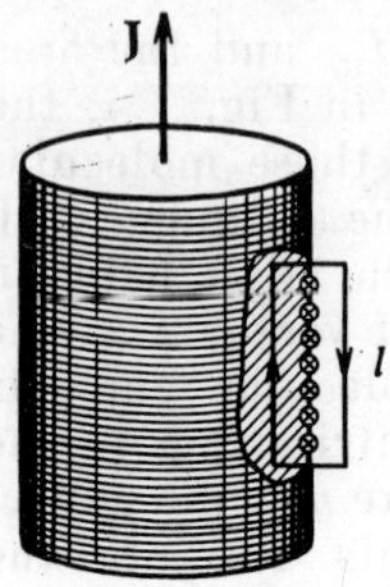

Fig. 7.5

Example. Find the surface magnetization current per unit length of a cylinder made of a homogeneous magnetic, if its magnetization is **J**, vector **J** being directed everywhere along the cylinder axis.

Let us apply Eq. (7.5) to the contour chosen as is shown in Fig. 7.5. It can be easily seen that the circulation of vector **J** around this contour is equal to the product Jl. In this case, the magnetization current is the surface current. If we denote its linear density by i', the contour under consideration embraces the magnetization current $i'l$. It follows from the equality $Jl = i'l$ that

$$i' = J. \tag{7.7}$$

It should be noted, by the way, that vectors $\mathbf{i}'$ and **J** are mutually perpendicular: $\mathbf{i}' \perp \mathbf{J}$.

7.3. Vector H

Theorem on Circulation of Vector H (for magnetic fields of *direct* currents). In magnetics placed into an external magnetic field magnetization currents are induced. Hence circulation of vector **B** will now be determined not only by conduction currents but by magnetization currents as well:

$$\oint \mathbf{B}\, d\mathbf{l} = \mu_0 (I + I'), \tag{7.8}$$

where I and I' are the conduction and magnetization currents embraced by a given contour Γ.

Since the determination of currents I' in the general case is a complicated problem, formula (7.8) becomes inapplicable in practical respect. It turns out, however, that we can find a certain auxiliary vector whose circulation will be determined only by the conduction currents embraced by the contour Γ. Indeed, as we know, the current I' is related to the circulation of the magnetization:

$$\oint \mathbf{J}\, d\mathbf{l} = I'. \tag{7.9}$$

Assuming that the circulation of vectors **B** and **J** is taken over the same contour Γ, we express I' in Eq. (7.8) through formula (7.9). Then

$$\oint \left(\frac{\mathbf{B}}{\mu_0} - \mathbf{J}\right) d\mathbf{l} = I. \tag{7.10}$$

Let us denote the integrand of this expression by **H**. Thus, we have found an auxiliary *vector* **H**,

$$\boxed{\mathbf{H} = \frac{\mathbf{B}}{\mu_0} - \mathbf{J},} \tag{7.11}$$

whose circulation around an arbitrary contour Γ is equal to the algebraic sum of the conduction currents I embraced by this contour:

$$\boxed{\oint \mathbf{H}\, d\mathbf{l} = I.} \tag{7.12}$$

This formula expresses the *theorem on circulation of vector* **H**: *the circulation of vector* **H** *around an arbitrary closed contour is equal to the algebraic sum of the conduction currents embraced by this contour.*

The sign rule for currents is the same as in the case of circulation of vector **B** (see p. 149).

It should be noted that vector **H** is a combination of two quite different quantities, $\mathbf{B}/\mu_0$ and **J**. Hence, vector **H** is indeed an auxiliary vector which does not have any deep physical meaning.* However, the important property of vector **H** expressed in the theorem on its circulation justifies the introduction of this vector: in many cases, it considerably simplifies the investigation of the field in magnetics.

Relations (7.11) and (7.12) are valid for any magnetics, including anisotropic ones.

---

* The quantity **H** is often called *magnetic field intensity*, but we shall not be using this term to emphasize again its auxiliary nature.

It follows from formula (7.12) that the magnitude of vector **H** has the dimensions of the current per unit length. Hence, **H** is measured in *amperes per metre* (A/m).

Differential form of the theorem on circulation of vector H:

$$\boxed{\nabla \times \mathbf{H} = \mathbf{j},} \tag{7.13}$$

i.e. the curl of vector **H** is equal to the density of the conduction current at the same point of the substance.

Relation Between Vectors J and H. It was shown above that magnetization **J** depends on the magnetic induction **B** at a given point of the substance. However, customarily vector **J** is related not with **B** but with **H**. We shall confine ourselves to the analysis of only those magnetics for which the dependence between **J** and **H** is linear, viz.

$$\mathbf{J} = \chi \mathbf{H}, \tag{7.14}$$

where χ is the *magnetic susceptibility*. It is a dimensionless quantity typical of each magnetic (that χ is dimensionless follows from the fact that, according to (7.11), the dimensions of **H** and **J** are the same).

Unlike electric susceptibility $\varkappa$ which is always positive, magnetic susceptibility χ may be either positive or negative. Accordingly, the magnetics satisfying formula (7.14) are divided into *paramagnetics* ($\chi > 0$) and *diamagnetics* ($\chi < 0$). In paramagnetics, $\mathbf{J} \uparrow\uparrow \mathbf{H}$, while in diamagnetic $\mathbf{J} \downarrow\uparrow \mathbf{H}$.

It should be noted that in addition to these magnetics, there exist *ferromagnetics* for which the dependence **J** (**H**) has a rather complicated form: it is not linear, and besides, it has a *hysteresis*, i.e. **J** depends on the prehistory of a magnetic. (Ferromagnetics will be described in greater detail in Sec. 7.6.)

Relation Between B and H. For magnetics obeying (7.14), expression (7.11) acquires the form $(1 + \chi)\,\mathbf{H} = \mathbf{B}/\mu_0$. Hence

$$\mathbf{B} = \mu\mu_0 \mathbf{H}, \tag{7.15}$$

where μ is the *permeability* of the medium,

$$\mu = 1 + \chi. \tag{7.16}$$

For paramagnetics $\mu > 1$, while for diamagnetics $\mu < 1$. It should be noted that in both cases μ differs from unity only slightly, i.e. the magnetic properties of these magnetics are manifested very weakly.

A Remark about the Field of H. Let us consider a question which often involves misunderstanding: which currents determine the field of vector **H**? Generally, the field of **H** (just as the field of **B**) depends on all currents, both conduction and magnetization. This can be seen even from (7.15). However, in some cases the field **H** is determined only by conduction currents. Vector **H** is very helpful for these cases in particular. At the same time, this makes grounds for the erroneous conclusion that the field of **H** always depends on conduction currents and for incorrect interpretations of the theorem on circulation of vector **H** and Eq. (7.13). This theorem expresses only a certain property of the field of **H** without defining the field itself.

Example. A system consists of a long straight wire with current I and an arbitrary piece of a paramagnetic (Fig. 7.6). Let us find the changes in the fields of vectors **B** and **H** and in the circulation of vector **H** around a certain fixed contour Γ upon the removal of the magnetic.

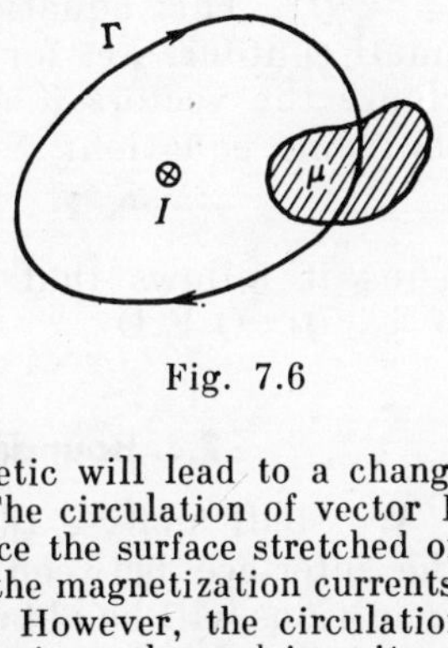

Fig. 7.6

The field **B** at each point of space is determined by the conduction current I as well as by the magnetization currents in the paramagnetic. Since in our case, in accordance with (7.15), $\mathbf{H} = \mathbf{B}/\mu\mu_0$, this also applies to the field of vector **H**, i.e. it also depends on the conduction current I and the magnetization currents.

The removal of the piece of paramagnetic will lead to a change in the field **B**, and hence in the field **H**. The circulation of vector **B** around the contour Γ will also change, since the surface stretched on the contour Γ will no longer be pierced by the magnetization currents. Only the conduction current will remain. However, the circulation of vector **H** around the contour Γ will remain unchanged in spite of the variation of the field **H** itself.

When j′ = 0 Inside a Magnetic? We shall show now that magnetization currents inside a magnetic are absent if (1) the magnetic is homogeneous and (2) there are no conduction currents in it ($\mathbf{j} = 0$). In this case, for any shape of a magnetic and any configuration of the magnetic field, we can be sure that volume magnetization currents are equal to zero, and only surface magnetization currents remain.

In order to prove this, we shall use the theorem on circulation of vector **J** around an arbitrary contour Γ completely lying inside the magnetic. In the case of a homogeneous magnetic, we can, substituting $\chi\mathbf{H}$ for **J**, take χ in Eq. (7.5) out of the integral and write

$$I' = \chi \oint \mathbf{H}\, d\mathbf{l}.$$

In accordance with (7.12), the remaining integral is equal to the algebraic sum of the conduction currents I embraced by the contour Γ. Hence for a homogeneous magnetic, we have

$$I' = \chi I. \qquad (7.17)$$

This relationship between the currents I' and I is valid for any contour inside the magnetic, in particular, for a very small contour, when $I' \to dI' = j'_n\, dS$ and $I \to dI = j_n\, dS$. Then $j'_n\, dS = \chi j_n\, dS$, and after cancelling dS, we obtain $j'_n = \chi j_n$. This equation holds for any orientation of the small contour, i.e. for any direction of the normal **n** to it. Hence the vectors **j**′ and **j** themselves are related through the same equation:

$$\mathbf{j}' = \chi \mathbf{j}. \qquad (7.18)$$

Thus it follows that in a homogeneous magnetic $\mathbf{j}' = 0$ if $\mathbf{j} = 0$, Q.E.D.

7.4. Boundary Conditions for B and H

We shall analyse the conditions for vectors **B** and **H** at the interface between two homogeneous magnetics. These conditions will be obtained, just as in the case of dielectrics, with the help of the Gauss theorem and the theorem on circulation. We recall that for vectors **B** and **H** these theorems have the form

$$\oint \mathbf{B}\, d\mathbf{S} = 0, \quad \oint \mathbf{H}\, d\mathbf{l} = I. \qquad (7.19)$$

Condition for Vector B. Let us imagine a very low cylinder located at the interface between two magnetics as is shown in Fig. 7.7. Then the flux of **B** in the outward direction (we ignore the flux through the lateral surface) can be written in the form

$$B_{2n}\,\Delta S + B_{1n'}\,\Delta S = 0.$$

Taking the two projections of vector **B** onto the normal **n** to the interface, we obtain $B_{1n'} = -B_{1n}$, and after cancelling ΔS, the previous equation becomes

$$\boxed{B_{2n} = B_{1n},} \tag{7.20}$$

i.e. the normal component of vector **B** turns out to be the

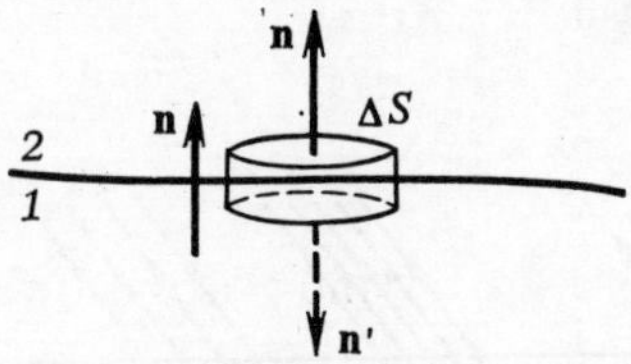

Fig. 7.7

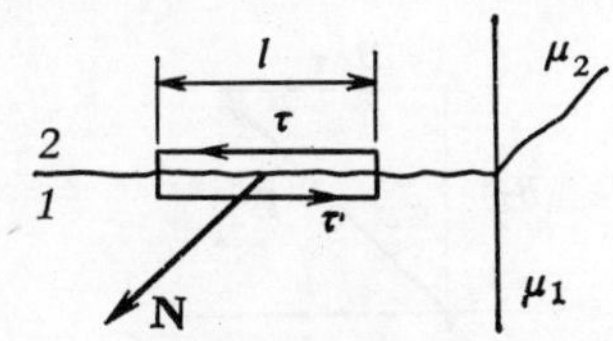

Fig. 7.8

same on both sides of the interface. This quantity does not have a discontinuity.

Condition for Vector H. We shall assume, for higher generality that a surface conduction current with linear densiy **i** flows over the interfacial surface of the magnetics. Let us apply the theorem on circulation of vector **H** to a very small rectangular contour whose height is negligibly small in comparison with its length l (the arrangement of the contour is shown in Fig. 7.8). Neglecting the contributions from smaller sides of the contour to the circulation, we can write for the entire contour

$$H_{2\tau}l + H_{1\tau'}l = i_N l,$$

where i_N is the projection of vector **i** onto the normal **N** to the contour (vector **N** forms a right-handed system with the direction of contour circumvention). Taking the two projections of vector **H** onto the common unit vector of the tangent τ (in medium *2*), we obtain $H_{1\tau'} = -H_{1\tau}$. Cancelling l out of the previous equation, we get

$$\boxed{H_{2\tau} - H_{1\tau} = i_N,} \tag{7.21}$$

i.e. the tangential component of vector **H** generally has a discontinuity upon a transition through the interface, which is due to the presence of conduction currents.

If, however, there are no conduction currents at the interface between magnetics (**i** = 0), the tangential component of vector **H** turns out to be the same on both sides of the interface:

$$\boxed{H_{2\tau}=H_{1\tau}.} \tag{7.22}$$

Thus, if there is no conduction current at the interface between two homogeneous magnetics, the components B_n

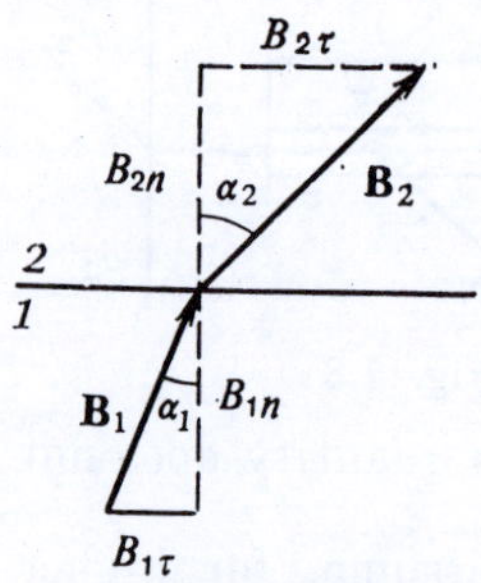

Fig. 7.9

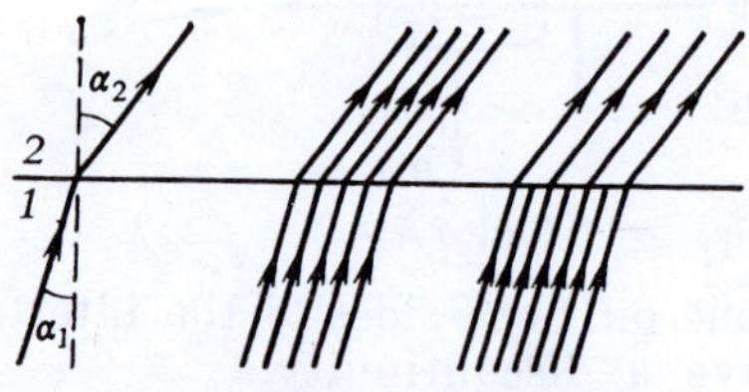

Fig. 7.10

and H_τ vary continuously (without a jump) upon a transition through this interface. On the other hand, the components B_τ and H_n in this case have discontinuities.

It should be noted that vector **B** behaves at the interface like vector **D**, while vector **H** behaves like **E**.

Refraction of Lines of B. The lines of **B** undergo refraction at the interface between two magnetics (Fig. 7.9). As in the case of dielectrics, we shall find the ratio of the tangents of angles α_1 and α_2:

$$\frac{\tan\alpha_2}{\tan\alpha_1}=\frac{B_{2\tau}/B_{2n}}{B_{1\tau}/B_{1n}}.$$

We shall confine ourselves to the case when there is no conduction current at the interface. According to (7.22) and (7.20), in this case we have

$$B_{2\tau}/\mu_2=B_{1\tau}/\mu_1 \quad \text{and} \quad B_{2n}=B_{1n}.$$

Taking these relations into account, we obtain the law of refraction of the lines of vector **B** (and hence of vector **H** as well) similar to (2.25):

$$\frac{\tan \alpha_2}{\tan \alpha_1} = \frac{\mu_2}{\mu_1}. \tag{7.23}$$

Figure 7.10 depicts the fields of vectors **B** and **H** near the interface between two magnetics (in the absence of conduction currents). Here $\mu_2 > \mu_1$. A comparison of the densities of the lines shows that $B_2 > B_1$, while $H_2 < H_1$. The lines of **B** do not have a discontinuity upon a transition through the interface, while the **H** lines do (due to the surface magnetization currents).

The refraction of magnetic field lines is used in *magnetic protection*. If, for example, a closed iron shell (layer) is introduced into an external magnetic field, the field lines will be concentrated (condensed) mainly on the shell itself. Inside the shell (in the cavity) the magnetic field turns out to be considerably weakened in comparison with the external field. In other words, the iron shell acts like a screen. This is used in order to protect sensitive devices from external magnetic fields.

7.5. Field in a Homogeneous Magnetic

It was mentioned in Sec. 7.1 that determination of the resultant magnetic field in the presence of arbitrary magnetics is generally a complicated problem. Indeed, for this purpose, in accordance with (7.1), the field $\mathbf{B}_0$ of conduction currents must be supplemented by the macroscopic field $\mathbf{B}'$ created by magnetization currents. The problem is that we do not know beforehand the configuration of magnetization currents. We can only state that the distribution of these currents depends on the nature and configuration of the magnetic as well as on the configuration of the external field $\mathbf{B}_0$, viz. the field of conduction currents. And since we do not know the distribution of magnetization currents, we cannot calculate the field $\mathbf{B}'$.

The only exception is the case when the entire space occupied by the field **B** is filled by a homogeneous isotropic dielectric. Let us consider this case in greater detail. But first of all, we shall analyse the phenomena observed when conduction current flows along a homogeneous conductor in a vacuum. Since each conductor is a magnetic, magnetization currents also flow through it, viz. volume currents given by (7.18) and surface currents. Let us take a contour embracing our current-carrying conductor. In accordance with the theorem on circulation of vector **J** (7.5), the algebraic sum of magnetization (volume and surface) currents is equal to zero everywhere since $\mathbf{J} = 0$ at all points of the contour, i.e. $I' = I'_v + I'_s = 0$. Hence $I'_v = -I'_s$, i.e. the volume and surface magnetization currents are equal in magnitude and opposite in direction.

Thus, it can be stated that in an ordinary case, when currents flow along sufficiently thin wires, the magnetic field in the surrounding space (in a vacuum) is determined only by conduction currents, since magnetization currents compensate each other (except for, perhaps, the points lying very close to the wire).

Let us now fill the space surrounding the conductor by a homogeneous nonconducting magnetic (for the sake of definiteness, we assume that it is a paramagnetic, $\chi > 0$). At the interface between *this* magnetic and the wire, surface magnetization current I' will appear. It can be easily seen that this current has the same direction as the conduction current I (when $\chi > 0$).

As a result, we shall have the conduction current I, the surface and volume magnetization currents in the conductor (the magnetic fields of these currents compensate one another and hence can be disregarded), and the surface magnetization current I' on the nonconducting magnetic. For sufficiently thin wires, the magnetic field **B** in the magnetic will be determined as the field of the current $I + I'$.

Thus, the problem is reduced to finding the current I'. For this purpose, we surround the conductor by a contour arranged in the surface layer of the nonconducting magnetic. Let the plane of the contour be perpendicular to the wire axis, i.e. to the magnetization currents. Then, taking (7.7) and (7.14) into consideration, we can write

$$I' = \oint i'\,dl = \oint J\,dl = \chi \oint H\,dl.$$

According to (7.12), it follows that $I' = \chi I$.

The configurations of the magnetization current I' and

of the conduction current I practically coincide (the wires are thin), and hence at all points the induction $\mathbf{B}'$ of the field of the magnetization currents differs from the induction $\mathbf{B}_0$ of the field of the conduction currents only in magnitude, these vectors being related in the same way as the corresponding currents, viz.

$$\mathbf{B}' = \chi \mathbf{B}_0. \tag{7.24}$$

Then the induction of the resultant field $\mathbf{B} = \mathbf{B}_0 + \mathbf{B}' = (1 + \chi) \mathbf{B}_0$ or

$$\mathbf{B} = \mu \mathbf{B}_0. \tag{7.25}$$

This means that when the entire space is filled with a homogeneous magnetic, $\mathbf{B}$ increases μ times. In other words, the quantity μ shows the increase in the magnetic field $\mathbf{B}$ upon filling the entire space occupied by the field with a magnetic.

If we divide both parts of (7.25) by $\mu\mu_0$, we obtain

$$\mathbf{H} = \mathbf{H}_0 \tag{7.26}$$

(in the case under consideration, field $\mathbf{H}$ turns out to be the same as in a vacuum).

Formulas (7.24)-(7.26) are also valid *when a homogeneous magnetic fills the entire volume bounded by the surfaces formed by the lines of* $\mathbf{B}_0$ (the field of the conduction current). In this case too the magnetic induction $\mathbf{B}$ inside the magnetic is μ times larger than $\mathbf{B}_0$.

In the cases considered above, the magnetic induction $\mathbf{B}'$ of the field of magnetization currents is connected with magnetization $\mathbf{J}$ of the magnetic through the simple relation

$$\mathbf{B}' = \mu_0 \mathbf{J}. \tag{7.27}$$

This expression can be easily obtained from the formula $\mathbf{B} = \mathbf{B}_0 + \mathbf{B}'$ if we take into account that $\mathbf{B}' = \chi \mathbf{B}_0$ and $\mathbf{B} = \mu\mu_0 \mathbf{H}$, where $\mathbf{H} = \mathbf{J}/\chi$.

As has already been mentioned above, in other cases the situation is much more complicated, and formulas (7.24)-(7.27) become inapplicable.

Concluding the section, let us consider two simple examples.

Example 1. Field B in a solenoid. A solenoid having nI ampere-turns per unit length is filled with a homogeneous magnetic of the permeability $\mu > 1$. Find the magnetic induction B of the field in the magnetic.

According to (6.20), in the absence of magnetic the magnetic induction $\mathbf{B}_0$ in the solenoid is equal to $\mu_0 nI$. Since the magnetic fills the entire space where the field differs from zero (we ignore the edge effects), the magnetic induction B must be μ times larger:

$$B = \mu\mu_0 nI. \tag{7.28}$$

In this case, the field of vector $\mathbf{H}$ remains the same as in the absence of the magnetic, i.e. $\mathbf{H} = \mathbf{H}_0$.

The change in field $\mathbf{B}$ is caused by the appearance of magnetization currents flowing over the surface of the magnetic in the same direction as the conduction currents in the solenoid winding when $\mu > 1$. If, however, $\mu < 1$, the directions of these currents will be opposite.

The obtained results are also valid when the magnetic has the form of a very long rod arranged inside the solenoid so that it is parallel to the solenoid's axis.

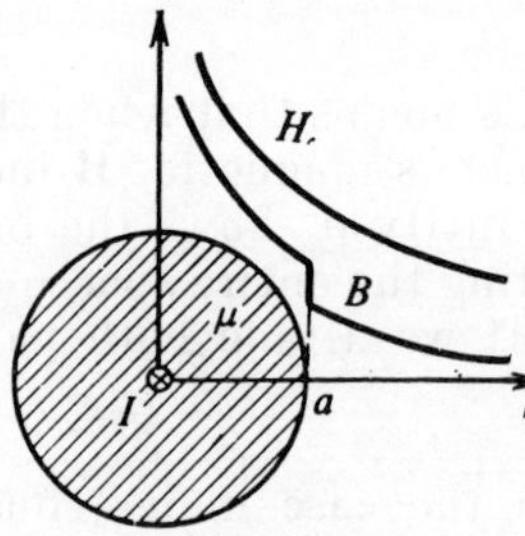

Fig. 7.11

Example 2. The field of straight current in the presence of a magnetic. Suppose that a magnetic fills a long cylinder of radius a along whose axis a given current I flows. The permeability of the magnetic $\mu > 1$. Find the magnetic induction $\mathbf{B}$ as a function of the distance r from the cylinder axis.

We cannot use directly the theorem on circulation of vector $\mathbf{B}$ since magnetization currents are unknown. In this situation, vector $\mathbf{H}$ is very helpful: its circulation is determined only by conduction currents. For a circle of radius r, we have $2\pi rH = I$, whence

$$B = \mu\mu_0 H = \mu\mu_0 I/2\pi r.$$

Upon a transition through the magnetic-vacuum interface, the magnetic induction $\mathbf{B}$, unlike $\mathbf{H}$, undergoes a discontinuity (Fig. 7.11).

An increase in B inside the magnetic is caused by surface magnetization currents. These currents coincide in direction with the current I in the wire on the axis of the system and hence "amplify" this current. On the other hand, outside the cylinder the surface magnetization current is directed oppositely, but it does not produce any effect on field $\mathbf{B}$ in the magnetic. Outside the magnetic, the magnetic fields of both currents compensate one another.

7.6. Ferromagnetism

Ferromagnetics. In magnetic respect, all substances can be divided into weakly magnetic (paramagnetics and diamagnetics) and strongly magnetic (ferromagnetics). It is well known that in the absence of magnetic field, para- and diamagnetics are not magnetized and are characterized by a one-to-one correspondence between magnetization **J** and vector **H**.

Ferromagnetics are the substances (solids) that may possess *spontaneous magnetization*, i.e. which are magnetized even

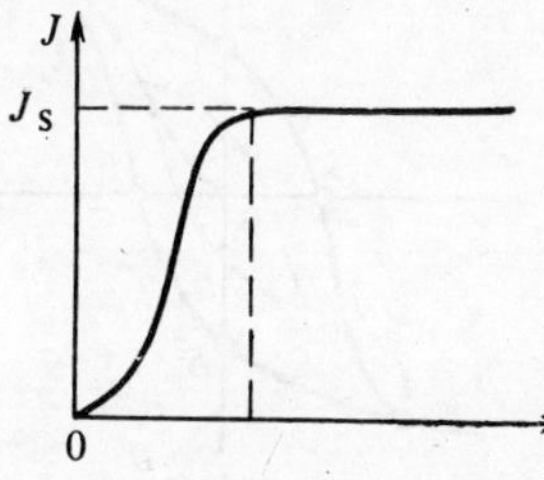

Fig. 7.12

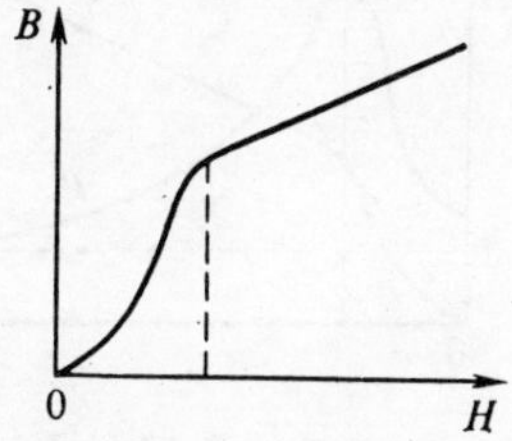

Fig. 7.13

in the absence of external magnetic field. Typical ferromagnetics are iron, cobalt, and many of their alloys.

Basic Magnetization Curve. A typical feature of ferromagnetics is the complex nonlinear dependence **J** (**H**) or **B** (**H**). Figure 7.12 shows the magnetization curve for a ferromagnetic whose magnetization for **H** = 0 is also zero. This curve is called the *basic magnetization curve*. Even at comparatively small values of H, magnetization J attains saturation (J_s). Magnetic induction $B = \mu_0 (H + J)$ also increases with H. After attaining saturation, B continues to grow with H according to the linear law $B = \mu_0 H + \text{const}$, where const $= \mu_0 J_s$. Figure 7.13 represents the basic magnetization curve on the B-H diagram.

In view of the nonlinear dependence $B(H)$, the permeability μ for ferromagnetics cannot be defined as a constant characterizing the magnetic properties of a specific ferromagnetic. However, as before, it is assumed that $\mu = B/\mu_0 H$, but here μ is a function of H (Fig. 7.14). The value μ_{max} of permeability for ferromagnetics may be very large. For example, for pure iron $\mu_{max} = 5{,}000$, while for supermalloy $\mu_{max} = 800{,}000$.

It should be noted that the concept of permeability is

applicable only to the basic magnetization curve since, as will be shown, the B (H) dependence is nonunique.

Magnetic Hysteresis. Besides the nonlinear dependence B (H) or J (H), ferromagnetics also exhibit magnetic *hysteresis*: the relation between B and H or J and H turns out to

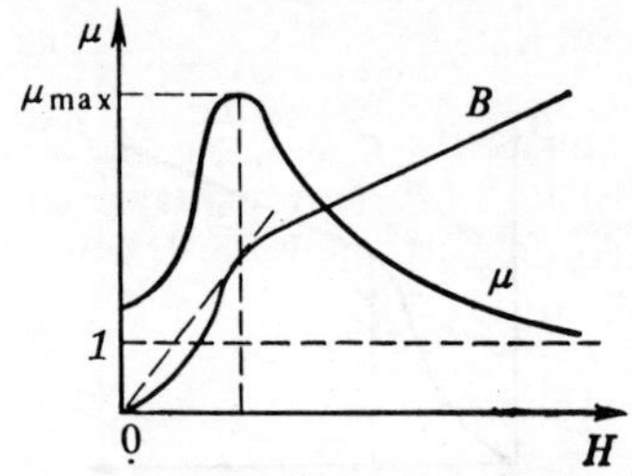

Fig. 7.14

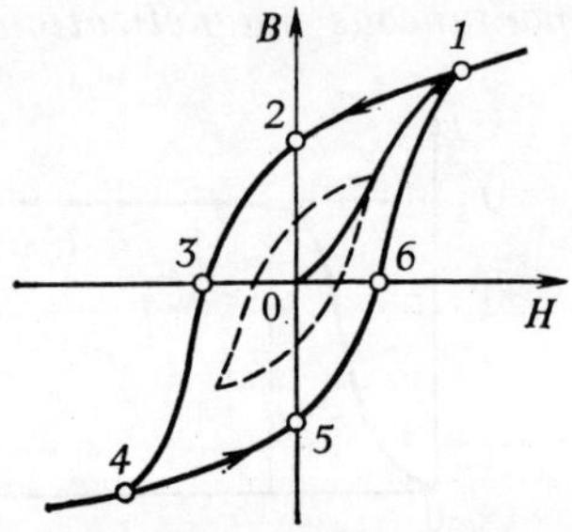

Fig. 7.15

be ambiguous and is determined by the history of the ferromagnetic's magnetization. If we magnetize an initially nonmagnetized ferromagnetic by increasing H from zero to the value at which saturation sets in (point *1* in Fig. 7.15), and then reduce H from $+H_1$ to $-H_1$, the magnetization curve will not follow path *10*, but will take path *1 2 3 4*. If we then change H from $-H_1$ to $+H_1$ the magnetization curve will lie below, and take path *4 5 6 1*.

The obtained closed curve is called the *hysteresis loop*. The *maximum* hysteresis loop is obtained when saturation is attained at points *1* and *4*. If, however, there is no saturation at extreme points, the hysteresis loops have similar form but smaller size, as if inscribed into the maximum hysteresis loop.

Figure 7.15 shows that for $H = 0$ magnetization does not vanish (point *2*) and is characterized by the quantity B_r called the *residual induction*. It corresponds to the *residual magnetization* J_r. The existence of permanent magnets is associated with this residual magnetization. The quantity B vanishes (point *3*) only under the action of the field H_c which has the direction opposite to that of the field causing the magnetization. The quantity H_c is called the *coercive force*.

The values of B_r and H_c for different ferromagnetics vary over broad ranges. For transformer steel, the hysteresis loop is narrow (H_c is small), while for ferromagnetics used for manufacturing permanent magnets it is wide (H_c is large). For example, for alnico alloy, $H_c = 50{,}000$ A/m and $B_r = 0.9$ T.

These peculiarities of magnetization curves are used in a convenient practical method for demagnetizing ferromagnetics. A magnetized sample is placed into a coil through which an alternating current is passed, its amplitude being gradually reduced to zero. In this case, the ferromagnetic is subjected to multiple cyclic reverse magnetizations in which hysteresis loops gradually decrease, contracting to the point where magnetization is zero.

Experiments show that a ferromagnetic is heated upon reverse magnetization. It can be shown that in this case the amount of heat Q_u liberated per unit volume of the ferromagnetic is numerically equal to the "area" S_n of the hysteresis loop:

$$Q_u = \oint H\,dB = S_n. \tag{7.29}$$

Curie Point. As the temperature increases, the ability of ferromagnetics to get magnetized becomes weaker, and in particular, saturation magnetization becomes lower. At a certain temperature called the *Curie point*, ferromagnetic properties of the material disappear.

At temperatures above the Curie point, a ferromagnetic becomes a paramagnetic.

On the Theory of Ferromagnetism. The physical nature of ferromagnetism could be explained only with the help of quantum mechanics. The so-called *exchange forces* may appear in crystals under certain conditions. These forces orient parallel to each other the magnetic moments of electrons. As a result, regions (1-10 μm in size) of spontaneous magnetization, which are called *domains*, appear in crystals. Within the limits of each domain, a ferromagnetic is magnetized to saturation and has a certain magnetic moment. The directions of these moments are different for different domains. Hence, in the absence of external field, the resultant magnetic moment of the sample is equal to zero, and the sample as a whole is macroscopically nonmagnetized.

When an external magnetic field is switched on, the

domains oriented along the field grow at the expense of domains oriented against the field. In weak fields, this growth is reversible. In stronger fields, the simultaneous reorientation of magnetic moments within the entire domain takes place. This process is irreversible, which explains hysteresis and residual magnetization.

Problems

● **7.1. Conditions at the interface.** In the vicinity of point A (Fig. 7.16) on a magnetic-vacuum interface, the magnetic induction

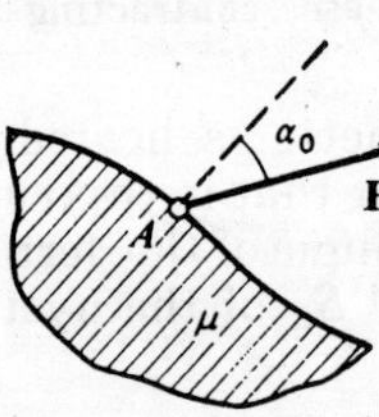

Fig. 7.16

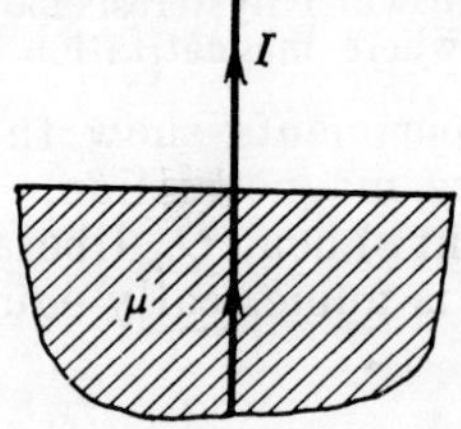

Fig. 7.17

in the vacuum is B_0, vector $\mathbf{B}_0$ forming an angle α_0 with the normal to the interface. The permeability of the magnetic is μ. Find the magnetic induction B in the magnetic in the vicinity of point A.

Solution. The required quantity

$$B = \sqrt{B_n^2 + B_\tau^2}. \quad (1)$$

Considering conditions (7.20) and (7.22) at the interface, we find

$$B_n = B_0 \cos \alpha_0,$$

$$B_\tau = \mu\mu_0 H_\tau = \mu\mu_0 H_{0\tau} = \mu B_{0\tau} = \mu B_0 \sin \alpha_0,$$

where $H_{0\tau}$ is the tangential component of vector $\mathbf{H}_0$ in the vacuum. Substituting these expressions into (1), we obtain

$$B = B_0 \sqrt{\cos^2 \alpha_0 + \mu^2 \sin^2 \alpha_0}.$$

● **7.2. Surface magnetization current.** A long thin current-carrying conductor with current I is arranged perpendicularly to a plane vacuum-magnetic interface (Fig. 7.17). The permeability of the magnetic is μ. Find the linear current density i' of the surface magnetization current on this interface as a function of the distance r from the conductor.

Solution. Let us first consider the configuration of the surface magnetization current. It is clear from Fig. 7.17 that this current has the radial direction. We shall use the theorem on circulation of magnetization vector **J**, taking as a contour a small rectangle whose plane is perpendicular to the magnetization current in this region. The arrangement of this contour is shown in Fig. 7.18, where the crosses indicate the direction of the surface magnetization current. From the equality $Jl = i'l$ we get $i' = J$.

Then we can write $J = \chi H$, where H can be determined from the circulation of vector **H** around the circle of radius r with the centre

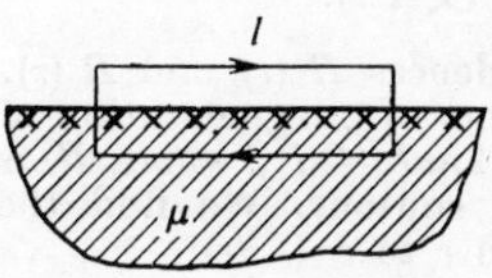

Fig. 7.18

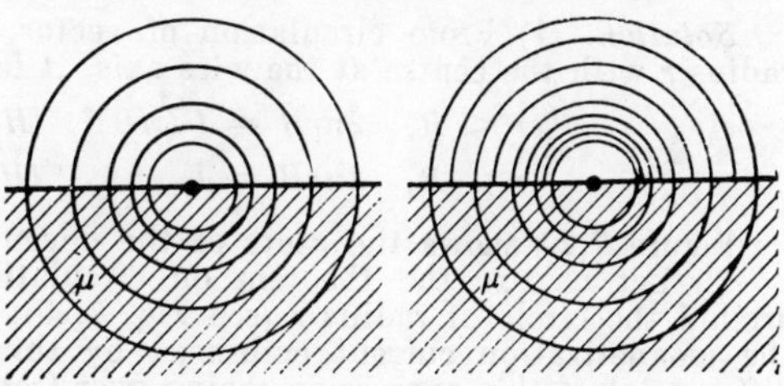

Field **B** Field **H**

Fig. 7.19

on the conductor axis: $2\pi r H = I$ (it is clear from symmetry considerations that the lines of **H** must be circles lying in the planes perpendicular to the conductor with the current I). This gives

$$i' = (\mu - 1)\, I/2\pi r.$$

● **7.3. Circulation of vector H.** A long thin conductor with current I lies in the plane separating the space filled with a nonconducting magnetic of permeability μ from a vacuum. Find the magnetic induction B in the entire space as a function of the distance r from the conductor. Bear in mind that the lines of **B** are circles with centres on the conductor axis.

Solution. The lines of **H** are clearly also circles. Unlike vector **B**, vector **H** undergoes a discontinuity at the vacuum-magnetic interface. We denote by **H** and $\mathbf{H}_0$ magnetic fields in the magnetic and vacuum respectively. Then, in accordance with the theorem on circultaion of vector **H** around the contour in the form of a circle of radius r with the centre on the conductor axis, we have

$$\pi r H + \pi r H_0 = I. \tag{1}$$

Besides, $B = B_0$ at the interface, or

$$\mu H = H_0. \tag{2}$$

Solving Eqs. (1) and (2) together, we obtain

$$H = \frac{I}{(1+\mu)\,\pi r}, \quad B = \mu\mu_0 H = \frac{\mu\mu_0 I}{(1+\mu)\,\pi r}.$$

The configurations of the fields **B** and **H** in this case are shown in Fig. 7.19. We advise the reader to convince himself that for $\mu = 1$, we arrive at the well-known formulas for B and H in a vacuum.

● **7.4. Circulation of vectors H and J.** A direct current I flows along a long homogeneous cylindrical wire with a circular cross section of radius R. The wire is made of a paramagnetic with the susceptibility χ. Find (1) field B as a function of the distance r from the axis of the wire and (2) magnetization current density j' inside the wire.

Solution. (1) From circulation of vector **H** around the circle of radius r with the centre at the wire axis, it follows that

$$r < R, \quad 2\pi r H = I\,(r/R)^2 \quad (H \propto r),$$
$$r > R, \quad 2\pi r H = I \quad (H \propto 1/r).$$

Figure 7.20 shows the plots of the dependences $H\,(r)$ and $B\,(r)$.

(2) Let us use the theorem on circulation of magnetization **J** around the circle of radius r (see Fig. 7.20): $2\pi r J = I'$, where I' is the magnetization current embraced by this contour. We find the differential of this expression (going over from r to $r + dr$):

$$2\pi\, d\,(rJ) = dI'.$$

Since $dI' = j' 2\pi r\, dr$, the above equation can be transformed as follows:

$$j' = \frac{J}{r} + \frac{dJ}{dr}.$$

Let us now consider that $J = \chi H = (\chi I/2\pi R^2)\, r$. This gives

$$j' = \chi I/\pi R^2.$$

It can be easily seen that this current has the same direction as the conduction current (unlike the surface magnetization current which has the opposite direction).

● **7.5.** A long solenoid is filled with a nonhomogeneous isotropic paramagnetic whose susceptibility depends only on the distance r from the solenoid axis as $\chi = ar^2$, where a is a constant. The magnetic induction on the solenoid axis is B_0. Find (1) magnetization J and (2) magnetization current density j' as functions of r.

Solution. (1) $J = \chi H$. In our case, H is independent of r (this directly follows from circulation of vector **H** around the contour shown in the left part of Fig. 7.21). Hence $H = H_0$ on the solenoid's axis, and we obtain

$$J = ar^2 H_0 = ar^2 B_0/\mu_0.$$

(2) From the theorem on circulation of magnetization **J** around an infinitely narrow contour shown in Fig. 7.21 on the right, it follows that

$$Jl - (J + dJ)\, l = j'_n l\, dr,$$

where l is the contour height and dr its width. Hence

$$j'_n = -\frac{dJ}{dr} = -\frac{2aB_0}{\mu_0} r.$$

The minus sign indicates that vector $\mathbf{j}'$ is directed against the normal vector $\mathbf{n}$ which forms the right-hand system with the direction of cir-

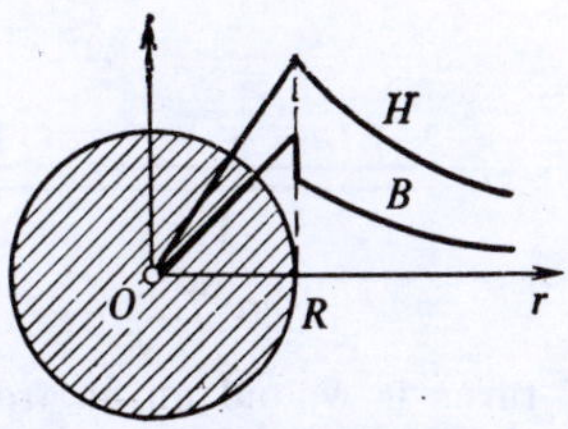

Fig. 7.20

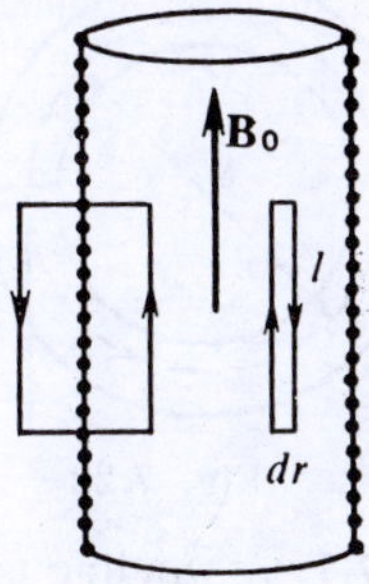

Fig. 7.21

cumvention of the contour. In other words, $\mathbf{j}'$ is directed towards us in the region of the right contour in the figure, i.e. volume magnetization currents form with vector $\mathbf{B}_0$ the left-hand system.

● 7.6. **A permanent magnet** has the shape of a ring with a narrow gap between the poles. The middle diameter of the ring is d. The gap width is b, the magnetic induction of the field in the gap is $\mathbf{B}$. Find the magnitudes of vectors $\mathbf{H}$ and $\mathbf{J}$ inside the magnet, ignoring the dissipation of the field at the edges of the gap.

Solution. Using the theorem on circulation of vector $\mathbf{H}$ around the dashed circle of diameter d (Fig. 7.22) and considering that there is no conduction current, we can write

$$(\pi d - b) H_\tau + bB/\mu_0 = 0,$$

where H_τ is the projection of vector $\mathbf{H}$ onto the direction of contour circumvention (which is taken so that it coincides with the direction of vector $\mathbf{B}$ in the gap). Hence

$$H_\tau = -\frac{bB}{\mu_0(\pi d - b)} \approx -\frac{bB}{\mu_0 \pi d}. \quad (1)$$

The minus sign indicates that the direction of vector $\mathbf{H}$ inside the magnetic material is opposite to the direction of vector $\mathbf{B}$ at the same point. It should be noted that as $b \to 0$, $H \to 0$ as well.

The magnitude of magnetization $\mathbf{J}$ can be found by formula (7.11), taking into account (1):

$$J = \frac{B/\mu_0}{1-b/\pi d} \approx \frac{B}{\mu_0}.$$

The relationship between vectors $\mathbf{B}/\mu_0$, $\mathbf{H}$, and $\mathbf{J}$ at any point inside the magnet is shown in Fig. 7.23.

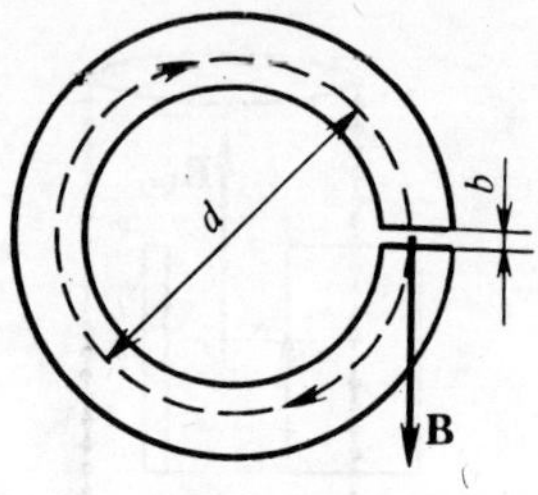

Fig. 7.22

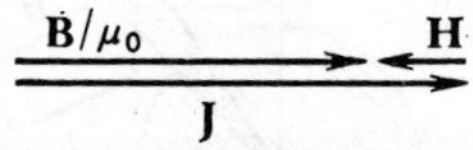

Fig. 7.23

● **7.7.** A winding containing N turns is wound on an iron core in the form of a torus with the middle diameter d. A narrow transverse slot of width b (Fig. 7.22) is cut in the core. When current I is passed through the winding, the magnetic induction in the slot is B. Find the permeability of iron under these conditions, neglecting the dissipation of the field at the edges of the slot.

Solution. In accordance with the theorem on circulation of vector $\mathbf{H}$ around the circle of diameter d (see Fig. 7.22), we have

$$(\pi d - b)\, H + bH_0 = NI,$$

where H and H_0 are the magnitudes of $\mathbf{H}$ in iron and in the slot respectively. Besides, the absence of field dissipation at the edges means that

$$B = B_0.$$

Using these two equations and taking into account that $B = \mu\mu_0 H$ and $b \ll d$, we obtain

$$\mu = \frac{\pi\, dB}{\mu_0 NI - bB}.$$

● **7.8. The force acting on a magnetic.** The device shown in Fig. 7.24 is used for measuring (with the help of a balance) a force with which a small paramagnetic sphere of volume V is attracted to the pole of magnet M. The magnetic induction on the axis of the pole shoe depends on the height as $B = B_0 e^{-ax^2}$, where B_0 and a are constants. Find (1) the height x_m at which the sphere must be placed so that the attractive force be maximum and (2) the magnetic susceptibility of the paramagnetic if the maximum force of attraction is F_m.

Solution. (1) Let, for the sake of definiteness, vector **B** on the axis be directed upwards and coincide with the X-axis. Then, in accordance with (6.34), $F_x = p_m\, \partial B/\partial x$, where we took into account that the magnetic moment $\mathbf{p}_m$ is directed as vector **B** (for paramagnetics) and hence replaced ∂n by ∂x.

Next, since $p_m = JV = \chi HV$ and $\partial B/\partial x = -2aB_0 x e^{-ax^2}$, we have

$$F_x = -Axe^{-2ax^2}, \quad (1)$$

where $A = 2aB_0^2\chi V/\mu\mu_0$.

Having calculated the derivative dF_x/dx and equating it to zero, we obtain the following equation for determining x_m: $1 - 4ax^2 = 0$, whence

$$x_m = 1/2\sqrt{a}. \quad (2)$$

(2) Substituting (2) into (1), we find

$$\chi = \frac{\mu_0 F_m}{B_0^2 V}\sqrt{\frac{e}{a}},$$

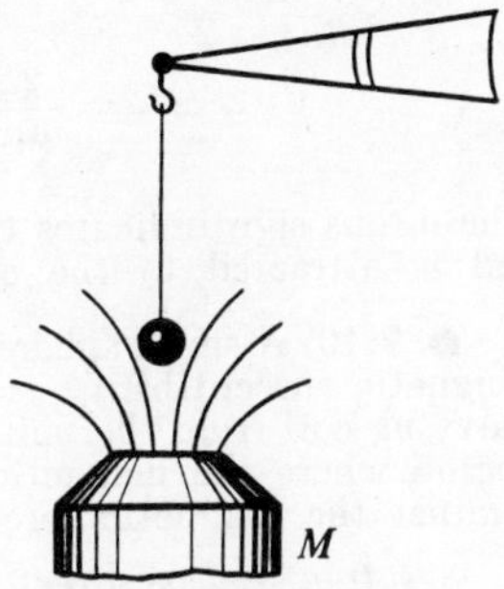

Fig. 7.24

where we took into account that $\mu \simeq 1$ for paramagnetics.

● **7.9.** A long thin cylindrical rod made of a paramagnetic with the magnetic susceptibility χ and cross-sectional area S is arranged

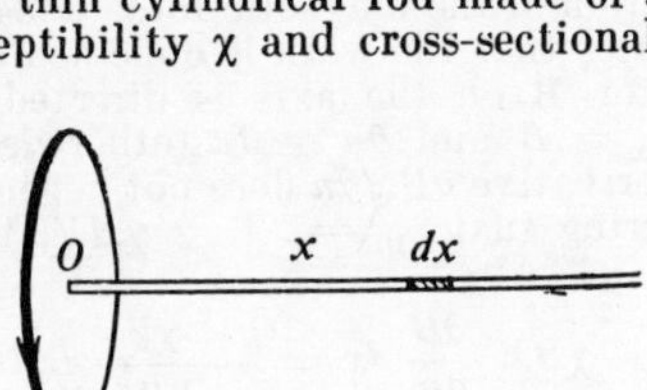

Fig. 7.25

along the axis of a current-carrying coil. One end of the rod is at the centre of the coil, where the magnetic field is B, while the other end lies in the region where the magnetic field is practically absent. Find the force with which the coil acts on the rod.

Solution. Let us mentally isolate an element of the rod of length dx (Fig. 7.25). The force acting on this element is given by

$$dF_x = dp_m \frac{\partial B_x}{\partial n}.$$

Suppose that vector **B** on the axis of the coil is directed to the right (see the figure), towards positive values of x. Then $B_x = B$, $\partial n = \partial x$, and since $dp_m = JS\,dx = \chi HS\,dx$, we get

$$dF_x = \chi HS\,dx\,\frac{\partial B}{\partial x} = \frac{\chi S}{\mu\mu_0}\,B\,dB.$$

Having integrated this equation, we obtain

$$F_x = \frac{\chi S}{\mu\mu_0}\int_B^0 B\,dB = -\frac{\chi SB^2}{2\mu\mu_0}.$$

The minus sign indicates that vector **F** is directed to the left, i.e. the rod is attracted to the current-carrying coil.

● **7.10.** A small sphere of volume V, made of a paramagnetic with magnetic susceptibility χ, was displaced along the axis of a current-carrying coil from the point where the magnetic induction is B to the region where the magnetic field is practically absent. Find the work against the magnetic forces accomplished in this case.

Solution. Let us direct the X-axis along the axis of the coil. Then the elementary work done against the magnetic forces upon the displacement of the sphere by dx is equal to

$$\delta A = -F_x\,dx = -p_m\,\frac{\partial B_x}{\partial n}\,dx, \tag{1}$$

where F_x is the projection of the magnetic force (6.34) onto the X-axis The minus sign indicates that the work is done against this force.

Suppose that vector **B** on the axis is directed towards positive values of x. Then $B_x = B$ and $\partial n = \partial x$ (otherwise, $B_x = -B$ and $\partial n = -\partial x$, i.e. the derivative $\partial B_x/\partial n$ does not depend on the direction of vector **B**). Considering that $p_m = JV = \chi HV$, we rewrite Eq. (1) in the form

$$\delta A = -\chi HV\,\frac{\partial B}{\partial x}\,dx = -\frac{\chi V}{\mu\mu_0}\,B\,dB.$$

Integrating this equation between B and 0, we obtain

$$A = -\frac{\chi V}{\mu\mu_0}\int_B^0 B\,dB = \frac{\chi B^2 V}{2\mu\mu_0}.$$

8. Relative Nature of Electric and Magnetic Fields

8.1. Electromagnetic Field. Charge Invariance

So far, we have considered electric and magnetic fields *separately*, without establishing any clear relation between them. This could be done only because the two fields were static. In other cases, however, it is impossible.

It will be shown that electric and magnetic fields must always be considered *together* as a single total electromagnetic field. In other words, it turns out that electric and magnetic fields are in a certain sense the components of a single physical object which we call the *electromagnetic field.*

The division of the electromagnetic field into electric and magnetic fields is of relative nature since it depends to a very large extent on the reference system in which the phenomena are considered. The field which is constant in one reference frame in the general case is found to vary in another reference frame.

Let us consider some examples.

A charge moves in an inertial system K at a constant velocity $\mathbf{v}$. In this system, we shall observe both the electric and magnetic fields of this charge, the two fields varying with time. If, however, we go over to an inertial system K' moving together with the charge, the charge will be at rest in this system, and we shall observe only the electric field.

Suppose two identical charges move in the system K towards each other at the same velocity v. In this system, we shall observe both electric and magnetic fields, both these fields varying. In this case it is impossible to find such a system K' where only one of the fields would be observed.

In the system K, there exists a permanent nonuniform magnetic field (e.g., the field of a fixed permanent magnet). Then in the system K' moving relative to the system K we shall observe a varying magnetic field and, as will be shown below, an electric field as well.

Thus, it is clear that in different reference frames different relations are observed between electric and magnetic fields.

Before considering the main points of this chapter, viz. the laws of transformation of fields upon a transition from one reference system to another, we must answer the question which is important for further discussion: how do an electric charge q itself and the Gauss theorem for vector $\mathbf{E}$ behave upon such transitions?

Charge Invariance. At present, there is exhaustive evidence that the total charge of an isolated system does not change with the change in the motion of charge carriers.

This may be proved by the neutrality of a gas consisting of hydrogen molecules. Electrons in these molecules move at much higher velocities than protons. Therefore, if the charge depended on the velocity, the charges of the electrons and protons would not compensate each other and the gas would be charged. However, no charge (to 20 decimal places!) has been observed in hydrogen gas.

Another proof can be obtained while observing the heating of a piece of a substance. Since the electron mass is considerably smaller than the mass of nuclei, the velocity of electrons upon heating must increase more than that of nuclei. And if the charge depended on the velocity, the substance would become charged upon heating. Nothing of this kind was ever observed.

Further, if the electron charge depended on its velocity, then the total charge of a substance would change as a result of chemical reactions, since the average velocities of electrons in a substance depend on its chemical composition. Calculations show that even a weak dependence of the charge on the velocity would result in extremely strong electric fields even in the simplest chemical reactions. But nothing of this kind was observed in this case either.

Finally, the design and operation of all modern particle accelerators are based on the assumption that a particle's charge is invariable when its velocity is changed.

Thus, we arrive at the conclusion that the charge of any particle is a relativistic invariant quantity and does not depend on the particle's velocity or on the choice of the reference system.

Invariance of the Gauss Theorem for Field E. It turns out, as follows from the generalization of experimental results, that the Gauss theorem $\oint \mathbf{E}\, d\mathbf{S} = q/\varepsilon_0$ is valid not only for fixed charges but for moving charges also. In the latter case, the surface integral must be taken for a definite instant of time in a given reference system.

Besides, since different inertial reference systems are physically equivalent (in accordance with the relativity principle), we can state that the Gauss theorem is valid for all inertial systems of reference.

8.2. Laws of Transformation for Fields E and B

While going over from one reference system to another, fields **E** and **B** are transformed in a certain way. The laws of this transformation are established in the special theory of relativity in a rather complicated way. For this reason, we shall not present here the corresponding derivations and will pay main attention to the content of these laws, the corollaries following from them, and the application of these laws for solving some specific problems.

Formulation of the Problem. Suppose we have two inertial systems of reference: the system K and the system K' moving relative to the first system at a velocity $\mathbf{v}_0$. We know the magnitudes of the fields **E** and **B** at a certain point in space and time in the system K. What are the magnitudes of the fields **E**′ and **B**′ at the same point in space and time in the system K'? (Recall that the same point in space and time is a point whose coordinates and time in the two reference systems are related through the Lorentz transformations*)

As was mentioned above, the answer to this question is given in the theory of relativity from which it follows that the *laws of transformation of fields* are expressed by the formulas

$$\boxed{\begin{gathered}\mathbf{E}'_{\parallel}=\mathbf{E}_{\parallel},\qquad \mathbf{B}'_{\parallel}=\mathbf{B}_{\parallel},\\ \mathbf{E}'_{\perp}=\frac{\mathbf{E}_{\perp}+[\mathbf{v}_0\times\mathbf{B}]}{\sqrt{1-\beta^2}},\qquad \mathbf{B}'_{\perp}=\frac{\mathbf{B}_{\perp}-[\mathbf{v}_0\times\mathbf{E}]/c^2}{\sqrt{1-\beta^2}}.\end{gathered}} \tag{8.1}$$

Here the symbols $\parallel$ and $\perp$ mark the *longitudinal* and *transverse* (relative to the vector $\mathbf{v}_0$) components of the electric and magnetic fields, $\beta=v_0/c$, and c is the velocity of light in a vacuum ($c^2=1/\varepsilon_0\mu_0$).

Written in terms of projections, these formulas have the form

$$\boxed{\begin{aligned}E'_x&=E_x, & B'_x&=B_x,\\ E'_y&=\frac{E_y-v_0B_z}{\sqrt{1-\beta_2}}, & B'_y&=\frac{B_y+v_0E_z/c^2}{\sqrt{1-\beta^2}},\\ E'_z&=\frac{E_z+v_0B_y}{\sqrt{1-\beta^2}}, & B'_z&=\frac{B_z-v_0E_y/c^2}{\sqrt{1-\beta^2}},\end{aligned}} \tag{8.2}$$

where it is assumed that the X- and X'-axes of coordinates are directed

---

* $x'=\dfrac{x-v_0t}{\sqrt{1-(v_0/c)^2}},\quad y'=y,\ z'=z,\quad t'=\dfrac{t-xv_0/c^2}{\sqrt{1-(v_0/c)^2}}.$

along the vector $\mathbf{v}_0$, the Y'-axis is parallel to the Y-axis and the Z'-axis to the Z-axis.

It follows from Eqs. (8.1) and (8.2) that each of the vectors $\mathbf{E}'$ and $\mathbf{B}'$ is expressed both in terms of $\mathbf{E}$ and $\mathbf{B}$. This is an evidence of a *common* nature of electric and magnetic fields. Taken separately, each of these fields has no absolute meaning: speaking of electric and magnetic fields, we must indicate the system of reference in which these fields are considered.

It should be emphasized that the properties of electromagnetic field expressed by the laws of its transformation are *local*; the values of $\mathbf{E}'$ and $\mathbf{B}'$ at a certain point in space and time in the system K' are uniquely determined only by the values of $\mathbf{E}$ and $\mathbf{B}$ at the same point in space and time in the system K.

The following features of the laws of field transformation are also very important.

1. Unlike the transverse components of $\mathbf{E}$ and $\mathbf{B}$, which change upon a transition from one reference system to another, their *longitudinal components do not change* and remain the same in all systems.

2. Vectors $\mathbf{E}$ and $\mathbf{B}$ in different systems are connected with each other through highly symmetric relations. This can be seen most clearly in the form of the laws of transformation for the projections of the fields [see (8.2)].

3. In order to obtain the formulas of the inverse transformation (from K' to K), it is sufficient to replace in formulas (8.1) and (8.2) all primed quantities by unprimed ones (and vice versa) as well as the sign of v_0.

Special Case of Field Transformation ($v_0 \ll c$). If the system K' moves relative to the system K at a velocity $v_0 \ll c$, the radicals in the denominators of formulas (8.1) can be replaced by unity, and we obtain

$$\mathbf{E}'_{\parallel} = \mathbf{E}_{\parallel}, \qquad \mathbf{B}'_{\parallel} = \mathbf{B}_{\parallel},$$
$$\mathbf{E}'_{\perp} = \mathbf{E}_{\perp} + [\mathbf{v}_0 \times \mathbf{B}], \quad \mathbf{B}'_{\perp} = \mathbf{B}_{\perp} - [\mathbf{v}_0 \times \mathbf{E}]/c^2. \tag{8.3}$$

Hence it follows that

$$\boxed{\mathbf{E}' = \mathbf{E} + [\mathbf{v}_0 \times \mathbf{B}], \quad \mathbf{B}' = \mathbf{B} - [\mathbf{v}_0 \times \mathbf{E}]/c^2.} \tag{8.4}$$

It should be noted that the first formula in (8.4) can be directly obtained in a very simple way. Let a charge q in the system K have a velocity $\mathbf{v}_0$ at a certain instant t. The Lorentz force acting on this charge is $\mathbf{F} = q\mathbf{E} + q[\mathbf{v}_0 \times \mathbf{B}]$. We go over to the inertial system K' moving relative to the system K at the same velocity as the charge q at the instant t, viz. the velocity $\mathbf{v}_0$. At this instant, the charge q is at rest in the system K', and the force acting on the immobile charge is purely electric: $\mathbf{F}' = q\mathbf{E}'$. If $v_0 \ll c$ (as in the case under consideration), this force is invariant ($\mathbf{F}' = \mathbf{F}$), whence we obtain the first formula in (8.4).

However, the transformation formulas for magnetic field can be obtained only with the help of the theory of relativity as a result of rather cumbersome calculations.

Let us consider a simple example of the application of formulas (8.4).

Example. A large metallic plate moves at a constant nonrelativistic velocity v in a uniform magnetic field B (Fig. 8.1). Find the surface density of charges induced on the surfaces of the plate as a result of its motion.

Let us go over to the reference system fixed to the plate. In accordance with the first of formulas (8.4), in this system of reference a constant uniform electric field

$$\mathbf{E}' = [\mathbf{v} \times \mathbf{B}]$$

will be observed. It is directed towards us. Under the action of this external field, the charges will be displaced so that positive charges will appear on the front surface of the plate and negative charges, on the hind surface.

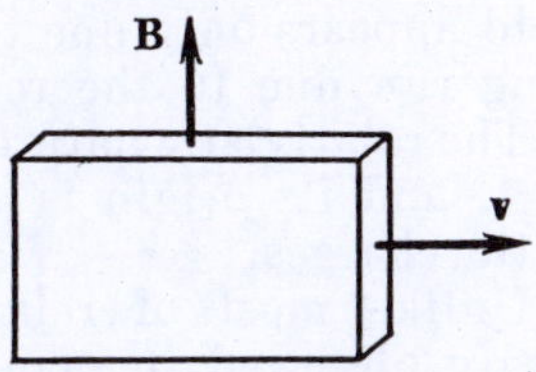

Fig. 8.1

The surface density σ of these charges will be such that the field created by these charges inside the plate completely compensates the external field $\mathbf{E}'$, since in equilibrium the resultant electric field inside the plate must be equal to zero. Taking into account relation (1.11), we obtain

$$\sigma = \varepsilon_0 E' = \varepsilon_0 vB.$$

It should be noted that while solving this problem, we could follow another line of reasoning, viz. from the point of view of the reference frame in which the plate moves at the velocity $\mathbf{v}$. In *this* system there will be an *electric field* inside the plate. It appears as a result of the action of the magnetic component of the Lorentz force causing the displacement of all the electrons in the plate behind the plane of Fig. 8.1. As a result, the front surface of the plate will be charged positively and the hind surface, negatively, and the electric field appearing in the plate will be such that the electric force $q\mathbf{E}$ compensates the magnetic component of the Lorentz force, $q\,[\mathbf{v} \times \mathbf{B}]$, whence $\mathbf{E} = -[\mathbf{v} \times \mathbf{B}]$. This field is connected with the surface charge density through the same formula $\sigma = \varepsilon_0 vB$.

Both approaches to the solution of the given problem are equally justified.

Relativistic Nature of Magnetism. Formulas (8.1) and (8.2) for the transformation of fields lead to a remarkable conclusion: the appearance of the magnetic field is a purely relativistic effect following from the existence in nature of the limiting velocity c equal to the velocity of light in a vacuum.

If this velocity (and hence the velocity of propagation of interactions) were infinite, no magnetism would exist at all. Indeed, let us consider a free electric charge. Only electric field exists in the system K where this charge is at rest. Consequently, according to (8.1), no magnetic field would appear in any other system K' *if c tended to infinity*. This field appears only due to the fact that c is finite, i.e. in the long run due to the relativistic effect.

The relativistic nature of magnetism is a universal physical fact, and its origin is associated with the absence of magnetic charges.

Unlike most of relativistic effects, magnetism can be easily observed in many cases, for example, the magnetic field of a current-carrying conductor. The reason behind such favourable circumstances is that the magnetic field may be created by a very large number of moving electric charges under the conditions of the almost complete vanishing of the electric field due to practically ideal balance of the numbers of electrons and protons in conductors. In this case the magnetic interaction is predominant.

The almost complete compensation of electric charges made it possible for physicists to study relativistic effects (i.e. magnetism) and discover *correct* laws. For this reason, the laws of electromagnetism, unlike Newton's laws, have not been refined after the creation of the theory of relativity.

Field Varies Rather than Moves. Since the relations between electric and magnetic fields vary upon a transition to another reference system, the fields **E** and **B** must be treated with care. For instance, even the concept of the force exerted on a moving charge by a *moving* magnetic field does not have any precise meaning. The force is determined by the values of the quantities **E** and **B** at the point where the charge is located. If as a result of motion of the *sources* of the fields **E** and **B** their values at this point change, the force will change as well. Otherwise the motion of the sources does not affect the magnitude of the force.

Thus, while solving the problem about the force acting on a charge, we must know **E** and **B** at the point of location of the charge as well as its velocity **v**. All these quantities must be taken relative to the inertial reference frame we are interested in.

The expression "moving field" (if it is used) should be interpreted as a convenient way of the verbal description of a varying field under certain conditions, and nothing more.

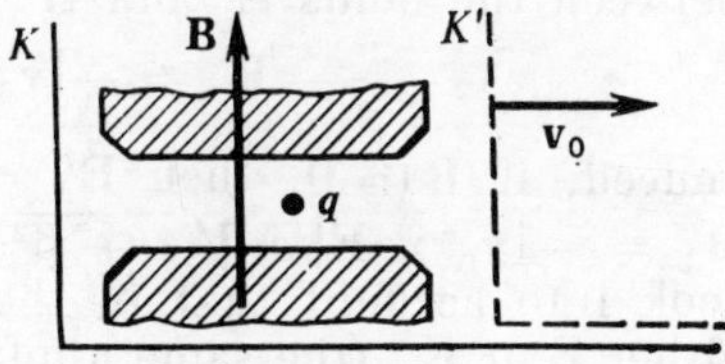

Fig. 8.2

The following simple example will illustrate the fact that upon a transition from one reference frame to another the field must be treated with care.

Example. A charged particle is at rest between the poles of a magnet fixed in the system K. Let us go over to the system K' that moves to the right (Fig. 8.2) at a nonrelativistic velocity $\mathbf{v}_0$ relative to the system K. (1) Can we state that the charged particle moves in the magnetic field in the system K'? (2) Find the force acting on this particle in the system K'.

(1) Yes, the particle *moves in the magnetic field*. It should be noted, however, that it moves in the magnetic field and not relative to the magnetic field. It is meaningful to say that a particle moves relative to the reference system, a magnet, or other objects. However, we cannot state that a particle moves relative to the magnetic field. The latter statement has no physical meaning. All this refers not only to magnetic but to electric fields as well.

(2) In order to find the force, we must take into account that the electric field $\mathbf{E}' = [\mathbf{v}_0 \times \mathbf{B}]$ directed towards us (Fig. 8.2), will appear in the system K'. In this system, the charge will move to the left at the velocity $-\mathbf{v}_0$, and this motion will occur in crossed electric and magnetic fields. For the sake of definiteness, we suppose that the

charge of the particle is $q > 0$. Then the Lorentz force in the system K' is

$$\mathbf{F}' = q\mathbf{E}' + q[-\mathbf{v}_0 \times \mathbf{B}] = q[\mathbf{v}_0 \times \mathbf{B}] - q[\mathbf{v}_0 \times \mathbf{B}] = 0$$

which, however, directly follows from the invariance of the force upon nonrelativistic transformations from one reference frame to the other.

8.3. Corollaries of the Laws of Field Transformation

Some Simple Corollaries. Several simple and useful relations follow in some cases from transformation formulas (8.1).

1. If the *electric* field $\mathbf{E}$ alone is present in the system K (the magnetic field $\mathbf{B} = 0$), the following relation exists between the fields $\mathbf{E}'$ and $\mathbf{B}'$ in the system K':

$$\mathbf{B}' = -[\mathbf{v}_0 \times \mathbf{E}']/c^2. \tag{8.5}$$

Indeed, if $\mathbf{B} = 0$, then $\mathbf{E}'_\perp = \mathbf{E}_\perp/\sqrt{1-\beta^2}$ and $\mathbf{B}'_\parallel = 0$, $\mathbf{B}'_\perp = -[\mathbf{v}_0 \times \mathbf{E}]/c^2\sqrt{1-\beta^2} = -[\mathbf{v}_0 \times \mathbf{E}']/c^2$, where we took into account that in the cross product we can write either $\mathbf{E}$ or $\mathbf{E}_\perp$ (the same applies to the primed quantities). Considering that $\mathbf{B}' = \mathbf{B}'_\parallel + \mathbf{B}'_\perp = \mathbf{B}'_\perp$, we arrive at formula (8.5).

2. If the *magnetic* field $\mathbf{B}$ alone is present in the system K (while the electric field $\mathbf{E} = 0$), then for the system K' we have

$$\mathbf{E}' = [\mathbf{v}_0 \times \mathbf{B}']. \tag{8.6}$$

Indeed, if $\mathbf{E} = 0$, then $\mathbf{B}'_\perp = \mathbf{B}_\perp/\sqrt{1-\beta^2}$ and $\mathbf{E}'_\parallel = 0$, $\mathbf{E}'_\perp = [\mathbf{v}_0 \times \mathbf{B}]/\sqrt{1-\beta^2}$. Substituting $\mathbf{B}_\perp$ for $\mathbf{B}$ and then $\mathbf{B}'$ for $\mathbf{B}'_\perp$ in the last cross product, we arrive at formula (8.6).

Formulas (8.5) and (8.6) lead to the following important conclusion: *If only one of the fields* ($\mathbf{E}$ *or* $\mathbf{B}$) *is present in the system* K, *the electric and magnetic fields in the system* K' *are mutually perpendicular* ($\mathbf{E}' \perp \mathbf{B}'$). It should be noted that the opposite statement is true only under certain additional limitations imposed on the magnitudes of the vectors $\mathbf{E}$ and $\mathbf{B}$.

Finally, since Eqs. (8.5) and (8.6) contain only the quantities referring to the same system of reference, these equations can be easily applied to the fields varying in space and time. An illustrative example is the field of a uniformly moving point charge.

Field of Freely Moving Relativistic Charge. The formulas for field transformation are interesting above all by that they express remarkable properties of electromagnetic field. Besides, they are important from a purely practical point of view since they allow us to solve some problems in a simpler way.

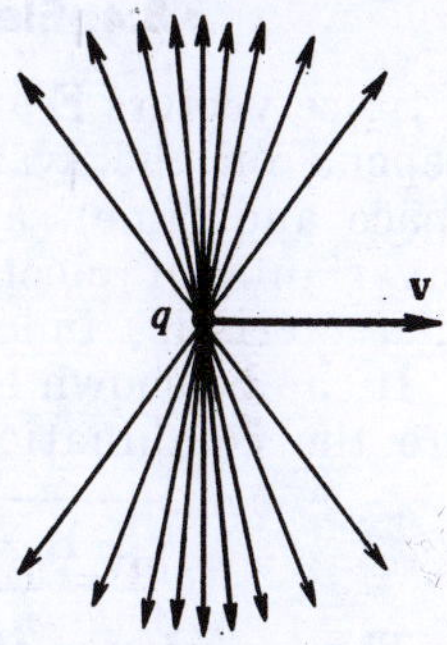

Fig. 8.3

For example, the problem of determining the field of a uniformly moving point charge can be solved as a result of transformation of a purely Coulomb field which is observed in the reference system fixed to the charge. Calculations show (see Problem 8.10) that the lines of the field **E** created by a freely moving charge q have the form shown in Fig. 8.3, where **v** is the velocity of the charge. This pattern corresponds to the instant "picture" of the electric field configuration. At an arbitrary point P of the reference system, vector **E** is directed *along* the radius vector **r** drawn from the point of location of the charge q to point P.

The magnitude of vector **E** is determined by the formula

$$E = \frac{1}{4\pi\varepsilon_0} \frac{q}{r^2} \frac{1-\beta^2}{(1-\beta^2 \sin^2 \vartheta)^{3/2}}, \tag{8.7}$$

where $\beta = v/c$, ϑ is the angle between the radius vector **r** and the vector **v** corresponding to the velocity of the charge.

The electric field is "flattened" in the direction of motion of the charge (see Fig. 8.3), the degree of flattening being the larger the closer the velocity **v** of the charge to the velocity c of light. It should be also borne in mind that the field shown in the figure "moves" together with the charge. That is why the field **E** in the reference frame relative to which the charge moves varies with time.

Knowing the field **E**, we can find the field **B** in the same reference system:

$$\mathbf{B} = \frac{1}{c^2} [\mathbf{v} \times \mathbf{E}] = \frac{\mu_0}{4\pi} \frac{q [\mathbf{v} \times \mathbf{r}]}{r^3} \frac{1-\beta^2}{(1-\beta^2 \sin^2 \vartheta)^{3/2}}. \tag{8.8}$$

This formula is a corollary of relation (8.5) in which we have replaced the primed quantities by unprimed ones and **v** by —**v**.

When $v \ll c$ ($\beta \ll 1$), expressions (8.7) and (8.8) are transformed to (1.2) and (6.3) respectively.

8.4. Electromagnetic Field Invariants

Since vectors **E** and **B** characterizing electromagnetic field depend on the system of reference (at the same point in space and time), a natural question arises concerning the invariants of electromagnetic field, i.e. the quantitative characteristics independent of the reference system.

It can be shown that there exist two such invariants which are the combinations of vectors **E** and **B**, viz.

$$\boxed{\mathbf{E}\cdot\mathbf{B}=\text{inv.} \quad \text{and} \quad E^2-c^2B^2=\text{inv.}} \qquad (8.9)$$

The invariance of these quantities (relative to the Lorentz transformations) is a consequence of field transformation formulas (8.1) or (8.2). This question is considered in greater detail in Problem 8.9.

The application of these invariants allows us in some cases to find the solution quickly and simply and make the corresponding conclusions and predictions. We shall list the most important conclusions.

1. The invariance of $\mathbf{E}\cdot\mathbf{B}$ immediately implies that if $\mathbf{E} \perp \mathbf{B}$ in some reference system, i.e. $\mathbf{E}\cdot\mathbf{B} = 0$, in all other inertial systems $\mathbf{E}' \perp \mathbf{B}'$ as well.

2. It follows from the invariance of $E^2 - c^2B^2$ that if $E = cB$ (i.e. $E^2 - c^2B^2 = 0$), in any other inertial system $E' = cB'$ as well.

3. If in a certain reference system the angle between vectors **E** and **B** is acute (obtuse) (this means that $\mathbf{E}\cdot\mathbf{B}$ is greater (less) than zero), in any other reference system the angle between vectors $\mathbf{E}'$ and $\mathbf{B}'$ will be acute (obtuse) as well.

4. If in a certain reference system $E > cB$ (or $E < cB$), (which means that $E^2 - c^2B^2$ is greater (or less) than zero), in any other reference system $E' > cB'$ (or $E' < cB'$) as well.

5. If both invariants are equal to zero, then $\mathbf{E} \perp \mathbf{B}$ and $E = cB$ in all reference systems. It will be shown later that this is precisely observed in electromagnetic waves.

6. If the invariant $\mathbf{E}\cdot\mathbf{B}$ alone is equal to zero, we can find a reference system in which either $\mathbf{E}' = 0$ or $\mathbf{B}' = 0$. The sign of the other invariant determines which of these vectors is equal to zero. The

opposite is also true: if $\mathbf{E} = 0$ or $\mathbf{B} = 0$ in some reference system, then $\mathbf{E}' \perp \mathbf{B}'$ in any other system. (This conclusion has already been drawn in Sec. 8.3.)

And finally one more remark: it should be borne in mind that generally fields $\mathbf{E}$ and $\mathbf{B}$ depend both on coordinates

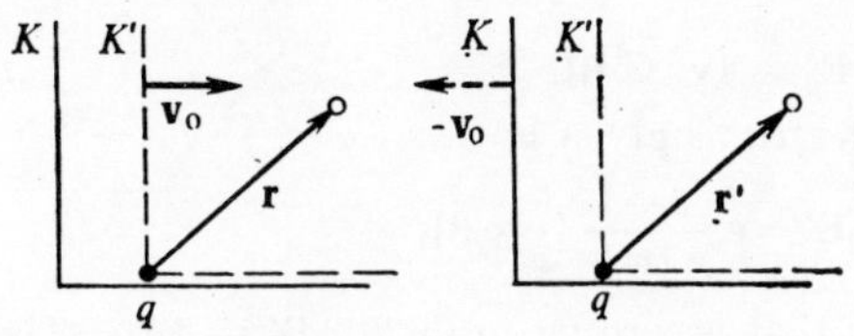

Fig. 8.4

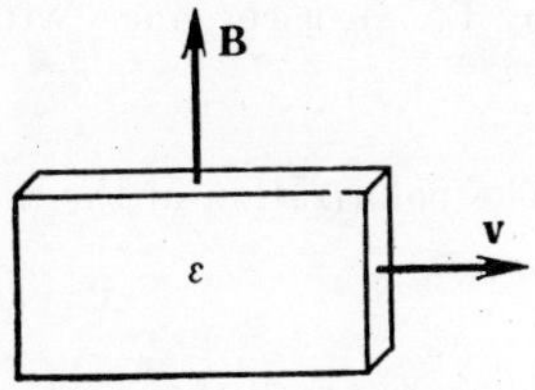

Fig 8.5

and on time. Consequently, each invariant of (8.9) refers to the same point in space and time of the field, the coordinates and time of the point in different systems of reference being connected through the Lorentz transformations.

Problems

● **8.1. Special case of field transformation.** A nonrelativistic point charge q moves at a constant velocity $\mathbf{v}$. Using the transformation formulas, find the magnetic field $\mathbf{B}$ of this charge at a point whose position relative to the charge is defined by the radius vector $\mathbf{r}$.

Solution. Let us go over to the system of reference K' fixed to the charge. In this system, only the Coulomb field with the intensity

$$\mathbf{E}' = \frac{1}{4\pi\varepsilon_0} \frac{q}{r^3} \mathbf{r},$$

is present, where we took into account that the radius vector $\mathbf{r}' = \mathbf{r}$ in the system K' (nonrelativistic case). Let us now return from the system K' to the system K that moves relative to the system K' at the velocity $-\mathbf{v}$. For this purpose, we shall use the formula for the field $\mathbf{B}$ from (8.4), in which the role of primed quantities will be played by unprimed quantities (and vice versa), and replace the velocity $\mathbf{v}_0$ by $-\mathbf{v}_0$ (Fig. 8.4). In the case under consideration, $\mathbf{v}_0 = \mathbf{v}$, and hence $\mathbf{B} = \mathbf{B}' + [\mathbf{v} \times \mathbf{E}']/c^2$. Considering that in the system K' $\mathbf{B}' = 0$ and that $c^2 = 1/\varepsilon_0\mu_0$, we find

$$\mathbf{B} = \frac{\mu_0}{4\pi} \frac{q\,[\mathbf{v} \times \mathbf{r}]}{r^3}.$$

We have obtained formula (6.3) which was earlier postulated as a result of the generalization of experimental data.

14—0181

● **8.2.** A large plate made of homogeneous dielectric with the permittivity ε moves at a constant nonrelativistic velocity $\mathbf{v}$ in a uniform magnetic field $\mathbf{B}$ as shown in Fig. 8.5. Find the polarization $\mathbf{P}$ of the dielectric and the surface density σ' of bound charges.

Solution. In the reference system fixed to the plate, in addition to the magnetic field, there will be an electric field which we shall denote by $\mathbf{E}_0'$. In accordance with formulas (8.4) of field transformation, we have

$$\mathbf{E}_0' = [\mathbf{v} \times \mathbf{B}].$$

The polarization of the dielectric is given by

$$\mathbf{P} = \varkappa\varepsilon_0\mathbf{E}' = \varepsilon_0\frac{\varepsilon - 1}{\varepsilon}[\mathbf{v} \times \mathbf{B}],$$

where we took into account that according to (3.29), $\mathbf{E}' = \mathbf{E}_0'/\varepsilon$. The surface density of bound charges is

$$|\sigma'| = P = \varepsilon_0\frac{\varepsilon - 1}{\varepsilon}vB.$$

At the front surface of the plate (Fig. 8.5), $\sigma' > 0$, while on the opposite face $\sigma' < 0$.

● **8.3.** Suppose we have an uncharged long straight wire with a current I. Find the charge per unit length of this wire in the reference system moving translationally with a nonrelativistic velocity v_0 along the conductor in the direction of the current I.

Solution. In accordance with the transformation formulas (8.4), the electric field $\mathbf{E}' = [\mathbf{v}_0 \times \mathbf{B}]$ will appear in the moving reference system, or

$$E_r' = -v_0\mu_0 I/2\pi r \tag{1}$$

where r is the distance from the wire axis.

The expression for B in this formula was obtained with the help of the theorem on circulation.

On the other hand, according to the Gauss theorem (in the moving system) we have

$$E_r' = \lambda'/2\pi\varepsilon_0 r, \tag{2}$$

where λ' is the charge per unit length of the wire.

A comparison of (1) and (2) gives

$$\lambda' = -v_0 I/c^2,$$

where $c^2 = 1/\varepsilon_0\mu_0$. The origin of this charge is associated with the Lorentz contraction experienced by the "chains" of positive and negative charges (they have different velocities!).

● **8.4.** There is a narrow beam of protons moving with a relativistic velocity v in the system of reference K. At a certain distance from the beam, the intensity of the electric field is equal to E. Find the induction B' of the magnetic field at the same distance from the beam in the system K' that moves relative to the system K at the velocity v_0 in the direction of the proton beam.

Solution. This problem can be solved in the most simple way with the help of formulas (8.1). But first we must find the induction B in the system K at the same distance from the beam where the intensity E is given.

Using the theorem on circulation of vector **B** and the Gauss theorem for vector **E**, we find

$$B = \mu_0 I/2\pi r, \quad E = \lambda/2\pi\varepsilon_0 r,$$

where r is the distance from the beam, $I = \lambda v$ is the current, and λ is the charge per unit length of the beam. It follows from these formulas that

$$B/E = \varepsilon_0\mu_0 I/\lambda = v/c^2,$$

where $c^2 = 1/\varepsilon_0\mu_0$. Substituting the expression for B from this formula into the last of transformation formulas (8.1), we obtain

$$B' = \frac{E\,|v - v_0|}{c^2\sqrt{1-(v_0/c)^2}}.$$

If in this case $v_0 < v$, the lines of $\mathbf{B}'$ form the right-handed system with the vector $\mathbf{v}_0$. If $v_0 > v$, they form the left-handed system (since the current I' in the system K' in this case will flow in the opposite direction).

● **8.5.** A relativistic charged particle moves in the space occupied by uniform mutually perpendicular electric and magnetic fields **E** and **B**. The particle moves rectilinearly in the direction perpendicular to the vectors **E** and **B**. Find **E**′ and **B**′ in the reference system moving translationally with the particle.

Solution. It follows from the description of motion of the particle that its velocity must satisfy the condition

$$vB = E. \tag{1}$$

In accordance with the transformation formulas (8.1), we have

$$\mathbf{E}' = \frac{\mathbf{E} + [\mathbf{v} \times \mathbf{B}]}{\sqrt{1-\beta^2}} = 0,$$

since in the case under consideration the Lorentz force (and hence the quantity $\mathbf{E} + [\mathbf{v} \times \mathbf{B}]$) is equal to zero.

According to the same transformation formulas, the magnetic field is given by

$$\mathbf{B}' = \frac{\mathbf{B} - [\mathbf{v} \times \mathbf{E}]/c^2}{\sqrt{1-\beta^2}}.$$

The arrangement of vectors is shown in Fig. 8.6, from which it follows that $[\mathbf{v} \times \mathbf{E}] \uparrow\uparrow \mathbf{B}$. Consequently, considering that according to (1)

$v = E/B$, we can write

$$B' = \frac{B - E^2/Bc^2}{\sqrt{1-\beta^2}} = \frac{B(1-\beta^2)}{\sqrt{1-\beta^2}},$$

or in the vector form

$$\mathbf{B}' = \mathbf{B}\sqrt{1-(E/cB)^2}.$$

We advise the reader to verify that the obtained expressions satisfy both field invariants.

● **8.6. The motion of a charge in crossed fields E and B.** A nonrelativistic particle with a specific charge q/m moves in a region where

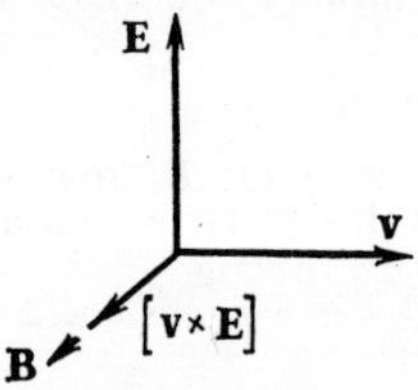

Fig. 8.6

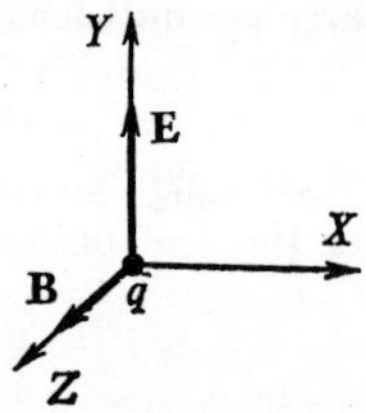

Fig. 8.7

uniform, mutually perpendicular fields **E** and **B** have been created (Fig. 8.7). At the moment $t = 0$ the particle was at point O and its velocity was equal to zero. Find the law of motion of the particle, $x(t)$ and $y(t)$.

Solution. The particle moves under the action of the Lorentz force. It can be easily seen that the particle always remains in the plane XY. Its motion can be described most easily in a certain system K' where only the magnetic field is present. Let us find this reference system.

It follows from transformations (8.4) that $E' = 0$ in a reference system that moves with a velocity $\mathbf{v}_0$ satisfying the relation $\mathbf{E} = -[\mathbf{v}_0 \times \mathbf{B}]$. It is more convenient to choose the system K' whose velocity $\mathbf{v}_0$ is directed towards the positive values on the X-axis (Fig. 8.7), since in such a system the particle will move perpendicularly to vector $\mathbf{B}'$ and its motion will be the simplest.

Thus, in the system K' that moves to the right at the velocity $v_0 = E/B$, the field $\mathbf{E}' = 0$ and only the field $\mathbf{B}'$ is observed. In accordance with (8.4) and Fig. 8.7, we have

$$\mathbf{B}' = \mathbf{B} - [\mathbf{v}_0 \times \mathbf{E}]/c^2 = \mathbf{B}(1 - v_0^2/c^2).$$

Since for a nonrelativistic particle $v_0 \ll c$, we can assume that $\mathbf{B}' = \mathbf{B}$.

In the system K', the particle will move only in the magnetic field, its velocity being perpendicular to this field. The equation of motion for this particle in the system K' will have the form

$$mv_0^2/R = qv_0B. \qquad (1)$$

This equation is written for the instant $t = 0$, when the particle moved in the system K' as is shown in Fig. 8.8. Since the Lorentz force **F** is always perpendicular to the velocity of the particle, $v_0 = \text{const}$, and it follows from (1) that in the system K' the particle will move in a circle of radius

$$R = mv_0/qB.$$

Thus, the particle moves uniformly with the velocity v_0 in a circle in the system K' which, in turn, moves uniformly to the right with the same velocity $v_0 = E/B$. This motion can be compared with the motion of the point q at the rim of a wheel (Fig. 8.9) rolling with the angular velocity $\omega = v_0/R = qB/m$.

Figure 8.9 readily shows that the coordinates of the particle q at the instant t are given by

$$x = v_0 t - R \sin \omega t = a\,(\omega t - \sin \omega t),$$

$$y = R - R \cos \omega t = a\,(1 - \cos \omega t),$$

where $a = mE/qB^2$ and $\omega = qB/m$.

● **8.7.** There is only a uniform electric field **E** in an inertial system K. Find the magnitudes and directions of vectors **E′** and **B′** in the system K' moving relative to the system K at a constant relativistic velocity $\mathbf{v}_0$ at an angle α to vector **E**.

Solution. According to transformation formulas (8.1) and taking into account that $\mathbf{B} = 0$ in the system K, we obtain

$$E'_{\parallel} = E \cos \alpha,\quad E'_{\perp} = E \sin \alpha / \sqrt{1-\beta^2},\quad \beta = v_0/c.$$

Hence the magnitude of vector **E′** is given by

$$E' = \sqrt{E'^2_{\parallel} + E'^2_{\perp}} = E \sqrt{(1-\beta^2 \cos^2 \alpha)/(1-\beta^2)}$$

and the angle α' between vectors **E′** and $\mathbf{v}_0$ can be determined from the formula

$$\tan \alpha' = E'_{\perp}/E'_{\parallel} = \tan \alpha / \sqrt{1-\beta^2}.$$

The magnitude and direction of vector **B′** can be found in a similar way:

$$\mathbf{B}'_{\parallel} = 0,\qquad \mathbf{B}'_{\perp} = -[\mathbf{v}_0 \times \mathbf{E}]/(c^2 \sqrt{1-\beta^2}),\qquad \mathbf{B}' = \mathbf{B}'_{\perp}.$$

This means that $\mathbf{B}' \perp \mathbf{v}_0$, and

$$B' = v_0 E \sin \alpha / (c^2 \sqrt{1-\beta^2}).$$

● **8.8.** Uniform electric and magnetic fields **E** and **B** of the same direction exist in a system of reference K. Find the magnitudes of vectors **E′** and **B′** and the angle between these vectors in the system K' moving at a constant relativistic velocity $\mathbf{v}_0$ in the direction perpendicular to vectors **E** and **B**.

Solution. In accordance with formulas (8.1), in the system K'

the two vectors $\mathbf{E}'$ and $\mathbf{B}'$ will be also perpendicular to the vector $\mathbf{v}_0$ (Fig. 8.10). The magnitudes of the vectors $\mathbf{E}'$ and $\mathbf{B}'$ can be found by the formulas

$$E' = \sqrt{\frac{E^2+(v_0B)^2}{1-(v_0/c)^2}}, \quad B' = \sqrt{\frac{B^2+(v_0E/c^2)^2}{1-(v_0/c)^2}}.$$

The angle between the vectors $\mathbf{E}'$ and $\mathbf{B}'$ can be found from its tangent through the formula

$$\tan\alpha' = \tan(\alpha'_E+\alpha'_B) = (\tan\alpha'_E+\tan\alpha'_B)/(1-\tan\alpha'_E\tan\alpha'_B).$$

Since $\tan\alpha'_E = v_0B/E$ and $\tan\alpha'_B = v_0E/c^2B$ (Fig. 8.10), we obtain

$$\tan\alpha' = \frac{v_0(B^2+E^2/c^2)}{(1-\beta^2)EB}.$$

This formula shows that as $v_0 \to c$ ($\beta \to 1$), the angle $\alpha' \to \pi/2$. The

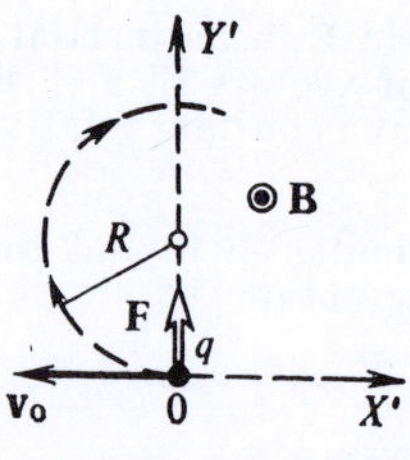

Fig. 8.8

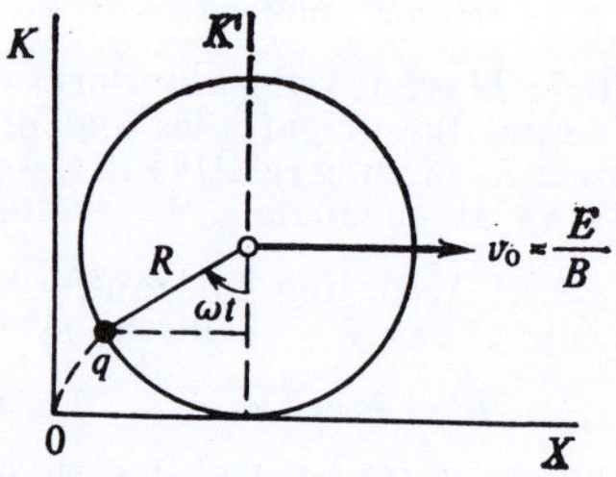

Fig. 8.9

opposite is also true: if we know $\mathbf{E}$ and $\mathbf{B}$ in one reference system, and the angle between these vectors is less than 90°, there exist reference systems where the two vectors $\mathbf{E}'$ and $\mathbf{B}'$ are parallel to one another.

● 8.9. **Invariant E·B.** Using transformation formulas (8.1) show that the quantity $\mathbf{E}\cdot\mathbf{B}$ is an invariant.

Solution. In the system K', this product will be

$$\mathbf{E}'\cdot\mathbf{B}' = (\mathbf{E}'_\parallel+\mathbf{E}'_\perp)\cdot(\mathbf{B}'_\parallel+\mathbf{B}'_\perp) = \mathbf{E}'_\parallel\cdot\mathbf{B}'_\parallel+\mathbf{E}'_\perp\cdot\mathbf{B}'_\perp. \quad (1)$$

Let us write the last term with the help of formulas (8.1):

$$\mathbf{E}'_\perp\cdot\mathbf{B}'_\perp = \frac{(\mathbf{E}_\perp+[\mathbf{v}_0\times\mathbf{B}_\perp])\cdot(\mathbf{B}_\perp-[\mathbf{v}_0\times\mathbf{E}_\perp]/c^2)}{1-\beta^2}. \quad (2)$$

Considering that the vectors $\mathbf{E}_\perp$ and $\mathbf{B}_\perp$ are perpendicular to the vector $\mathbf{v}_0$, we transform the numerator of (2) to the form

$$\mathbf{E}_\perp\cdot\mathbf{B}_\perp-(v_0/c)^2\,\mathbf{E}_\perp\cdot\mathbf{B}_\perp = \mathbf{E}_\perp\cdot\mathbf{B}_\perp(1-\beta^2), \quad (3)$$

where we used the fact that $[\mathbf{v}_0 \times \mathbf{B}_\perp]\cdot[\mathbf{v}_0 \times \mathbf{E}_\perp] = v_0^2 B_\perp E_\perp \times \times \cos\alpha = v_0^2 \mathbf{B}_\perp \cdot \mathbf{E}_\perp$ (Fig. 8.11). The remaining two scalar products in (2) are equal to zero since the vectors are mutually perpendicular.

Thus, the right-hand side of (1) becomes

$$\mathbf{E}'_{\parallel}\cdot\mathbf{B}'_{\parallel}+\mathbf{E}'_{\perp}\cdot\mathbf{B}'_{\perp}=\mathbf{E}_{\parallel}\cdot\mathbf{B}_{\parallel}+\mathbf{E}_{\perp}\cdot\mathbf{B}_{\perp}=\mathbf{E}\cdot\mathbf{B}$$

Q.E.D.

● **8.10. Field E of a uniformly moving charge.** A point charge q moves uniformly and rectilinearly with a relativistic velocity **v**.

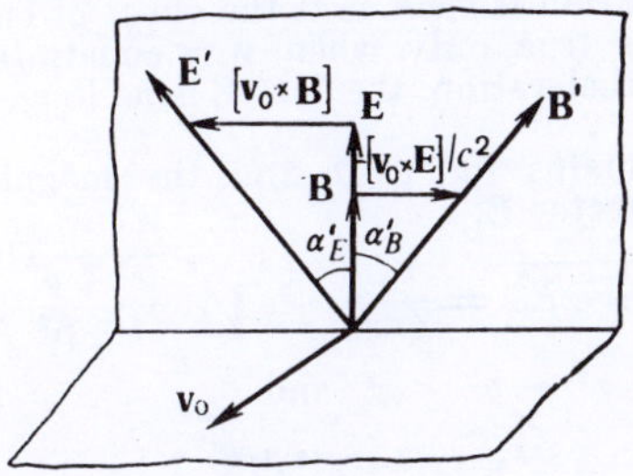

Fig. 8.10

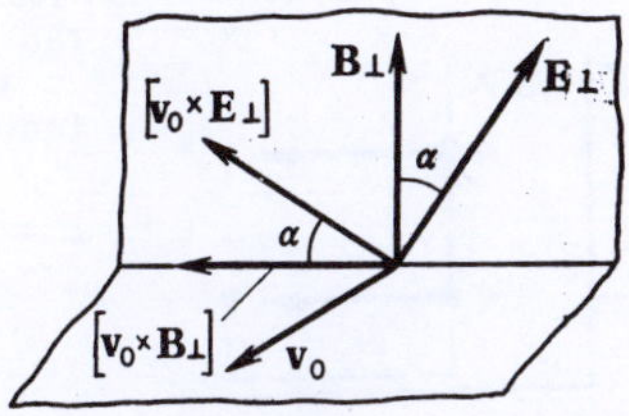

Fig. 8.11

Find the intensity **E** of the field created by this charge at a point whose radius vector relative to the charge is equal to **r** and forms an angle ϑ with the vector **v**.

Solution. Suppose that the charge moves in the positive direction of the X-axis in the system of reference K. Let us go over to the system K' at whose origin this charge is at rest (the axes X' and X of these systems coincide, while the Y'- and Y-axes are parallel). In the system K' the field $\mathbf{E}'$ of the charge has a simpler form:

$$\mathbf{E}' = \frac{1}{4\pi\varepsilon_0}\frac{q}{r'^3}\mathbf{r}',$$

and in the $X'Y'$ plane we have

$$E'_x = \frac{1}{4\pi\varepsilon_0}\frac{q}{r'^3}x', \quad E'_y = \frac{1}{4\pi\varepsilon_0}\frac{q}{r'^3}y'. \tag{1}$$

Let us now perform the reverse transition to the initial system K, which moves relative to the system K' with the velocity $-\mathbf{v}$. At the moment when the charge passes through the origin of the system K, the projections x and y of vector **r** are connected with the projections x' and y' of vector **r**′ through the following relations:

$$x = r\cos\vartheta = x'\sqrt{1-\beta^2} \quad y = r\sin\vartheta = y' \tag{2}$$

where $\beta = v/c$. Here we took into account that the longitudinal dimensions undergo the Lorentz contraction, while the transverse dimensions do not change. Besides, in accordance with the transformations inverse to (8.2), we obtain

$$E_x = E'_x, \quad E_y = E'_y/\sqrt{1-\beta^2}.$$

Substituting into these expressions (1) where x' and y' are replaced by the corresponding expressions from formulas (2), we get

$$E_x = \frac{1}{4\pi\varepsilon_0}\frac{q}{r'^3}\frac{x}{\sqrt{1-\beta^2}}, \qquad E_y = \frac{1}{4\pi\varepsilon_0}\frac{q}{r'^3}\frac{y}{\sqrt{1-\beta^2}}.$$

It should be noted that $E_x/E_y = x/y$, i.e. vector **E** is directed *radially* along the vector **r**. The situation is such as if the effect of lag were absent altogether. But this is true only when **v** = const. If, however, the charge moves with an acceleration, the field **E** is no longer radial.

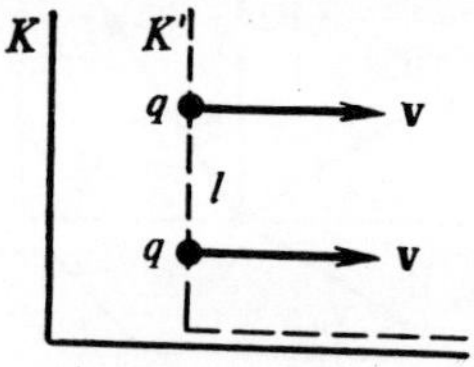

Fig. 8.12

It remains for us to find the magnitude of vector **E**:

$$E = \sqrt{E_x^2 + E_y^2} = \frac{1}{4\pi\varepsilon_0}\frac{q}{r'^3}\sqrt{\frac{x^2+y^2}{1-\beta^2}}.$$

Since $x^2 + y^2 = r^2$ and

$$r'^3 = (x'^2 + y'^2)^{3/2} = r^3\left(\frac{1-\beta^2\sin^2\vartheta}{1-\beta^2}\right)^{3/2},$$

in accordance with (2), the electric field intensity will be given by

$$E = \frac{1}{4\pi\varepsilon_0}\frac{q}{r^2}\frac{1-\beta^2}{(1-\beta^2\sin^2\vartheta)^{3/2}}.$$

● **8.11. Interaction between two moving charges.** The relativistic particles having the same charge q move parallel to each other at the same velocity v as is shown in Fig. 8.12. The separation between the particles is l. Using expression (8.7), find the force of interaction between the particles.

Solution. In this case, the angle between the vector **v** of one of the particles and the direction towards the other particle is $\vartheta = 90°$. Hence, in accordance with formula (8.7), the electric component of the Lorentz force is given by

$$F_e = qE = \frac{1}{4\pi\varepsilon_0}\frac{q^2}{l^2\sqrt{1-\beta^2}} \tag{1}$$

while the magnetic component of the Lorentz force is

$$F_m = qvB = \frac{\mu_0}{4\pi}\frac{q^2v^2}{l^2\sqrt{1-\beta^2}}, \tag{2}$$

where we took into account that in the case under consideration **B** is related to **E** through formula (8.5) from which it follows that $B =$

$= vE/c^2$, $c^2 = 1/\varepsilon_0\mu_0$. It should be noted that the ratio

$$F_m/F_e = \varepsilon_0\mu_0 v^2 = (v/c)^2,$$

just as in the nonrelativistic case (6.5). It can be seen that as $v \to c$, the magnetic component of the force F_m tends to F_e.

The resultant force of interaction (repulsive force) is given by

$$F = F_e - F_m = \frac{1}{4\pi\varepsilon_0} \frac{q^2}{l^2} \sqrt{1-\beta^2}.$$

9. Electromagnetic Induction

9.1. Faraday's Law of Electromagnetic Induction. Lenz's Law

In the previous chapter it was established that the relation between the "components" of an electromagnetic field, viz. electric and magnetic fields, is mainly determined by the choice of the reference system. In other words, the two components of electromagnetic field are interrelated. We shall show here that there exists a still closer connection between the **E**- and **B**-fields, which is exhibited in phenomena of electromagnetic induction.

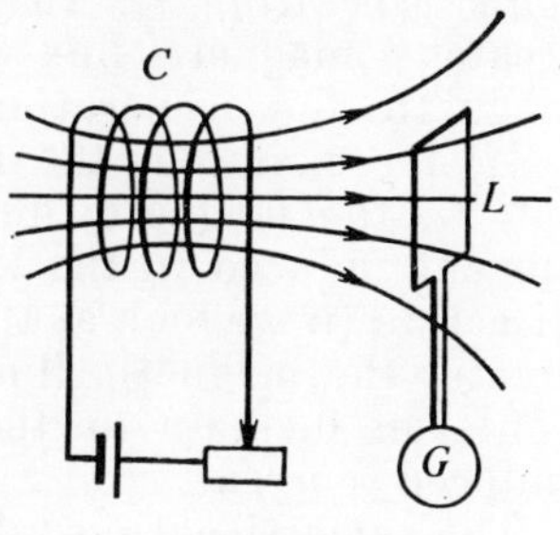

Fig. 9.1

Faraday's Discovery. In 1831 Faraday made one of the most fundamental discoveries in electrodynamics — the phenomenon of *electromagnetic induction.* It consists in that *an electric current* (called the *induced current*) *appears in a closed conducting loop upon a change in the magnetic flux* (i.e. the flux of **B**) *enclosed by this loop.*

The appearance of the induced current indicates that the *induced electromotive force* $\mathscr{E}_i$ appears in the loop as a result of a change of the magnetic flux. It is important here that $\mathscr{E}_i$ does not at all depend on how the change in the magnetic flux Φ was realized and is determined only by the rate of its variation, i.e. by $d\Phi/dt$. Besides, a change in the sign of the derivative $d\Phi/dt$ leads to a change in the sign or "direction" of $\mathscr{E}_i$.

Faraday found out that the induced current may be generated in two different ways. This is illustrated in Fig. 9.1, which shows the coil C with current I (the coil creates the magnetic field) and the loop L connected to the galvanometer G which indicates the induced current.

The first way consists in the displacement of the loop L (or its separate parts) in the field of the fixed coil C.

In the second case, the loop is fixed but the magnetic field varies either due to the motion of the coil C or as a result of a change in the current I in it, or due to both reasons.

In all these cases the galvanometer G indicates the presence of induced current in the loop L.

Lenz's Law. The direction of the induced current (and hence the sign of the induced e.m.f.) is determined by *Lenz's law*: ***the induced current is directed so that it counteracts the cause generating it.*** In other words, the induced current creates a magnetic flux which prevents the variation of the magnetic flux generating the induced e.m.f.

If, for example, the loop L (Fig. 9.1) is drawn to the coil C, the magnetic flux through the loop increases. The current induced in the loop in this case is in the clockwise direction (if we look at the loop from the right). This current creates the magnetic flux "directed" to the left, which prevents an increase in the magnetic flux, which causes the induced current.

The same situation takes place if we increase the current in the coil C, keeping fixed the coil and the loop L. On the other hand, if we decrease the current in the coil C, the current induced in the loop L will reverse its direction (it will now be directed counterclockwise if we look from the right).

Lenz's law expresses an important physical fact, viz. the tendency of a system to counteract the change in its state (electromagnetic inertia).

Faraday's Law of Electromagnetic Induction. According to this law, the e.m.f. induced in a loop is defined by the formula

$$\boxed{\mathscr{E}_i = -\frac{d\Phi}{dt},} \tag{9.1}$$

regardless of the cause of the variation of the magnetic flux embraced by the closed conducting loop. The minus sign in this equation is connected with a certain sign rule. The sign of the magnetic flux Φ is determined by the choice of the normal to the surface S bounded by the loop under consideration, while the sign of $\mathscr{E}_i$ depends on the choice of the positive direction of circumvention of the loop.

As before, we assume that the direction of the normal $\mathbf{n}$ to the surface S and the positive direction of the loop circumvention are connected through the *right-hand* screw rule* (Fig. 9.2). Consequently, when we choose (arbitrarily) the direction of the normal, we define the sign of the flux Φ as well as the sign (and hence the "direction") of the induced e.m.f. $\mathscr{E}_i$.

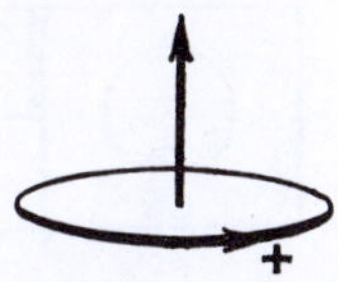

Fig. 9.2

With such a choice of positive directions (in accordance with the right-hand screw rule), the quantities $\mathscr{E}_i$ and $d\Phi/dt$ have opposite signs.

The unit of the magnetic flux is the *weber* (Wb). If the rate of variation of the magnetic flux is 1 Wb/s, the e.m.f. induced in the loop is equal to 1 V [see (9.1)].

Total Magnetic Flux (Magnetic-flux Linkage). If a closed loop in which an e.m.f. is induced contains not one but N turns (like a coil), $\mathscr{E}_i$ will be equal to the sum of the e.m.f.s induced in each turn. And if the magnetic flux embraced by each turn is the same and equal to Φ_1, the total flux Φ through the surface stretched over such a complex loop can be represented as

$$\Phi = N\Phi_1. \tag{9.2}$$

This quantity is called the *total magnetic flux*, or the *magnetic-flux linkage*. In this case, the e.m.f. induced in the loop is defined, in accordance with (9.1), by the formula

$$\mathscr{E}_i = -N\frac{d\Phi_1}{dt}. \tag{9.3}$$

* If these two directions were connected through the left-hand screw rule, Eq. (9.1) would not contain the minus sign.

9.2. Origin of Electromagnetic Induction

Let us now consider the physical reasons behind the appearance of induced e.m.f. and try to derive the law of induction (9.1) from what we already know. Let us consider two cases.

A Loop Moves in a Permanent Magnetic Field. Let us first take a loop with a movable jumper of length l (Fig. 9.3).

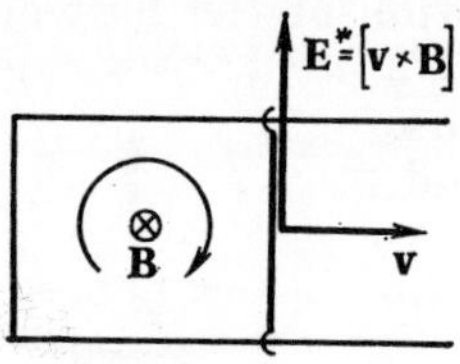

Fig. 9.3

Suppose that it is in a uniform magnetic field perpendicular to the plane of the loop and directed behind the plane of the figure. We start moving the jumper to the right with velocity **v**. The charge carriers in the jumper, viz. the electrons, will also start to move with the same velocity. As a result, each electron will be acted upon by a magnetic force $\mathbf{F} = -e\,[\mathbf{v} \times \mathbf{B}]$ directed along the jumper. The electrons will start moving downwards along the jumper, which is equivalent to electric current directed upwards. This is just the induced current. The charges redistributed over the surfaces of the conductors will create an electric field which will induce a current in the remaining parts of the loop.

The magnetic force **F** plays the role of an extraneous force. The field corresponding to it is $\mathbf{E}^* = \mathbf{F}/(-e) = [\mathbf{v} \times \mathbf{B}]$. It should be noted that this expression can also be obtained with the help of field transformation formulas (8.4).

Circulation of vector $\mathbf{E}^*$ around a closed contour gives, by definition, the magnitude of the induced e.m.f. In the case under consideration, we have

$$\mathscr{E}_i = -vBl, \tag{9.4}$$

where the minus sign is taken in accordance with the adopted sign rule: the normal **n** to the surface stretched over the loop was chosen so that it is directed behind the plane of Fig. 9.3 (towards the field **B**), and hence, according to the right-hand screw rule, the positive direction of circumvention is the clockwise direction, as is shown in the figure. In this case, the extraneous field $\mathbf{E}^*$ is directed against the positive direction of loop circumvention, and hence $\mathscr{E}_i$ is a negative quantity.

The product vl in (9.4) is the increment of the area bounded by the loop per unit time (dS/dt). Hence $vBl = B\ dS/dt = d\Phi/dt$, where $d\Phi$ is the increment of the magnetic flux through the area of the loop (in our case, $d\Phi > 0$). Thus,

$$\mathscr{E}_i = -d\Phi/dt. \tag{9.5}$$

It can be proved in the general form that law (9.1) is valid for any loop moving arbitrarily in a permanent non-uniform magnetic field (see Problem 9.2).

Thus, the excitation of induced e.m.f. during the motion of a loop in a permanent magnetic field is explained by the action of the magnetic force proportional to $[\mathbf{v} \times \mathbf{B}]$, which appears during the motion of the conductor.

It should be noted by the way that the idea of the circuit shown in Fig. 9.3 forms the basis of all induction generators in which a rotor with a winding rotates in an external magnetic field.

A Loop Is at Rest in a Varying Magnetic Field. In this case also, the induced current is an evidence of the fact that the magnetic field varying with time creates extraneous forces in the loop. But what are these forces? What is their origin? Clearly, they cannot be magnetic forces proportional to $[\mathbf{v} \times \mathbf{B}]$, since these forces cannot set in motion the charges that have been at rest ($\mathbf{v} = 0$). But there are no other forces besides $q\mathbf{E}$ and $q\,[\mathbf{v} \times \mathbf{B}]$! This leaves the only conclusion that the induced current is due to the appearance of an electric field $\mathbf{E}$ in the wire. It is this field which is responsible for the induced e.m.f. in a fixed loop placed into a magnetic field varying with time.

Maxwell assumed that a magnetic field varying with time leads to the appearance in space of an electric field regardless of the presence of a conducting loop. The latter just allows us to reveal the existence of this electric field due to the current induced in the loop.

Thus, according to Maxwell, a magnetic field varying with time generates an electric field. Circulation of vector $\mathbf{E}$ of this field around any fixed loop is defined as

$$\oint \mathbf{E}\,d\mathbf{l} = -\frac{\partial \Phi}{\partial t}. \tag{9.6}$$

Here the symbol of partial derivative with respect to time ($\partial/\partial t$) emphasizes the fact that the loop and the surface stretched on it are fixed. Since the flux $\Phi = \int \mathbf{B}\, d\mathbf{S}$ (the integration is performed around an arbitrary surface stretched over the loop we are interested in), we have

$$\frac{\partial}{\partial t} \int \mathbf{B}\, d\mathbf{S} = \int \frac{\partial \mathbf{B}}{\partial t}\, d\mathbf{S}.$$

In this equality, we exchanged the order of differentiation with respect to time and integration over the surface, which is possible since the loop and the surface are fixed. Then Eq. (9.6) can be represented in the form

$$\oint \mathbf{E}\, d\mathbf{l} = -\int \frac{\partial \mathbf{B}}{\partial t}\, d\mathbf{S}. \tag{9.7}$$

This equation has the same structure as Eq. (6.17) the role of the vector j being played by the vector $-\partial \mathbf{B}/\partial t$. Consequently, it can be written in the differential form the same as Eq. (6.26), i.e.

$$\nabla \times \mathbf{E} = -\partial \mathbf{B}/\partial t. \tag{9.8}$$

This equation expresses a *local* relation between electric and magnetic fields: *the time variation of magnetic field* B *at a given point determines the curl of vector* E *at the same point*. The fact that $\nabla \times \mathbf{E}$ differs from zero is an evidence of the presence of the electric field itself.

The fact that the circulation of the electric field induced by a varying magnetic field differs from zero indicates that this electric field is not a potential field. Like magnetic field, it is a *vortex* field. Thus, electric field can be either a potential field (in electrostatics) or a vortex field.

In the general case, electric field $\mathbf{E}$ may be the sum of the electrostatic field and the field induced by a magnetic field varying in time. Since the circulation of the electrostatic field is equal to zero, Eqs. (9.6)-(9.8) turn out to be valid for the general case as well, when the field $\mathbf{E}$ is the vector sum of these two fields.

Betatron. **The vortex electric field has found a brilliant application 'n the induction electron accelerator, viz. *betatron*. This accelerator onsists of a toroidal evacuated chamber arranged between the poles an electromagnet (Fig. 9.4). The variation of the current in the tromagnet winding creates a varying magnetic field which induces rtex electric field. The latter accelerates the electrons and simul-usly retains them on an equilibrium circular orbit of a certain**

radius (see Problem 9.5). Since the electric field is a vortex field, the direction of the force acting on the electrons always coincides with the direction of motion, and the electrons continuously increase their energy. During the time when the magnetic field increases (~ 1 ms), the electrons manage to make about a million turns and acquire an energy up to 400 MeV (at such energies, the electrons' velocity is almost equal to the velocity of light c in vacuum).

The induction accelerator (betatron) resembles a transformer in which the role of the secondary winding consisting of a single turn is played by an electron beam.

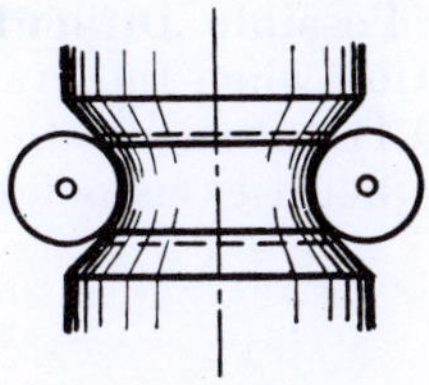

Fig. 9.4

Conclusion. Thus, the law of electromagnetic induction (9.1) is valid when the magnetic flux through a loop varies either due to the motion of the loop or the time variation of the magnetic field (or due to both reasons). However, we had to resort to two quite *different* phenomena for explaining the law in these two cases: for the moving loop we used the action of the magnetic force proportional to $[\mathbf{v} \times \mathbf{B}]$ while for the magnetic field varying with time, $\partial \mathbf{B}/\partial t$, the concept of vortex electric field $\mathbf{E}$ was employed.

Since there is no unique and profound principle which would combine these phenomena, we must interpret the law of electromagnetic induction as the combined effect of two entirely different phenomena. These phenomena are generally independent and nevertheless (which is astonishing) the e.m.f. induced in a loop is always equal to the rate of variation of the magnetic flux through this loop.

In other words, when the field $\mathbf{B}$ varies with time as well as the configuration or arrangement of the loop in the field, the induced e.m.f. should be calculated by formula (9.1) containing on its right-hand side the total time derivative $d\Phi/dt$ that automatically takes into consideration the two factors. In this connection, law (9.1) can be represented in the form

$$\oint \mathbf{E}\, d\mathbf{l} = -\frac{\partial \Phi}{\partial t} + \oint [\mathbf{v} \times \mathbf{B}]\, d\mathbf{l}. \tag{9.9}$$

The expression on the right-hand side of this equation is the total derivative $d\Phi/dt$. Here the first term is due to the time variation of the magnetic field, while the second is due to the motion of the loop. The origin of the second term is explained in greater detail in Problem 9.2.

Possible Difficulties. Sometimes we come across a situation when the law of electromagnetic induction in the form (9.1) turns out to be inapplicable (mainly because of the difficulties associated with the choice of the contour itself).

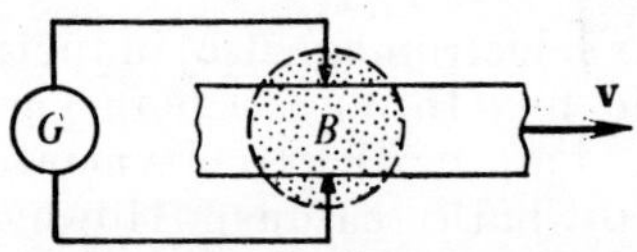

Fig. 9.5

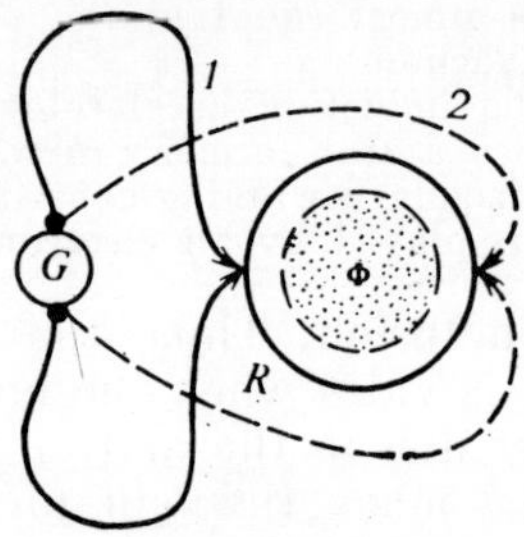

Fig. 9.6

In such cases it is necessary to resort to the basic laws, viz. the Lorentz force $q\mathbf{E} + q[\mathbf{v} \times \mathbf{B}]$ and the law $\nabla \times \mathbf{E} = -\partial\mathbf{B}/\partial t$. These laws express the physical meaning of the law of electromagnetic induction in all cases.

Let us consider two instructive examples.

Example 1. A conducting strip is displaced at a velocity v through a region in which a magnetic field $\mathbf{B}$ is applied (Fig. 9.5). In the figure, this region is shown by the dotted circle where the field $\mathbf{B}$ is directed towards us. A galvanometer G is connected to fixed contacts (arrows) touching the moving strip. Will the galvanometer indicate the presence of a current?

At first sight, the question seems not so simple, since in this case it is difficult to choose the contour itself: it is not clear where it should be "closed" in the strip and how this part of the contour will behave during the motion of the strip. If, however, we resort to the Lorentz force, it becomes immediately clear that the electrons in the moving strip will be displaced upwards, which will create an electric current in the galvanometer circuit, directed clockwise.

It should be noted that the idea of this experiment formed the basis of *MHD generator* in which the internal (thermal) energy is directly converted into electric energy. Instead of a conducting strip, a plasma (consisting of electrons and positive ions) is blown at a high speed through the generator. In all other respects, the situation is the same as in the case of the conducting strip.

Example 2. The dotted circle in Fig. 9.6. shows the region in which a permanent magnetic field **B** is localized (it is directed perpendicularly to the plane of the figure). This region is encircled by a fixed metallic ring R. By moving the sliding contacts to the other side of the ring (see Fig. 9.6), we introduce the magnetic flux Φ into the closed contour containing a galvanometer G (1—initial position, 2—final position). Will the galvanometer show a current pulse?

Applying formally the law (9.1), we should conclude that there will be an induced current. But it is not so. There is no current since in this case both $\partial \mathbf{B}/\partial t$ and the Lorentz force are equal to zero: the magnetic field **B** is constant and the closed loop moves in the region where there is no magnetic field. Thus, we have none of the physical reasons underlying the law of electromagnetic induction.

On an Apparent Paradox. We know that the force acting on an electric charge in a magnetic field is perpendicular to its velocity and hence accomplishes no work. Meanwhile, during the motion of a current-carrying conductor (moving charges!) Ampère's forces undoubtedly accomplish some work (electric motor!). What is the matter?

This seeming contradiction disappears if we take into account that the motion of the current-carrying conductor in the magnetic field is inevitably accompanied by electromagnetic induction. And since the e.m.f. induced in the conductor performs work on the charges, the total work of the forces of the magnetic field (the work done by Ampère's forces and by the induced e.m.f.) is equal to zero. Indeed, upon an elementary displacement of a current loop in the magnetic field, Ampère's forces perform the work (see Sec. 6.8)

$$\delta A_A = I\, d\Phi \tag{9.10}$$

while the induced e.m.f. accomplishes during the same time the work

$$\delta A_i = \mathscr{E}_i I\, dt = -I\, d\Phi \tag{9.11}$$

where we took into account that $\mathscr{E}_i = -d\Phi/dt$. It follows from these two formulas that the total work is

$$\delta A_A + \delta A_i = 0. \tag{9.12}$$

Thus, the work of the forces of the magnetic field includes not only the mechanical work (due to Ampère's forces) but also the work done by the e.m.f. induced during the motion of the loop. These works are equal in magnitude and opposite in sign, and hence their sum is equal to zero.

15–0181

The work of Ampère's forces is done not at the expense of the energy of the external magnetic field but at the expense of the power supply which maintains the current in the loop. In this case, the source performs additional work *against* the induced e.m.f., equal to $\delta A^{ad} = -\mathcal{E}_i I\, dt = I\, d\Phi$, which turns out to be the same as the work δA_A of Ampère's forces.

The work δA which is done during the displacement of the loop against the impeding Ampère's forces (which appear due to the current induced in accordance with Lenz's law) is transformed into the work of the induced e.m.f.:

$$\delta A = -\delta A_A = \delta A_i. \tag{9.13}$$

From the point of view of energy, this is the essence of the operation of all induction generators.

9.3. Self-induction

Electromagnetic induction appears in all cases when the magnetic flux through a loop changes. It is not important at all what causes the variation of this flux. If in a certain loop there is a time-varying current the magnetic field of this current will also change. And this leads to the variation of the magnetic flux through the contour, and hence to the appearance of induced e.m.f.

Thus, the variation of current in a circuit leads to the appearance of an induced e.m.f. in this circuit. This phenomenon is called *self-induction.*

Inductance. If there are no ferromagnetics in the space where a loop with current I is located, field $\mathbf{B}$ and hence total magnetic flux Φ through the contour are proportional to the current I, and we can write

$$\Phi = LI \tag{9.14}$$

where L is the coefficient called the *inductance* of the circuit. In accordance with the adopted sign rule for the quantities Φ and I, it turns out that Φ and I always have the same signs. This means that the inductance L is essentially a positive quantity.

The inductance L depends on the shape and size of the loop as well as on the magnetic properties of the surrounding medium. If the loop is rigid and there are no ferromagnetics in its neighbourhood, the inductance is a constant quantity independent of the current I.

The unit of inductance is *henry* (H). In accordance with (9.14) the inductance of 1 H corresponds to the loop the magnetic flux through which is equal to 1 Wb at the current of 1 A, which means that 1 H = 1 Wb/A.

Example. Find the inductance of a solenoid, neglecting edge effects. Let the volume of the solenoid be V, the number of turns per unit length n, and the magnetic permeability of a substance inside the solenoid μ.

According to (9.14), $L = \Phi/I$. Consequently, the problem is reduced to the determination of the total magnetic flux Φ, assuming that the current I is given. For the current I, the magnetic field in the solenoid is $B = \mu\mu_0 nI$. The magnetic flux through a turn of the solenoid is $\Phi_1 = BS = \mu\mu_0 nIS$, while the total magnetic flux piercing N turns is

$$\Phi = N\Phi_1 = nl \cdot BS = \mu\mu_0 n^2 VI,$$

where $V = Sl$. Hence the inductance of the solenoid is

$$L = \mu\mu_0 n^2 V. \tag{9.15}$$

On Certain Difficulties. It should be noted that the determination of inductance with the help of formula $L = \Phi/I$ is fraught with certain difficulties. No matter how thin the wire may be, its cross-sectional area is finite, and we simply do not know how to draw in the body of the conductor a geometrical contour required for calculating Φ. The result becomes ambiguous. For a sufficiently thin wire this ambiguity is immaterial. However, for thick wires the situation is entirely different: in this case, the result of calculation of L may contain a gross error due to the indeterminacy in the choice of the geometrical contour. This should be always kept in mind. We shall later show (see Sec. 9.5) that there exists another method of determining L, which is not subject to the indicated drawbacks.

E.M.F. of Self-induction. According to (9.1), a variation of the current in the circuit leads to the appearance of the e.m.f. of self-induction $\mathscr{E}_s$

$$\mathscr{E}_s = -\frac{d\Phi}{dt} = -\frac{d}{dt}(LI). \tag{9.16}$$

If the inductance L remains constant upon a variation of the current (the configuration of the circuit does not change and there are no ferromagnetics), then

$$\mathscr{E}_s = -L\frac{dI}{dt} \quad (L = \text{const}). \tag{9.17}$$

The minus sign here indicates that $\mathscr{E}_s$ is always directed so that it prevents a change in the current (in accordance with the Lenz's law). This e.m.f. tends to keep the current constant: it counteracts the current when it increases and sustains it when it decreases. In self-induction phenomena, the current has "inertia". Therefore, the induction effects strive to retain the magnetic flux constant just in the same way as mechanical inertia strives to preserve the velocity of a body.

Examples of Self-induction Effects. Typical manifestations of self-induction are observed at the moments of connection or disconnection of an electric circuit. The increase in the current before it reaches the steady-state value after closing the circuit and the decrease in the current when the circuit is disconnected occur gradually and not instantaneously. These effects of lag are the more pronounced the larger the inductance of the circuit.

Any large electromagnet has a large inductance. If its winding is disconnected from the source, the current rapidly diminishes to zero and creates during this process a huge e.m.f. This often leads to the appearance of a Voltaic arc between the contacts of the switch which is not only *very dangerous* for the electromagnet winding but may even prove fatal. For this reason, a bulb with the resistance of the same order of magnitude as that of the wire is normally connected in parallel to the electromagnet winding. In this case, the current in the winding decreases slowly and is not a hazard.

Let us now consider in greater detail the mode of vanishing and establishment of the current in a circuit.

Example 1. Current collapse upon the disconnection of a circuit. Suppose that a circuit consists of a coil of constant inductance L, a resistor R, an ammeter A, an e.m.f. source $\mathscr{E}$ and a special key K (Fig. 9.7a). Initially, the key K is in the lower position (Fig. 9.7b) and a current $I_0 = \mathscr{E}/R$ flows in the circuit (we assume that the resistance of the source of e.m.f. $\mathscr{E}$ is negligibly small).

At the instant $t = 0$, we rapidly turn key K clockwise from the lower to the upper position (Fig. 9.7a). This leads to the following:

the key short-circuits the source for a very short time and then disconnects it from the circuit without breaking the latter.

The current through the inductance coil L starts to decrease, which leads to the appearance of self-induction e.m.f. $\mathscr{E}_s = -L\, dI/dt$

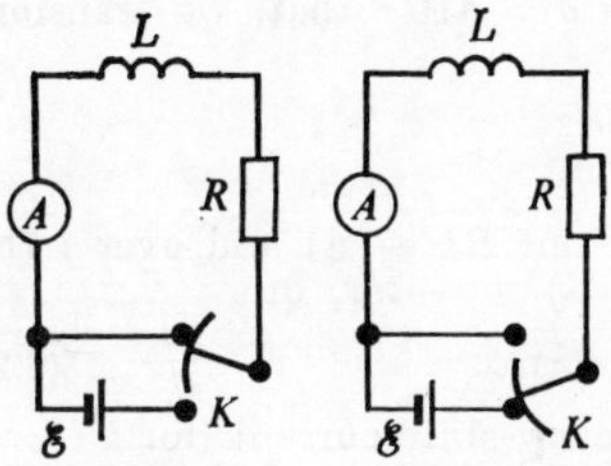

(a) (b)

Fig. 9.7

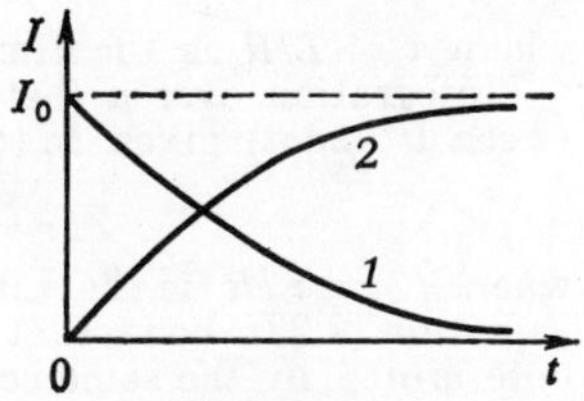

Fig. 9.8

which, in accordance with Lenz's law, counteracts the decrease in the current. At each instant of time, the current in the circuit will be determined by Ohm's law $I = \mathscr{E}_s/R$, or

$$RI = -L\frac{dI}{dt}. \tag{9.18}$$

Separating the variables, we obtain

$$\frac{dI}{I} = -\frac{R}{L}\,dt.$$

Integration of this equation over I (between I_0 and I) and over t (between 0 and t) gives $\ln(I/I_0) = -RL/t$, or

$$I = I_0 e^{-t/\tau} \tag{9.19}$$

where τ is a constant having the dimension of time:

$$\tau = L/R. \tag{9.20}$$

It is called the *time constant* (or the *relaxaton time*). This quantity characterizes the rate of decrease in the current: it follows from (9.19) that τ is the time during which the current decreases to (1/e) times its initial value. The larger the value of τ, the slower the decrease in the current. Figure 9.8 shows the curve of the dependence $I(t)$ describing the decrease of current with time (curve *1*).

Example 2. Stabilization of current upon closure of a circuit. At the instant $t = 0$, we rapidly turn the switch S counterclockwise from the upper to the lower position (Fig. 9.7*b*). We thus connected the source $\mathscr{E}$ to the inductance coil L. The current in the circuit starts to grow, and a self-induction e.m.f., counteracting this increase, will again appear. In accordance with Ohm's law, $RI =$

$= \mathscr{E} + \mathscr{E}_s$, or

$$RI = \mathscr{E} - L\frac{dI}{dt}. \tag{9.21}$$

We transpose $\mathscr{E}$ to the left-hand side of this equation and introduce a new variable $u = RI - \mathscr{E}$, $du = R\,dI$. After that, we transform the equation thus obtained to

$$du/u = -dt/\tau,$$

where $\tau = L/R$ is the time constant.

Integration over u (between $-\mathscr{E}$ and $RI - \mathscr{E}$) and over t (between 0 and t) gives $\ln[(RI - \mathscr{E})/(-\mathscr{E})] = -t/\tau$, or

$$I = I_0(1 - e^{-t/\tau}), \tag{9.22}$$

where $I_0 = \mathscr{E}/R$ is the value of the steady-state current (for $t \to \infty$). Equation (9.22) shows that the rate of stabilization of the currentis determined by the same constant τ. The curve $I(t)$ characterizing the increase in the current with time is shown in Fig. 9.8 (curve *2*).

On the Conservation of Magnetic Flux. Let a current loop move and be deformed in an arbitrary external magnetic field (permanent or varying). The current induced in the loop in this case is given by

$$I = \frac{\mathscr{E}_i + \mathscr{E}_s}{R} = -\frac{1}{R}\frac{d\Phi}{dt}.$$

If the resistance of the circuit $R = 0$, $d\Phi/dt$ must also be equal to zero, since the current I cannot be infinitely large. Hence it follows that $\Phi = \text{const}$.

Thus, when a superconducting loop moves in a magnetic field, the magnetic flux through its contour remains constant. This conservation of the flux is ensured by induced currents which, according to Lenz's law, prevent any change in the magnetic flux through the contour.

The tendency to conserve the magnetic flux through a contour always exists but is exhibited in the clearest form in the circuits of superconductors.

Example. A superconducting ring of radius a and inductance L is in a uniform magnetic field $\mathbf{B}$. In the initial position, the plane of the ring is parallel to vector $\mathbf{B}$, and the current in the ring is equal to zero. The ring is turned to the position perpendicular to vector $\mathbf{B}$. Find the current in the ring in the final position and the magnetic induction at its centre.

The magnetic flux through the ring does not change upon its rotation and remains equal to zero. This means that the magnetic fluxes of the field of the induced current and of the external current through the ring are equal in magnitude and opposite in sign. Hence $LI = \pi a^2 B$, whence

$$I = \pi a^2 B/L.$$

This current, in accordance with (6.13), creates a field $B_I = \pi\mu_0 aB/2L$ at the centre of the ring. The resultant magnetic induction at this point is given by

$$B_{res} = B - B_I = B\,(1 - \pi\mu_0 a/2L).$$

9.4. Mutual Induction

Mutual Inductance. Let us consider two fixed loops *1* and *2* (Fig. 9.9) arranged sufficiently close to each other. If current I_1 flows in loop *1*, it creates through loop *2* the total magnetic flux Φ_2 proportional (in the absence of ferromagnetics) to the current I_1:

$$\Phi_2 = L_{21} I_1. \tag{9.23}$$

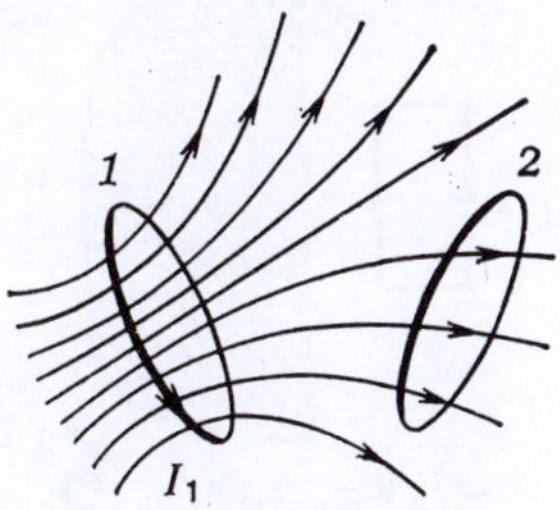

Fig. 9.9

Similarly, if current I_2 flows in loop *2*, it creates through the contour *1* the total magnetic flux

$$\Phi_1 = L_{12} I_2. \tag{9.24}$$

The proportionality factors L_{12} and L_{21} are called *mutual inductances* of the loops. Clearly, mutual inductance is numerically equal to the magnetic flux through one of the loops created by a unit current in the other loop. The coefficients L_{21} and L_{12} depend on the shape, size, and mutual arrangement of the loops, as well as on the magnetic permeability of the medium surrounding the loops. These coefficients are measured in the same units as the inductance L.

Reciprocity Theorem. Calculations show (and experiments confirm) that in the absence of ferromagnetics, the coefficients L_{12} and L_{21} are equal:

$$\boxed{L_{12} = L_{21}.} \tag{9.25}$$

This remarkable property of mutual inductance is usually called the *reciprocity theorem*. Owing to this theorem, we do not have to distinguish between L_{12} and L_{21} and can simply speak of the mutual inductance of two circuits.

The meaning of equality (9.25) is that in any case the magnetic flux Φ_1 through loop *1*, created by current I in loop *2*, is equal to the magnetic flux Φ_2 through loop *2*, created by the *same* current I in loop *1*. This circumstance often allows us to considerably simplify the calculation, for example, of magnetic fluxes. Here are two examples.

Fig. 9.10

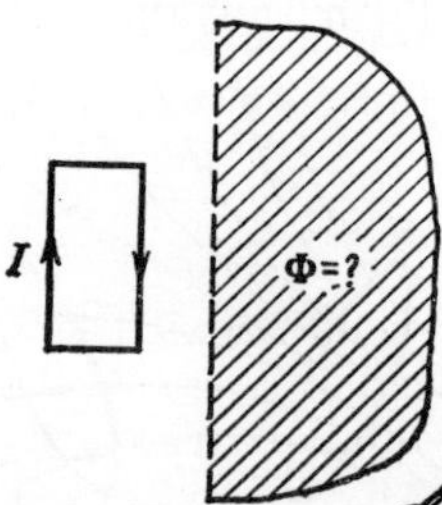

Fig. 9.1

Example 1. Two circular loops *1* and *2* whose centres coincide lie in a plane (Fig. 9.10). The radii of the loops are a_1 and a_2. Current I flows in loop *1*. Find the magnetic flux Φ_2 embraced by loop *2*, if $a_1 \ll a_2$.

The direct calculation of the flux Φ_2 is clearly a rather complicated problem since the configuration of the field itself is complicated. However, the application of the reciprocity theorem greatly simplifies the solution of the problem. Indeed, let us pass the same current I through loop *2*. Then the magnetic flux Φ_1 created by this current through loop *1* can be easily found, provided that $a_1 \ll a_2$: it is sufficient to multiply the magnetic induction B at the centre of the loop ($B = \mu_0 I/2a_2$) by the area πa_1^2 of the circle and take into account that $\Phi_2 = \Phi_1$ in accordance with the reciprocity theorem.

Example 2. A loop with current I has the shape of a rectangle. Find the magnetic flux Φ through the hatched half-plane (Fig. 9.11) whose boundary is at a given distance from the contour. Assume that this half-plane and the loop are in the same plane.

In this case too, the magnetic field of current I has a complex configuration, and hence it is very difficult to directly calculate the flux Φ in which we are interested. However, the solution can be considerably simplified by using the reciprocity theorem.

Suppose that current I flows not around the rectangular contour but along the boundary of the half-plane, enveloping it at infinity. The magnetic field created by this current in the region of the rectangular loop has a simple configuration—this is the field of a straight current. Hence we can easily find the magnetic flux Φ' through the rectangular contour (with the help of simple integration). In accordance with the reciprocity theorem, the required flux $\Phi = \Phi'$ and the problem is thus solved.

However, the presence of ferromagnetics changes the

situation, and the reciprocity theorem becomes inapplicable. Let us verify this with the help of the following concrete example.

Example. A long ferromagnetic cylinder of volume V has two windings (one over the other). One winding contains n_1 turns per unit length, while the other contains n_2 turns. Find their mutual inductance ignoring the edge effects.

According to (9.23), $L_{21} = \Phi_2/I_1$. This means that we must create current I_1 in winding *1* and calculate the total magnetic flux through all the turns of winding *2*. If winding *2* contains N_2 turns, we obtain

$$\Phi_2 = N_2 B_1 S,$$

where S is the cross-sectional area of the cylinder. Considering that $N_2 = n_2 l$, where l is the cylinder length, and $B_1 = \mu_1\mu_0 n_1 I_1$, where μ_1 is the magnetic permeability for current I_1, we write $\Phi_2 = \mu_1\mu_0 n_1 n_2 V I_1$, where $V = lS$. Hence

$$L_{21} = \mu_1\mu_0 n_1 n_2 V.$$

Similarly, we can find L_{12}:

$$L_{12} = \mu_2\mu_0 n_1 n_2 V.$$

Since the values of μ_1 and μ_2 in the last two expressions are generally different (they depend on currents I_1 and I_2 in ferromagnetics), the values of L_{21} and L_{12} do not coincide.

Mutual Induction. The presence of a magnetic coupling between circuits is manifested in that any variation of current in one of the circuits leads to the appearance of an induced e.m.f. in the other circuit. This phenomenon is called *mutual induction*.

According to Faraday's law, the e.m.f.'s appearing in contours *1* and *2* are given by

$$\mathscr{E}_1 = -\frac{d\Phi_1}{dt} = -L_{12}\frac{dI_2}{dt},$$

$$\mathscr{E}_2 = -\frac{d\Phi_2}{dt} = -L_{21}\frac{dI_1}{dt}, \tag{9.26}$$

respectively. Here we assume that the contours are fixed and there are no ferromagnetics in the surroundings.

Taking into account self-induction, the current appearing, say, in contour *1* upon a change in currents of both circuits is determined according to Ohm's law through the formula

$$R_1 I_1 = \mathscr{E}_1 - L_1\frac{dI_1}{dt} - L_{12}\frac{dI_2}{dt},$$

where $\mathscr{E}_1$ is the extraneous e.m.f. in contour *1* (besides induced e.m.f.s) and L_1 is the inductance of contour *1*. A similar equation can be written for the current I_2 in contour *2*.

It should be noted that transformers, devices intended for transforming currents and voltages, are based on the principle of induction.

On the Sign of L_{12}. Unlike inductance L which, as was mentioned earlier, is essentially a positive quantity, mutual inductance L_{12} is an *algebraic* quantity (which may, in particular, be equal to zero). This is due to the fact that, for instance, in (9.23) the quantities Φ_2 and I_1 refer to *different* contours. It immediately follows from Fig. 9.9 that the sign of the magnetic flux Φ_2 for a given direction of current I_1 will depend on the choice of the normal to the surface bounded by contour *2* (or on the choice of the positive direction of circumvention of this contour).

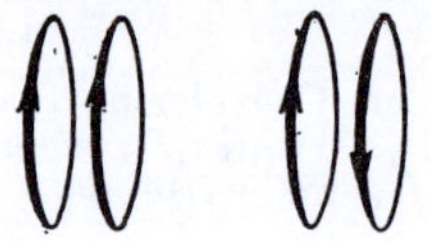

$L_{12} > 0$ (a) $L_{12} < 0$ (b)

Fig. 9.12

The positive directions for currents (and e.m.f.s) in both contours can always be chosen arbitrarily (and the positive direction of contour circumvention is uniquely (through the right-hand screw rule) related to the direction of the normal **n** to the surface bounded by the contour, i.e. in the long run with the sign of the magnetic flux). As soon as these directions are specified, the quantity L_{12} should be assumed to be positive if the fluxes of mutual induction through the contours are positive for positive currents, i.e. when they coincide in sign with self-induction fluxes.

In other words, $L_{12} > 0$ if for positive currents the contours "magnetize" each other. Otherwise, $L_{12} < 0$. In special cases, the positive directions of circumvention of the contours can be established beforehand so that the required sign of the quantity L_{12} can be obtained (Fig. 9.12).

9.5. Magnetic Field Energy

Magnetic Energy of Current. Let us connect a fixed circuit containing induction coil L and resistor R to a source of e.m.f. $\mathscr{E}_0$. As we already know, the current in the circuit will start increasing. This leads to the appearance of a self-induced e.m.f. $\mathscr{E}_s$. In accordance with Ohm's law, $RI = \mathscr{E}_0 + \mathscr{E}_s$, whence

$$\mathscr{E}_0 = RI - \mathscr{E}_s.$$

Let us find the elementary work done by extraneous forces (i.e. the source of $\mathscr{E}_0$) during time dt. For this purpose, we multiply the above equality by $I\,dt$:

$$\mathscr{E}_0 I\,dt = RI^2\,dt - \mathscr{E}_s I\,dt.$$

Taking into account the meaning of each term and the relation $\mathscr{E}_s = -d\Phi/dt$, we can write

$$\delta A_{extr} = \delta Q + I\,d\Phi.$$

It can be seen that in the process of current stabilization, when the flux Φ varies so that $d\Phi > 0$ (if $I > 0$), the work done by the source of $\mathscr{E}_0$ turns out to be *greater* than the Joule heat liberated in the circuit. A part of this work (additional work) is performed *against* the self-induced e.m.f. It should be noted that after the current has been stabilized, $d\Phi = 0$, and the entire work of the source of $\mathscr{E}_0$ will be spent for the liberation of the Joule heat.

Thus, the additional work accomplished by extraneous forces against the self-induced e.m.f. in the process of current stabilization is given by

$$\boxed{\delta A^{add} = I\,d\Phi.} \qquad (9.27)$$

This relation is of a general nature. It is valid in the presence of ferromagnetics as well, since while deriving it no assumptions have been made concerning the magnetic properties of the surroundings.

Here (and below) we shall assume that there are no ferromagnetics. Then $d\Phi = L\,dI$, and

$$\delta A^{add} = LI\,dI. \qquad (9.28)$$

Having integrated this equation, we obtain $A^{add} = LI^2/2$. According to the law of conservation of energy, any work is equal to the increment of any kind of energy. It is clear that a part of the work done by extraneous forces

($\mathscr{E}_0$) is spent on increasing the internal energy of conductors (it is associated with the liberation of the Joule heat), while another part (during current stabilization) is spent on something else. This "something" is just the magnetic field, since its appearance is associated with the appearance of the current.

Thus, we arrive at the conclusion that in the absence of ferromagnetics, the contour with the induction coil L and current I has the energy

$$\boxed{W = \frac{1}{2} LI^2 = \frac{1}{2} I\Phi = \frac{\Phi^2}{2L}.} \tag{9.29}$$

This energy is called the *magnetic*, or ***intrinsic***, *energy of current*. It can be completely converted into the internal energy of conductors if we disconnect the source of $\mathscr{E}_0$ from the circuit as shown in Fig. 9.7, i.e. if we rapidly turn the key K from position b to position a.

Magnetic Field Energy. Formula (9.29) expresses the magnetic energy of current in terms of inductance and current (in the absence of ferromagnetics). In this case, however, the energy like the electric energy of charged bodies, can be directly expressed in terms of magnetic induction $\mathbf{B}$. Let us show this first for a simple case of a long solenoid, ignoring field distortions at its ends (edge effects). Substituting the expression $L = \mu\mu_0 n^2 V$ into (9.29) we obtain

$$W = (1/2)LI^2 = (1/2)\mu\mu_0 n^2 I^2 V.$$

And since $nI = H = B/\mu\mu_0$, we have

$$W = \frac{B^2}{2\mu\mu_0} V = \frac{\mathbf{B}\cdot\mathbf{H}}{2} V. \tag{9.30}$$

This formula is valid for a uniform field filling the volume V (as in the case of the solenoid under consideration).

The general theory shows that the energy W can be expressed in terms of $\mathbf{B}$ and $\mathbf{H}$ in any case (but in the absence of ferromagnetics) through the formula

$$\boxed{W = \int \frac{\mathbf{B}\cdot\mathbf{H}}{2} dV.} \tag{9.31}$$

The integrand in this equation has the meaning of the energy contained in the volume element dV.

Thus, as in the case of the electric field energy, we arrive at the conclusion that the magnetic energy is also localized in the space occupied by a magnetic field.

It follows from formulas (9.30) and (9.31) that the magnetic energy is distributed in space with the volume density

$$\boxed{w = \frac{\mathbf{B}\cdot\mathbf{H}}{2} = \frac{B^2}{2\mu\mu_0}}. \tag{9.32}$$

It should be noted that this expression can be applied only to the media for which the **B** vs. **H** dependence is *linear*, i.e. μ in the relation $\mathbf{B} = \mu\mu_0\mathbf{H}$ is independent of **H**. In other words, expressions (9.31) and (9.32) are applicable only to paramagnetics and diamagnetics. In the case of ferromagnetics, these expressions are inapplicable*

It should also be noted that the magnetic energy is an essentially positive quantity. This can be easily seen from the last two formulas.

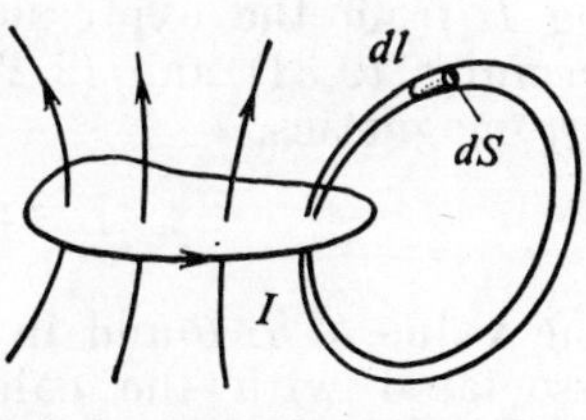

Fig. 9.13

Another Approach to the Substantiation of Formula (9.32). Let us prove the validity of this formula by proceeding "the other way round", i.e. let us show that if formula (9.32) is valid, the magnetic energy of the current loop is $W = LI^2/2$.

For this purpose, we consider the magnetic field of an arbitrary loop with current I (Fig. 9.13). Let us imagine that the entire field is divided into elementary tubes whose generatrices are the field lines of **B**. We isolate in one such tube a volume element $dV = dl\, dS$. According to formula (9.32), the energy $\frac{1}{2}BH\, dl\, dS$ is localized in this volume.

Let us now find the energy dW in the volume of the entire elementary tube. For this purpose, we integrate the latter expression along the tube axis. The flux $d\Phi = B\, dS$ through the tube cross-section is

* This is due to the fact that in the long run expressions (9.31) and (9.32) are the consequences of the formula $\delta A^{\text{add}} = I\, d\Phi$ and of the fact that, in the absence of hysteresis, the work δA^{add} is spent *only* on increasing the magnetic energy dW. For a ferromagnetic medium, the situation is different: the work δA^{add} is also spent on increasing the internal energy of the medium, i.e. on its heating.

constant along the tube, and hence $d\Phi$ can be taken out of the integral:

$$dW = \frac{d\Phi}{2} \oint H\, dl = I\, \frac{d\Phi}{2}.$$

Here we used the theorem on circulation of vector **H** (in our case, projection $H_l = H$).

Finally, we sum up the energy of all elementary tubes:

$$W = \frac{1}{2} I \int d\Phi = I\Phi/2 = LI^2/2,$$

where $\Phi = LI$ is the total magnetic flux enveloped by the current loop, Q.E.D.

Determination of Inductance from the Expression for Energy. We have introduced the inductance L as the proportionality factor between the total magnetic flux Φ and current I. There is, however, another possibility of calculating L from the expression for energy. Indeed, comparing formulas (9.31) and (9.29), we see that in the absence of ferromagnetics,

$$L = \frac{1}{I^2} \int \frac{B^2}{\mu\mu_0}\, dV. \tag{9.33}$$

The value of L found in this way is free of indeterminacy associated with the calculation of the magnetic flux Φ in formula (9.14) (see p. 227). The discrepancy appearing sometimes in determining L through formula (9.33) and from the expression (9.14) for the flux is illustrated in Problem 9.9 about a coaxial cable.

9.6. Magnetic Energy of Two Current Loops

Intrinsic and Mutual Energies. Let us consider two fixed loops *1* and *2*, arranged near one another at a sufficiently small distance (so that there is a magnetic coupling between them). We assume that each loop includes its own source of constant e.m.f. Let us close each loop at the instant $t = 0$. In each loop, its own current will start being established, and hence self-induced e.m.f. $\mathscr{E}_s$ and the e.m.f. $\mathscr{E}_i$ of mutual induction will appear. The additional work done in this case by the sources of constant e.m.f. *against* $\mathscr{E}_s$ and $\mathscr{E}_i$ is spent, as was shown above, for creating the magnetic energy.

Let us find this work over the time dt:

$$\delta A^{\text{add}} = -(\mathscr{E}_{s1} + \mathscr{E}_{i1}) I_1\, dt - (\mathscr{E}_{s2} + \mathscr{E}_{i2}) I_2\, dt = dW.$$

Considering that $\mathscr{E}_{s1} = -L_1 dI_1/dt$, $\mathscr{E}_{i1} = -L_{12}\, dI_2/dt$, etc., we transform this formula as follows:

$$dW = L_1 I_1\, dI_1 + L_{12} I_1\, dI_2 + L_2 I_2\, dI_2 + L_{21} I_2\, dI_1.$$

Taking into account the fact that $L_{12} = L_{21}$, we represent this expression in the form

$$dW = d\,(L_1 I_1^2/2) + d\,(L_2 I_2^2/2) + d\,(L_{12} I_1 I_2),$$

whence

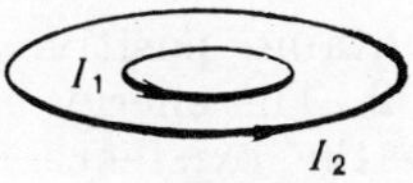

Fig. 9.14

$$W = \frac{L_1 I_1^2}{2} + \frac{L_2 I_2^2}{2} + L_{12} I_1 I_2. \qquad (9.34)$$

The first two terms in this expression are called the *intrinsic energies* of currents I_1 and I_2, while the last term, the *mutual energy* of the currents. Unlike the intrinsic energies of currents, the mutual energy is an algebraic quantity. A change in the direction of one of the currents leads to the reversal of the sign of the last term in (9.34), viz. the mutual energy.

Example. Suppose that we have two concentric loops with currents I_1 and I_2 whose directions are shown in Fig. 9.14. The mutual energy of these currents ($W_{12} = L_{12} I_1 I_2$) depends on three algebraic quantities whose signs are determined by the choice of positive directions of circumvention of the two loops. It is useful to verify, however, that the sign of W_{12} (in this case $W_{12} > 0$) depends only on the mutual orientation of the currents themselves and is independent of the choice of the positive directions of circumvention of the loops. We recall that the sign of the quantity L_{12} was considered in Sec. 9.4.

Field Treatment of Energy (9.34). There are some more important problems that can be solved by calculating the magnetic energy of two loops in a different way, viz. from the point of view of localization of energy in the field.

Let $\mathbf{B}_1$ be the magnetic field of current I_1, and $\mathbf{B}_2$ the field of current I_2. Then, in accordance with the principle of superposition, the field at each point is $\mathbf{B} = \mathbf{B}_1 + \mathbf{B}_2$, and according to (9.31), the field energy of this system of currents is $W = \int (B^2/2\mu\mu_0)\, dV$. Substituting into this formula $B^2 = B_1^2 + B_2^2 + 2\mathbf{B}_1\mathbf{B}_2$, we obtain

$$W = \int \frac{B_1^2}{2\mu\mu_0} dV + \int \frac{B_2^2}{2\mu\mu_0} dV + \int \frac{\mathbf{B}_1 \cdot \mathbf{B}_2}{\mu\mu_0} dV. \tag{9.35}$$

The correspondence between individual terms in formulas (9.35) and (9.34) is beyond doubt.

Formulas (9.34) and (9.35) lead to the following important sequences.

1. The magnetic energy of a system of two (or more) currents is an essentially positive quantity ($W > 0$). This follows from the fact that $W \propto \int B^2 \, dV$, where the integrand contains positive quantities.

2. The energy of currents is a nonadditive quantity (due to the existence of mutual energy).

3. The last integral in (9.35) is proportional to the product $I_1 I_2$ of currents since $B_1 \propto I_1$ and $B_2 \propto I_2$. The proportionality factor (i.e. the remaining integral) turns out to be symmetric with respect to indices *1* and *2* and hence can be denoted by L_{12} or L_{21} in accordance with formula (9.34). Thus, $L_{12} = L_{21}$ indeed.

4. Expression (9.35) leads to another definition of mutual inductance L_{12}. Indeed, a comparison of (9.35) and (9.34) gives

$$L_{12} = \frac{1}{I_1 I_2} \int \frac{\mathbf{B}_1 \cdot \mathbf{B}_2}{\mu\mu_0} dV. \tag{9.36}$$

9.7. Energy and Forces in Magnetic Field

The most general method for determining the forces acting in a magnetic field is the *energy method*, in which the expression for the magnetic field energy is used.

We shall confine ourselves to the case when the system consists of two loops with currents I_1 and I_2. The magnetic energy of such a system can be represented in the form

$$W = \frac{1}{2}(I_1 \Phi_1 + I_2 \Phi_2), \tag{9.37}$$

where Φ_1 and Φ_2 are the total magnetic fluxes piercing loops *1* and *2* respectively. This expression can be easily obtained from formula (9.34) by representing the last term in it as the sum $(1/2)\, L_{12} I_1 I_2 + (1/2)\, L_{21} I_2 I_1$ and then taking into account that

$$\Phi_1 = L_1 I_1 + L_{12} I_2, \quad \Phi_2 = L_2 I_2 + L_{21} I_1. \tag{9.38}$$

According to the law of conservation of energy, the work δA^* done by the sources of e.m.f. included into circuits *1* and *2* is converted into the heat δQ and is spent to increase the magnetic energy of the system by dW (due to the motion of the loops or the variation of currents in them as well as to accomplish the mechanical work δA_m (as a result of a displacement or deformation of the loops):

$$\delta A^* = \delta Q + dW + \delta A_m. \tag{9.39}$$

We assumed that the capacitance of the loops is negligibly small and hence the electric energy can be ignored.

Henceforth, we shall be interested not in the entire work δA^* of the source of e.m.f. but only in its part which is done *against* the induced and self-induced e.m.f.s (in each loop). This work (which was called the additional work) is equal to $\delta A^{add} = -(\mathscr{E}_{i1} + \mathscr{E}_{s1}) I_1\, dt - (\mathscr{E}_{i2} + \mathscr{E}_{s2}) I_2\, dt$. Considering that $\mathscr{E}_i + \mathscr{E}_s = -d\Phi/dt$ for each loop, we can write the expression for the additional work in the form

$$\delta A^{add} = I_1\, d\Phi_1 + I_2\, d\Phi_2. \tag{9.40}$$

It is just this part of the work of the e.m.f. sources (the work against the induced and self-induced e.m.f.s) that is associated with the variation of the fluxes Φ_1 and Φ_2 and spent for increasing the magnetic energy of the system as well as for accomplishing the mechanical work:

$$I_1\, d\Phi_1 + I_2\, d\Phi_2 = dW + \delta A_m. \tag{9.41}$$

This formula forms the *basis* for calculating the mechanical work δA_m and then the forces acting in a magnetic field.

Formula (9.41) can be used for obtaining simpler expressions for δA_m by assuming that either all the magnetic fluxes through the loops, or currents flowing in them remain unchanged during a displacement. Let us consider this in greater detail.

1. If the fluxes are constant (Φ_k = const), it immediately

follows from (9.41) that

$$\boxed{\delta A_{\mathrm{m}} = -dW\,|_{\Phi},} \tag{9.42}$$

where the subscript Φ indicates that the increment of the magnetic energy of the system must be calculated at constant fluxes through the loops. The obtained formula is similar to the corresponding formula (4.15) for the work in an electric field.

2. If the currents are constant (I_k = const), we have

$$\boxed{\delta A_{\mathrm{m}} = dW\,|_{I}.} \tag{9.43}$$

Indeed, for I_k = const, it follows from formula (9.37) that

$$dW|_I = \frac{1}{2}(I_1\,d\Phi_1 + I_2\,d\Phi_2),$$

i.e. in this case, in accordance with (9.40), the increment of the magnetic energy of the system is equal to half of the additional work done by the sources of e.m.f. Another half of this work is spent for doing mechanical work. In other words, when the currents are constant, $dW|_I = \delta A_{\mathrm{m}}$, Q.E.D.

It should be emphasized that the two expressions (9.42) and (9.43) obtained by us define the mechanical work of the same force, i.e. we can write

$$\mathbf{F}\cdot d\mathbf{l} = -dW|_{\Phi} = dW|_{I}. \tag{9.44}$$

In order to calculate the force with the help of these formulas, there is no need, of course, to choose a regime in which either magnetic fluxes or currents would necessarily remain constant. We must simply find the increment dW of magnetic energy of the system provided that either Φ_k = const or I_k = const, which is a purely mathematical operation.

The value of the obtained expressions (9.42) and (9.43) lies in their *generality*: they are applicable to systems consisting of any number of contours—one, two, or more.

Let us consider several examples illustrating the application of these formulas.

Example 1. Force in the case of one current loop. Suppose that we have a current loop, where AB is a movable jumper (Fig. 9.15). The inductance of this loop depends in a certain way on the coordinate x, i.e. $L(x)$ is known. Find the Ampère's force acting on the jumper, in two ways: for I = const and for Φ = const.

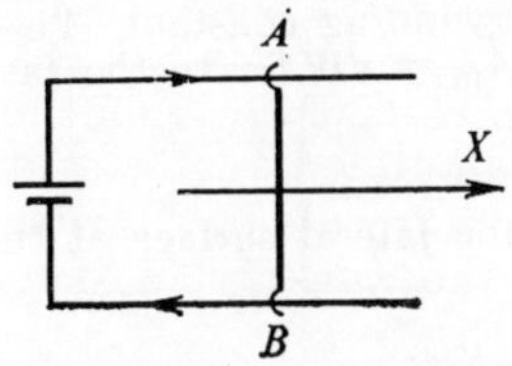

Fig. 9.15

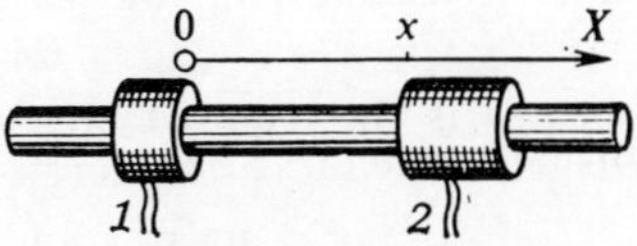

Fig. 9.16

In the case under consideration, the magnetic energy of the system can be represented, in accordance with (9.29), as follows:

$$W = LI^2/2 = \Phi^2/2L,$$

where $\Phi = LI$. Let us displace the jumper, for example, to the right by dx. Since $\delta A_m = F_x\, dx$, we have

$$F_x = \left.\frac{\partial W}{\partial x}\right|_I = \frac{I^2}{2}\frac{\partial L}{\partial x},$$

or

$$F_x = -\left.\frac{\partial W}{\partial x}\right|_\Phi = \frac{\Phi^2}{2L^2}\frac{\partial L}{\partial x} = \frac{I^2}{2}\frac{\partial L}{\partial x},$$

i.e. the calculations carried out with the help of the two formulas give, in accordance with (9.44), the same result.

Example 2. Interaction between two current-carrying coils. Two coils *1* and *2* with currents I_1 and I_2 are mounted on a magnetic core (Fig. 9.16). Suppose that the mutual inductance of the coils depends on their separation x according to a known law $L_{12}(x)$. Find the force of interaction between the coils.

The magnetic energy of the system of two coils is given by formula (9.34). For determining the force of interaction, we shall use expression (9.43). Let us displace coil *2* through a distance dx at constant currents I_1 and I_2. The corresponding increment of the magnetic energy of the system is

$$dW\,|_I = I_1 I_2\, dL_{12}(x).$$

Since the elementary mechanical work $\delta A_m = F_{2x}\, dx$, according to (9.43) we obtain

$$F_{2x} = I_1 I_2 \frac{\partial L_{12}(x)}{\partial x}.$$

Let currents I_1 and I_2 magnetize each other. Then $L_{12} > 0$, and for $dx > 0$ the increment $dL_{12} < 0$, i.e. $F_{2x} < 0$. Consequently, the force exerted on coil *2* by coil *1* is the attractive force: vector $\mathbf{F}_2$ is directed to the left in the figure.

Example 3. Magnetic pressure acting on solenoid winding. Let us mentally increase the radius of the solenoid cross section by dr, retaining the current I through the winding constant. Then Ampère's forces will accomplish the work $\delta A_m = dW|_I$. In the case under consideration, we have

$$\delta A_m = pS\,dr,$$

where p is the required pressure and S is the lateral surface of the solenoid, and

$$dW|_I = d\left(\frac{B^2}{2\mu_0}V\right) = \frac{B^2}{2\mu_0}S\,dr.$$

Here we took into account the fact that when $I = \text{const}$, $B = \text{const}$ as well. Equating these two expressions, we obtain

$$p = B^2/2\mu_0.$$

Magnetic Pressure. The expression for pressure obtained in Example 3 can be generalized for the case when the magnetic field is different ($\mathbf{B}_1$ and $\mathbf{B}_2$) on different sides of a surface with current (conduction current or magnetization current). In this case, the magnetic pressure is given by

$$p = \left|\frac{\mathbf{B}_1\cdot\mathbf{H}_1}{2} - \frac{\mathbf{B}_2\cdot\mathbf{H}_2}{2}\right|. \tag{9.45}$$

The situation is such as if the region with a higher magnetic energy density were the region of higher pressure.

Relation (9.45) is one of the basic relations in *magnetohydrodynamics* which studies the behaviour of electroconducting liquids (in electrical engineering and astrophysics).

Problems

● **9.1. Induced e.m.f.** A wire in the shape of a parabola $y = kx^2$ is in a uniform magnetic field $\mathbf{B}$ perpendicular to the plane XY. A jumper translates without initial velocity and at a constant acceleration a from the apex of the parabola (Fig. 9.17). Find the e.m.f. induced in the formed contour as a function of the coordinate y.

Solution. By definition, $\mathscr{E}_i = d\Phi/dt$. Having chosen the normal $\mathbf{n}$ to the plane of the contour in the direction of vector $\mathbf{B}$, we write $d\Phi = B\,dS$, where $dS = 2x\,dy$. Considering now that $x = \sqrt{y/k}$, we obtain

$$\mathscr{E}_i = -B\cdot 2\sqrt{y/k}\,dy/dt.$$

During the motion with a constant acceleration, the velocity $dy/dt = \sqrt{2ay}$, and hence

$$\mathscr{E}_i = -By\sqrt{8a/k}.$$

This formula shows that $\mathscr{E}_i \propto y$. The minus sign indicates that $\mathscr{E}_i$ in the figure acts counterclockwise

● **9.2. A loop moves arbitrarily.** A closed conducting loop is moved arbitrarily (even with a deformation) in a constant nonuniform magnetic field. Show that Faraday's law (9.1) will be fulfilled in this case.

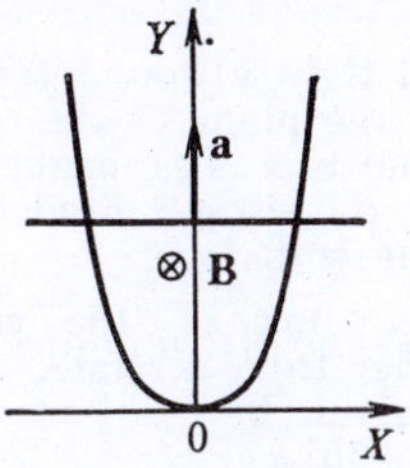

Fig. 9.17

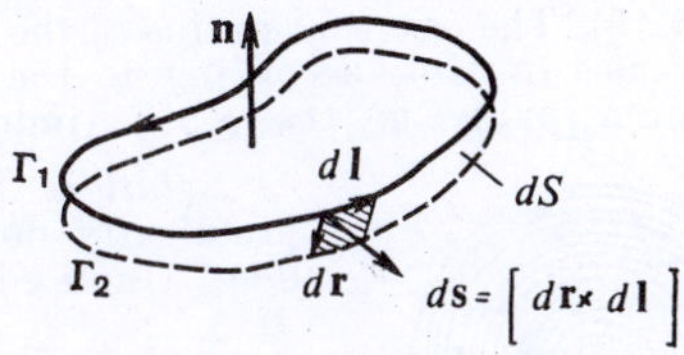

Fig. 9.18

Solution. Let us consider an element $d\mathbf{l}$ of loop length, which at a given instant moves at a velocity $\mathbf{v}$ in the magnetic field $\mathbf{B}$. In accordance with formulas (8.4) of field transformation, in the reference system fixed to the given element the electric field $\mathbf{E} = [\mathbf{v} \times \mathbf{B}]$ will be observed. It should be noted that this expression can be also obtained with the help of the Lorentz force, as it was done at the beginning of Sec. 9.2.

Circulation of vector $\mathbf{E}$ around the entire contour is, by definition, the induced e.m.f.:

$$\mathscr{E}_i = \oint [\mathbf{v} \times \mathbf{B}]\, d\mathbf{l}. \tag{1}$$

Let us now find the corresponding increment of the magnetic flux through the loop. For this purpose, we turn to Fig. 9.18. Suppose that the loop was displaced from position Γ_1 to Γ_2 during time dt. If in the first position the magnetic flux through the surface S_1 stretched over the loop was Φ_1, the corresponding magnetic flux in the second position can be represented as $\Phi_1 + d\Phi$, i.e. as the flux through the surface $S + dS$. Here $d\Phi$ is the required increment of the magnetic flux through the narrow strip dS between contours Γ_1 and Γ_2.

Using Fig. 9.18, we can write

$$d\Phi = \int \mathbf{B} \cdot d\mathbf{s} = \int \mathbf{B} \cdot [d\mathbf{r} \times d\mathbf{l}] = -\oint [d\mathbf{r} \times \mathbf{B}]\, d\mathbf{l}. \tag{2}$$

Here (1) the direction of normal $\mathbf{n}$ is matched with the direction of circumvention of the contour, viz. with vector $d\mathbf{l}$ (right-handed system); (2) the direction of vector $d\mathbf{s}$, viz. the area element of the strip, is matched with the choice of normals $\mathbf{n}$, and (3) the following cyclic transposition is used in the scalar triple product:

$$\mathbf{a} \cdot [\mathbf{b} \times \mathbf{c}] = \mathbf{b} \cdot [\mathbf{c} \times \mathbf{a}] = \mathbf{c} \cdot [\mathbf{a} \times \mathbf{b}] = -[\mathbf{b} \times \mathbf{a}] \cdot \mathbf{c},$$

Dividing expression (2) by dt, we find

$$d\Phi/dt = -\oint [\mathbf{v} \times \mathbf{B}]\, d\mathbf{l}, \tag{3}$$

where $\mathbf{v} = d\mathbf{r}/dt$. It remains for us to compare (3) with (1), which gives $\mathscr{E}_i = -d\Phi/dt$.

● **9.3.** A flat coil with a large number N of tightly wound turns is in a uniform magnetic field perpendicular to the plane of the coil (Fig. 9.19). The external radius of the coil is equal to a. The magnetic field varies in time according to the law $B = B_0 \sin \omega t$. Find the maximum value of the e.m.f. induced in the coil.

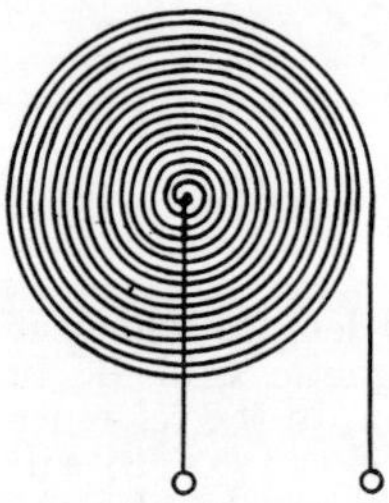

Fig. 9.19

Solution. Since each turn of the coil practically does not differ from a circle, the e.m.f. induced in it is given by

$$\varepsilon_i = -\,d\Phi/dt = -\pi r^2 B_0 \omega \cos \omega t,$$

where r is the radius of the turn under consideration. The number of turns corresponding to the interval dr of the values of the radius is $dN = (N/a)\, dr$. The turns are connected in series, hence the total e.m.f. induced in the coil is

$$\mathscr{E}_i = \int \varepsilon_i(r)\, dN.$$

Having integrated this expression, we obtain the following for the maximum value of the induced e.m.f.:

$$\mathscr{E}_{im} = (1/3)\pi a^2 N B_0 \omega.$$

● **9.4.** A coil of N turns with the cross-sectional area S is placed inside a long solenoid. The coil is rotated at a constant angular velocity ω around the axis coinciding with its diameter and perpendicular to the axis of the solenoid. The magnetic field in the solenoid varies according to the law $B = B_0 \sin \omega t$. Find the e.m.f. induced in the coil, if at the instant $t = 0$ the coil axis coincided with the axis of the solenoid.

Solution. At the instant t, the total magnetic flux through the coil is

$$\Phi = NBS \cos \omega t = NB_0 S \sin \omega t \cdot \cos \omega t = (1/2) NB_0 S \sin 2\omega t.$$

In accordance with Faraday's law, we have

$$\mathscr{E}_i = -d\Phi/dt = -(1/2) NB_0 S \cdot 2\omega \cos 2\omega t = -NB_0 S \omega \cos 2\omega t.$$

● **9.5. The betatron condition.** Show that electrons in a betatron will move in an orbit of a constant radius r_0 provided that the magnetic field B_0 on the orbit is equal to half of the value $\langle B \rangle$ of the magnetic field averaged over the area inside the orbit, i.e. $B_0 = \dfrac{\langle B \rangle}{2}$.

Solution. Let us write the relativistic equation for motion of an electron,

$$d\mathbf{p}/dt = e\mathbf{E} + e\,[\mathbf{v} \times \mathbf{B}_0], \tag{1}$$

where $\mathbf{E}$ is the vortex electric field, in terms of the projections onto the tangent $\boldsymbol{\tau}$ and the normal $\mathbf{n}$ to the trajectory. For this purpose,

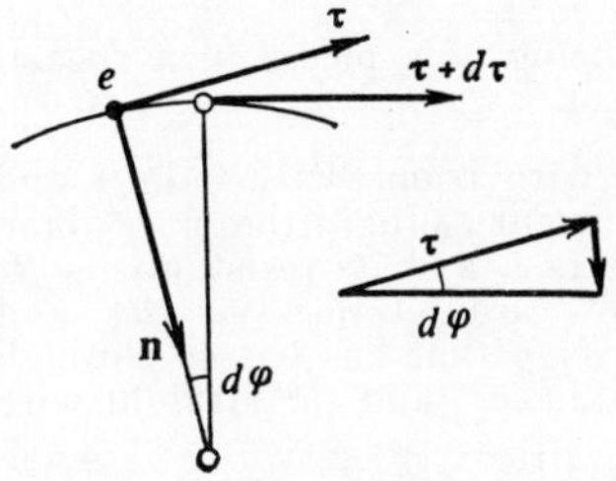

Fig. 9.20

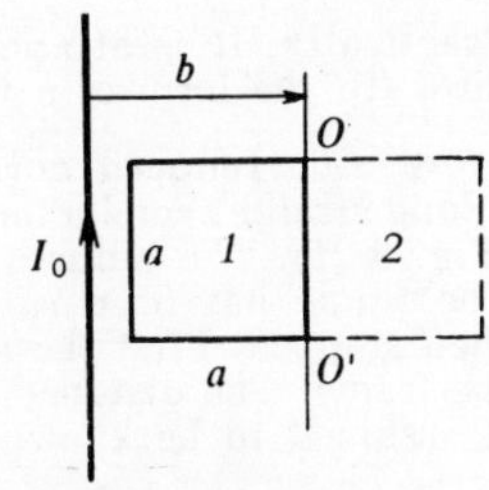

Fig. 9.21

we write the electron momentum as $\mathbf{p} = p\boldsymbol{\tau}$ and find its time derivative

$$\frac{d\mathbf{p}}{dt} = \frac{dp}{dt}\boldsymbol{\tau} + p\,\frac{d\boldsymbol{\tau}}{dt} = \frac{dp}{dt}\boldsymbol{\tau} + m\,\frac{v^2}{r_0}\,\mathbf{n}, \tag{2}$$

where we took into account that $\mathbf{p} = m\mathbf{v}$, m being the relativistic mass, and $d\boldsymbol{\tau}/dt = (v/r_0)\mathbf{n}$ (this can be easily seen from Fig. 9.20). Indeed, $d\boldsymbol{\tau} = d\varphi\cdot\mathbf{n} = (v\,dt/r_0)\mathbf{n}$, and the rest is obvious.

Besides, according to Faraday's law, $2\pi r_0 E = |d\Phi/dt|$, where $\Phi = \pi r_0^2 \langle B \rangle$. Hence

$$E = \frac{r_0}{2}\,\frac{d}{dt}\langle B\rangle. \tag{3}$$

Let us now write Eq. (1), taking into account (2) and (3), in terms of the projections onto the tangent and the normal to the trajectory:

$$\frac{dp}{dt} = eE = e\,\frac{r_0}{2}\,\frac{d}{dt}\langle B\rangle, \tag{4}$$

$$m\,\frac{v^2}{r_0} = evB_0.$$

The last equation can be written, after cancelling v, in the form

$$p = er_0B_0.$$

We differentiate this equation with respect to time, considering that $r_0 = \text{const}$:

$$\frac{dp}{dt} = er_0\,\frac{dB_0}{dt}. \tag{5}$$

A comparison of Eqs. (5) and (4) gives

$$\frac{d}{dt} B_0 = \frac{1}{2} \frac{d}{dt} \langle B \rangle.$$

In particular, the last condition will be satisfied when

$$B_0 = \frac{1}{2} \langle B \rangle.$$

Practically, it is attained by manufacturing pole pieces in a special form (in the form of a blunt-nosed cones).

● **9.6. Induced current.** A square wire frame with side a and a long straight conductor with a direct current I_0 lie in the same plane (Fig. 9.21). The inductance of the frame is L and its resistance is R. The frame has been rotated through 180° around the axis OO' and then stopped. Find the amount of electricity that has passed through the frame. The distance b between the axis OO' and the straight wire is assumed to be known.

Solution. According to Ohm's law, the current I appearing in the frame during its rotation is determined by the formula

$$RI = -\frac{d\Phi}{dt} - L\frac{dI}{dt}.$$

Hence the required amount of electricity (charge) is

$$q = \int I\,dt = -\frac{1}{R}\int (d\Phi + L\,dI) = -\frac{1}{R}(\Delta\Phi + L\,\Delta I).$$

Since the frame has been stopped after rotation, the current in it vanishes, and hence $\Delta I = 0$. It remains for us to find the increment of the flux $\Delta\Phi$ through the frame ($\Delta\Phi = \Phi_2 - \Phi_1$).

Let us choose the normal **n** to the plane of the frame, for instance, so that in the final position **n** is directed behind the plane of the figure (along **B**). Then it can be easily seen that in the final position $\Phi_2 > 0$, while in the initial position $\Phi_1 < 0$ (the normal is opposite to **B**), and $\Delta\Phi$ turns out to be simply equal to the flux through the surface bounded by the final and initial positions of the frame:

$$\Delta\Phi = \Phi_2 + |\Phi_1| = \int_{b-a}^{b+a} Ba\,dr,$$

where B is a function of r, whose form can be easily found with the help of the theorem on circulation.

Finally, omitting the minus sign, we obtain

$$q = \frac{\Delta\Phi}{R} = \frac{\mu_0 a I_0}{2\pi R} \ln\frac{b+a}{b-a}.$$

It can be seen that the obtained quantity is independent of the inductance of the circuit (the situation would be different if the circuit were superconducting).

● 9.7. A jumper *12* of mass m slides without friction along two long conducting rails separated by a distance l (Fig. 9.22). The system is in a uniform magnetic field perpendicular to the plane of the circuit.

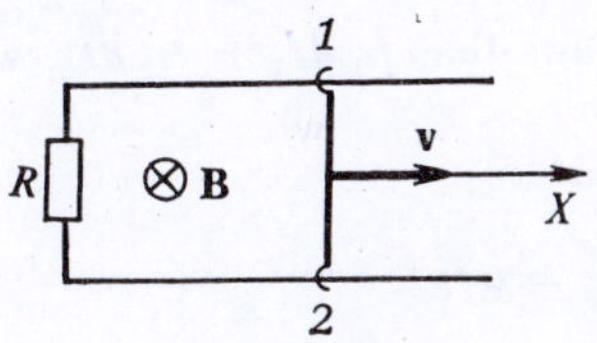

Fig. 9.22

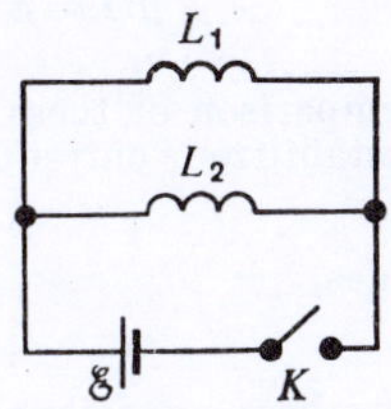

Fig. 9.23

The left ends of the rails are shunted through resistor R. At the instant $t = 0$, the jumper received the initial velocity v_0 directed to the right. Find the velocity of the jumper as a function of time, ignoring the resistances of the jumper and of the rails as well as self-inductance of the circuit.

Solution. Let us choose the positive direction of the normal **n** to the plane of the circuit away from us. This means that the positive direction of circumvention of the circuit (for induced e.m.f. and current) is chosen clockwise, in accordance with the right-hand screw rule. It follows from Ohm's law that

$$RI = -\frac{d\Phi}{dt} = -B\frac{dS}{dt} = -Blv, \tag{1}$$

where we took into account that when the jumper moves to the right, $d\Phi > 0$. According to Lenz's law, the induced current I causes the Ampère force counteracting the motion, directed to the left.

Having chosen the X-axis to the right, we write the equation of motion of the jumper

$$m\,dv/dt = IlB, \tag{2}$$

where the right-hand side is the projection of the Ampère force onto the X-axis (this quantity is negative, but we omit the minus sign since, as can be seen from (1), current $I < 0$).

Eliminating I from Eqs. (1) and (2), we obtain

$$dv/v = -a\,dt, \quad a = B^2l^2/mR.$$

The integration of this expression, taking into account the initial condition, gives

$$\ln(v/v_0) = -at, \quad v = v_0e^{-at}$$

● **9.8. The role of transient processes.** In the circuit shown in Fig. 9.23, the e.m.f. $\mathscr{E}$ of the source, its internal resistance R and the inductances L_1 and L_2 of superconducting coils are known. Find the currents established in the coils after key K has been closed.

Solution. Let us use Kirchhoff's laws for electric circuits $\mathscr{E}L_1$ and $\mathscr{E}L_2$:

$$RI=\mathscr{E}-L_1\frac{dI_1}{dt}, \quad RI=\mathscr{E}-L_2\frac{dI_2}{dt}.$$

A comparison of these expressions shows that $L_1\,dI_1 = L_2\,dI_2$, while for stabilized currents we have

$$L_1I_{10} = L_2I_{20}. \tag{1}$$

Besides,

$$I_{10} + I_{20} = I_0 = \mathscr{E}/R. \tag{2}$$

From these equations, we find

$$I_{10}=\frac{\mathscr{E}}{R}\frac{L_2}{L_1+L_2}, \quad I_{20}=\frac{\mathscr{E}}{R}\frac{L_1}{L_1+L_2}.$$

● **9.9. Calculation of inductance.** A coaxial cable consists of internal solid conductor of radius a and external thin-wall conducting tube of radius b. Find the inductance of a unit length of the cable, considering that the current distribution over the cross section of the internal conductor is uniform. The permeability is equal to unity everywhere.

Solution. In the case under consideration, the inner conductor is not thin, and hence the inductance should be determined in terms of the energy rather than through the magnetic flux. In accordance with (9.33), we can write

$$L_u=\frac{1}{I^2}\int_0^b \frac{B^2}{\mu_0}\,2\pi r\,dr, \tag{1}$$

where r is the distance from the cable axis. In order to evaluate this integral, we must find the dependence $B(r)$. Using the theorem on circulation, we have

$$B_{r<a}=\frac{\mu_0 I}{2\pi a^2}r, \quad B_{a<r<b}=\frac{\mu_0 I}{2\pi}\frac{1}{r}, \quad B_{r>b}=0. \tag{2}$$

The form of these dependences is shown in Fig. 9.24. On account of (2), integral (1) is split into two parts. As a result of integration, we obtain

$$L_u=\frac{\mu_0}{2\pi}\left(\frac{1}{4}+\ln\frac{b}{a}\right).$$

It should be noted that an attempt to determine this quantity in terms of magnetic flux by the formula $L_u = \Phi_u/I$ leads to a different (erroneous) result, namely, instead of 1/4, 1/2 is obtained in the parentheses. The thinner the central wire, i.e. the larger the ratio b/a, the smaller the relative difference in the results of calculation by these two methods, viz. through the energy and through the flux.

● **9.10. Mutual induction.** A long straight wire is arranged along the symmetry axis of a toroidal coil of rectangular cross section, whose

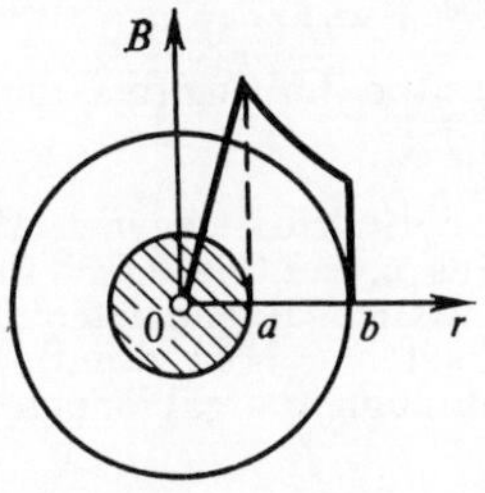

Fig. 9.24

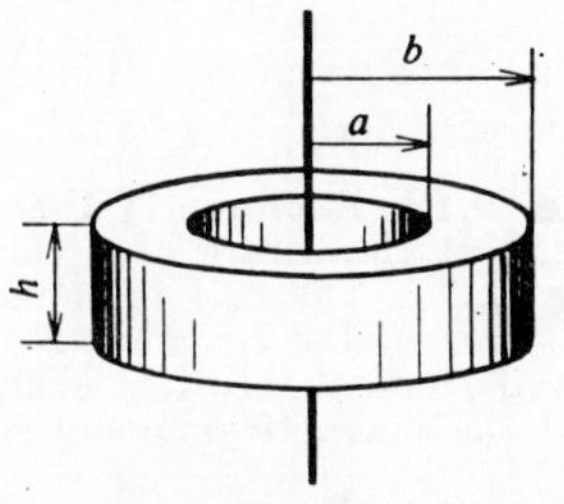

Fig. 9.25

dimensions are given in Fig. 9.25. The number of turns on the coil is N, and the permeability of the surrounding medium is unity. Find the amplitude of the e.m.f. induced in this coil if the current $I = I_m \cos \omega t$ flows along the straight wire.

Solution. The required e.m.f. $\mathscr{E}_i = -d\Phi/dt$, where $\Phi = N\Phi_1$ and Φ_1 is the magnetic flux through the cross section of the coil:

$$\Phi_1 = \int B_n \, dS = \int_a^b \frac{\mu_0}{2\pi r} Ih \, dr = \frac{\mu_0 h I}{2\pi} \ln \frac{b}{a},$$

where B_n is determined with the help of the theorem on circulation of vector **B**. Taking the time derivative of Φ_1 and multiplying the obtained result by N, we find the following expression for the amplitude value of induced e.m.f.:

$$\mathscr{E}_{im} = \frac{\mu_0 h \omega I_m N}{2\pi} \ln \frac{b}{a}.$$

● **9.11. Calculation of mutual inductance.** Two solenoids of the same length and of practically the same cross section are completely inserted one into the other. The inductances of solenoids are L_1 and L_2. Neglecting edge effects, find their mutual inductance (modulo).

Solution. By definition, the mutual inductance is

$$L_{12} = \Phi_1 / I_2, \tag{1}$$

where Φ_1 is the total magnetic flux through all the turns of solenoid *1* provided that current I_2 flows in solenoid *2*. The flux $\Phi_1 = N_1 B_2 S$, where N_1 is the number of turns in solenoid *1*, S is the cross-sectional area of the solenoid, and $B_2 = \mu\mu_0 n_2 I_2$. Hence formula (1) can be written (after cancelling I_2) as follows:

$$|L_{12}| = \mu\mu_0 n_2 N_1 S = \mu\mu_0 n_1 n_2 V, \tag{2}$$

where we took into account that $N_1 = n_1 l$, where l is the solenoid length and $lS = V$ is its volume. Expression (2) can be represented in terms of L_1 and L_2 as follows:

$$|L_{12}| = \sqrt{\mu\mu_0 n_1^2 V}\,\sqrt{\mu\mu_0 n_2^2 V} = \sqrt{L_1 L_2}.$$

It should be noted that this expression defines the limiting (maximum) value of $|L_{12}|$. In general, $|L_{12}| < \sqrt{L_1 L_2}$.

● **9.12. Reciprocity theorem.** A small cylindrical magnet M is placed at the centre of a thin coil of radius a, containing N turns (Fig. 9.26). The coil is connected to a ballistic galvanometer. The resistance of the circuit is R. After the magnet had been rapidly removed from the coil, a charge q passed through the galvanometer. Find the magnetic moment of the magnet.

Solution. In the process of removal of the magnet, the total magnetic flux through the coil was changing, which resulted in the emergence of induced current defined by the following equation:

$$RI = -\frac{d\Phi}{dt} - L\frac{dI}{dt}.$$

Fig. 9.26

We multiply both sides of this equation by dt and take into account that $I\,dt = dq$. This gives

$$R\,dq = -d\Phi - L\,dI.$$

Having integrated this expression, we obtain $Rq = -\Delta\Phi - L\,\Delta I$. Considering now that $\Delta I = 0$ (the current was equal to zero at the beginning as well as at the end of the process), we obtain

$$q = \Delta\Phi/R = \Phi/R, \tag{1}$$

where Φ is the magnetic flux through the coil at the beginning of the process (we omitted the minus sign since it is immaterial).

Thus, the problem is reduced to the calculation of the flux Φ through the coil. This quantity cannot be determined directly. This difficulty, however, can be overcome by using the reciprocity theorem. We mentally replace the magnet by a small current loop creating in the surrounding space the same magnetic field as the magnet. If the area of the loop is S and the current is I, their product will give the magnetic moment p_m of the magnet: $p_m = IS$. According to the reciprocity theorem, $L_{12}I = L_{21}I$ and the problem is reduced to determining the magnetic flux through the area S of the loop, which creates the same current I, but flowing in the coil. Assuming the field to be uniform within the loop, we obtain

$$\Phi = BS = \mu_0 NIS/2a. \tag{2}$$

It remains for us to substitute (2) into (1) and recall that $IS = p_m$. Then $q = \mu_0 N p_m / 2aR$, and

$$p_m = 2aRq/\mu_0 N.$$

10. Maxwell's Equations. Energy of Electromagnetic Field

10.1. Displacement Current

Maxwell's Discovery. The theory of electromagnetic field founded by Faraday was mathematically completed by Maxwell. One of the most important new ideas put forward by Maxwell was the idea of symmetry in the mutual dependence of electric and magnetic fields. Namely, since a time-varying magnetic field ($\partial \mathbf{B}/\partial t$) creates an electric field, it should be expected that a time-varying electric field ($\partial \mathbf{E}/\partial t$) creates a magnetic field.

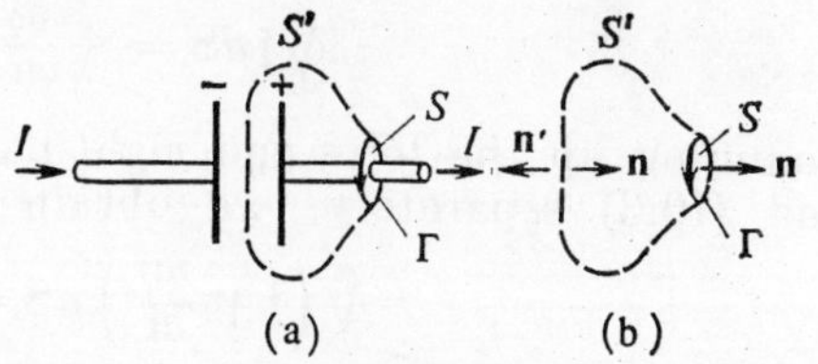

Fig. 10.1

This idea necessitating the existence of a new in principle phenomenon of induction can be approached, for example, by the following line of reasoning. We know that according to the theorem on circulation of vector **H**,

$$\oint \mathbf{H}\, d\mathbf{l} = \int \mathbf{j}\, d\mathbf{S}. \tag{10.1}$$

Let us apply this theorem to the case when a preliminarily charged parallel plate capacitor is being discharged through a certain external resistance (Fig. 10.1*a*). For the contour Γ we take a curve embracing the wire. We can stretch various surfaces on this contour, for example, S and S'. These two surfaces have "equal rights". However, current I flows through surface S, while through surface S' there is no current!

It turns out that the circulation of vector **H** depends on the surface stretched over a given contour (?!), which is obviously ruled out (and never happened in the case of direct currents).

Is it possible to change somehow the right-hand side of (10.1) to avoid this inconvenience? Fortunately, this can be done in the following way.

First, we note that surface S' "cuts" only the electric field. In accordance with the Gauss theorem, the flux of vector **D** through a closed surface is $\oint \mathbf{D}\, d\mathbf{S} = q$, whence

$$\oint \frac{\partial \mathbf{D}}{\partial t}\, d\mathbf{S} = \frac{\partial q}{\partial t}. \tag{10.2}$$

On the other hand, according to the continuity equation (5.4), we have

$$\oint \mathbf{j}\, d\mathbf{S} = -\frac{\partial q}{\partial t}. \tag{10.3}$$

Summing up the left- and right-hand sides of Eqs. (10.2) and (10.3) separately, we obtain

$$\oint \left(\mathbf{j} + \frac{\partial \mathbf{D}}{\partial t}\right) d\mathbf{S} = 0. \tag{10.4}$$

This equation is similar to the continuity equation for direct current. It can be seen that in addition to the density **j** of the conduction current, there is one more term $\partial \mathbf{D}/\partial t$ whose dimensions are the same as for current density. Maxwell termed this addend the density of *displacement current*:

$$\mathbf{j}_d = \partial \mathbf{D}/\partial t. \tag{10.5}$$

The sum of the conduction and displacement currents is called the *total current*. Its density is given by

$$\mathbf{j}_t = \mathbf{j} + \frac{\partial \mathbf{D}}{\partial t}. \tag{10.6}$$

According to (10.4), the lines of the total current are continuous in contrast to the lines of conduction current. If conduction currents are not closed, displacement currents close them.

We shall show now that the introduction of the concept of total current eliminates the difficulty associated with the dependence of circulation of vector **H** on the choice of the surface stretched over contour Γ. It turns out that for this it is sufficient to introduce on the right-hand side of Eq. (10.1) the total current instead of conduction current, viz. the quantity

$$I_t = \int \left(\mathbf{j} + \frac{\partial \mathbf{D}}{\partial t} \right) d\mathbf{S}. \tag{10.7}$$

Indeed, the right-hand side of (10.7) is the sum of conduction current I and displacement current $I_d : I_t = I + I_d$. Let us show that the total current I_t will be the same for surfaces S and S' stretched over the same contour Γ. For this purpose, we apply formula (10.4) to the closed surface composed of surfaces S and S' (Fig. 10.1*b*). Taking into account the fact that for a closed surface the normal $\mathbf{n}$ is directed outwards, we write

$$I_t(S') + I_t(S) = 0.$$

Now, if we direct the normal $\mathbf{n}'$ for surface S' to the same side as for surface S, the first term in this expression will change the sign, and we obtain

$$I_t(S') = I_t(S),$$

Q.E.D.

Thus, the theorem on circulation of vector $\mathbf{H}$, which was established for direct currents, can be generalized for an arbitrary case in the following form:

$$\oint \mathbf{H}\, d\mathbf{l} = \int \left(\mathbf{j} + \frac{\partial \mathbf{D}}{\partial t} \right) d\mathbf{S}. \tag{10.8}$$

In this form, the theorem on circulation of vector $\mathbf{H}$ is always valid, which is confirmed by the agreement of this equation with the results of experiments in all cases without any exception.

Differential form of Eg. (10.8):

$$\nabla \times \mathbf{H} = \mathbf{j} + \frac{\partial \mathbf{D}}{\partial t}, \tag{10.9}$$

where the curl of vector $\mathbf{H}$ is determined by the density $\mathbf{j}$ of conduction current and $\partial \mathbf{D}/\partial t$ of displacement current at the same point.

Several Remarks about Displacement Current. It should

be kept in mind that displacement current is equivalent to conduction current only from the point of view of its ability of creating a magnetic field.

Displacement currents exist only when an electric field varies with time. In dielectrics, displacement current consists of two essentially different components. Since vector $\mathbf{D} = \varepsilon_0\mathbf{E} + \mathbf{P}$, it follows that the density $\partial\mathbf{D}/\partial t$ of displacement current is the sum of the densities of the "true" displacement current, $\varepsilon_0\partial\mathbf{E}/\partial t$, and of the *polarization current* $\partial\mathbf{P}/\partial t$. The latter quantity appears due to the motion of bound charges. There is nothing unexpected in the fact that polarization currents excite a magnetic field, since these currents do not differ in nature from conduction currents. A new in principle statement consists in that the other component of the displacement current, $\varepsilon_0\partial\mathbf{E}/\partial t$, which is not connected with any motion of charges and is only due to the variation of the electric field, also excites a magnetic field. Even in a vacuum, any temporal change of an electric field excites in the surrounding space a magnetic field.

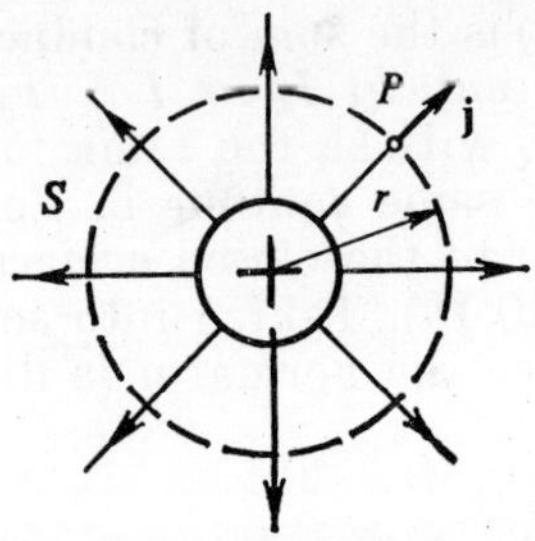

Fig. 10.2

The discovery of this phenomenon was the most essential and decisive step made by Maxwell while constructing the electromagnetic field theory. This discovery was as valuable as the discovery of electromagnetic induction, according to which a varying magnetic field excites a vortex electric field. It should also be noted that the discovery of displacement current by Maxwell was a purely theoretical discovery of primary importance.

Let us consider an example in which displacement currents are manifested.

Example. A metallic positively charged sphere is in an infinite homogeneous conducting medium (Fig. 10.2). Electric currents flowing in radial directions should excite a magnetic field. Let us find the direction of vector **B** at an arbitrary point P.

First of all, it is clear that vector **B** cannot have a radial component. Otherwise, the flux of **B** through the surface S of the sphere

(Fig. 10.2) would differ from zero, which is in contradiction with Eq. (7.2). Consequently, vector **B** must be perpendicular to the radial direction at the point P. But this is also impossible, since all directions perpendicular to the radial direction are absolutely equivalent. It remains for us to conclude that magnetic field is equal to zero everywhere.

The absence of magnetic field in the presence of electric current of density **j** indicates that in addition to conduction current **j**, displacement current $\mathbf{j}_d$ is also present in the system. This current is such that the total current is equal to zero everywhere, i.e. $\mathbf{j}_d = -\mathbf{j}$ at each point. This means that

$$j_d = j = \frac{I}{4\pi r^2} = \frac{\dot{q}}{4\pi r^2} = \frac{\partial D}{\partial t},$$

where we took into account that $D = q/4\pi r^2$ in accordance with the Gauss theorem.

10.2. Maxwell's Equations

Maxwell's Equations in the Integral Form. The introduction of displacement current brilliantly completed the macroscopic theory of electromagnetic field. The discovery of displacement current ($\partial \mathbf{D}/\partial t$) allowed Maxwell to create a *unified* theory of electric and magnetic phenomena. Maxwell's theory not only explained all individual phenomena of electricity and magnetism from a single point of view, but also predicted a number of new phenomena whose existence was subsequently confirmed.

Hitherto, we considered separate parts of this theory. Now we can represent the entire theory in the form of a system of fundamental equations in electrodynamics, called *Maxwell's equations* for stationary media. In all, there are four Maxwell's equations (we are already acquainted with each of these equations separately from previous sections; here we simply gather them together). In the integral form, the system of Maxwell's equations is written as follows:

$$\oint \mathbf{E}\, d\mathbf{l} = -\int \frac{\partial \mathbf{B}}{\partial t}\, d\mathbf{S}, \qquad \oint \mathbf{D}\, d\mathbf{S} = \int \rho\, dV, \tag{10.10}$$

$$\oint \mathbf{H}\, d\mathbf{l} = \int \left(\mathbf{j} + \frac{\partial \mathbf{D}}{\partial t}\right) d\mathbf{S}, \qquad \oint \mathbf{B}\, d\mathbf{S} = 0, \tag{10.11}$$

where ρ is the volume density of extraneous charges and **j** is the density of conduction current.

These equations express in a concise form all our knowledge about electromagnetic field. The content of these equations consists in the following.

1. Circulation of vector **E** around any closed contour is equal, with the minus sign, to the time derivative of the magnetic flux through an arbitrary surface bounded by the given contour. Here **E** is not only vortex electric field but also electrostatic field (whose circulation is known to be equal to zero).

2. The flux of **D** through any closed surface is equal to the algebraic sum of extraneous charges embraced by this surface.

3. Circulation of vector **H** around an arbitrary closed contour is equal to the total current (conduction current plus displacement current) through an arbitrary surface bounded by this contour.

4. The flux of **B** through an arbitrary closed surface is always equal to zero.

It follows from Maxwell's equations for circulations of vectors **E** and **H** that electric and magnetic fields cannot be treated as independent: a change in time in one of these fields leads to the emergence of the other. Hence it is only a combination of these fields, describing a unique electromagnetic field, that has a sense.

If the fields are stationary ($\mathbf{E}$ = const and $\mathbf{B}$ = const), Maxwell's equations split into two groups of *independent* equations

$$\oint \mathbf{E}\, d\mathbf{l} = 0, \quad \oint \mathbf{D}\, d\mathbf{S} = q, \qquad \oint \mathbf{H}\, d\mathbf{l} = I, \quad \oint \mathbf{B}\, d\mathbf{S} = 0. \tag{10.12}$$

In this case, the electric and magnetic fields are independent of one another, which allowed us to study first a constant electric field and then, independently, constant magnetic field.

It should be emphasized that the line of reasoning which has led us to Maxwell's equations can by no means be considered their proof. These equations cannot be "derived" since they are the fundamental axioms, or postulates of electrodynamics, obtained with the help of generalization of experimental results. These postulates play in electrodynamics the same role as Newton's laws in classical mechanics or the laws in thermodynamics.

Differential Form of Maxwell's Equations. Equations (10.10) and (11.11) can be represented in differential form, i.e. in the form of a system of differential equations, viz.

$$\nabla \times \mathbf{E} = -\frac{\partial \mathbf{B}}{\partial t}, \quad \nabla \cdot \mathbf{D} = \rho, \tag{10.13}$$

$$\nabla \times \mathbf{H} = \mathbf{j} + \frac{\partial \mathbf{D}}{\partial t}, \quad \nabla \cdot \mathbf{B} = 0. \tag{10.14}$$

Equations (10.13) show that there may be two causes for an electric field. First, its sources are electric charges, both extraneous and bound (which follows from the equation $\nabla \cdot \mathbf{D} = \rho$, if we take into account that $\mathbf{D} = \varepsilon_0 \mathbf{E} + \mathbf{P}$ and $\nabla \cdot \mathbf{P} = -\rho'$, then $\nabla \cdot \mathbf{E} \propto \rho + \rho'$). Second, the field **E** always emerges when a magnetic field varies in time (which expresses Faraday's law of electromagnetic induction).

On the other hand, Eqs. (10.14) indicate that the magnetic field **B** can be excited by moving electric charges (electric currents), by varying electric fields, or by both factors simultaneously (this follows from the equation $\nabla \times \mathbf{H} = \mathbf{j} + \partial \mathbf{D}/\partial t$, if we take into account that $\mathbf{H} = \mathbf{B}/\mu_0 - \mathbf{J}$ and $\nabla \times \mathbf{J} = \mathbf{j}'$; then $\nabla \times \mathbf{B} \propto \mathbf{j} + \mathbf{j}' + \partial \mathbf{P}/\partial t + \varepsilon_0 \partial \mathbf{E}/\partial t$, where $\mathbf{j}'$ is the magnetization current density and $\partial \mathbf{P}/\partial t$ is the *polarization current* density. The first three currents are associated with motion of charges, while the last current is due to a time-varying field **E**). As follows from the equation $\nabla \cdot \mathbf{B} = 0$, there are no sources of magnetic field in nature (called, by analogy, magnetic charges) that would be similar to electric charges.

The importance of Maxwell's equations in differential form consists not only in that they express the basic laws of electromagnetic field, but also in that the fields **E** and **B** themselves can be found by solving (integrating) these equations.

Together with the equation of motion of charged particles under the action of the Lorentz force,

$$dp/dt = q\mathbf{E} + q\,[\mathbf{v} \times \mathbf{B}], \tag{10.15}$$

Maxwell's equations form a fundamental system of equations. In principle, this system is sufficient for describing all electromagnetic phenomena in which quantum effects are not exhibited.

Boundary Conditions. Maxwell's equations in integral form are of a more general nature than their differential counterpart, since they remain valid in the presence of a *surface of discontinuity*, viz. the surface on which the properties of the medium or fields change abruptly. On the other hand, Maxwell's equations in the differential form presume that all quantities vary continuously in space and time.

However, the differential form of Maxwell's equations can be imparted the same generality by supplementing them with the *boundary conditions* which must be observed for an electromagnetic field at the interface between two media. These conditions are contained in the integral form of Maxwell's equations and are already familiar to us:

$$D_{1n} = D_{2n}, \quad E_{1\tau} = E_{2\tau}, \quad B_{1n} = B_{2n}, \ H_{1\tau} = H_{2\tau} \qquad (10.16)$$

(here the first and last conditions refer to the cases when there are neither extraneous charges nor conduction currents on the interface). It should also be noted that these boundary conditions are valid both for constant and for varying fields.

Material Equations. Maxwell's fundamental equations still do not form a complete system of equations for the electromagnetic field. These equations are insufficient for determining the fields from given distributions of charges and currents.

Maxwell's equations should be supplemented with relations which would contain the quantities characterizing individual properties of a medium. These relations are called *material equations*. Generally, these equations are rather complicated and are not as general and fundamental as Maxwell's equations.

Material equations have the simplest form for sufficiently weak electric fields which vary in time and space at a comparatively low rate. In this case, for isotropic media containing no ferroelectrics and ferromagnetics, the material equations have the following (already familiar to us) form:

$$\mathbf{D} = \varepsilon\varepsilon_0\mathbf{E}, \quad \mathbf{B} = \mu\mu_0\mathbf{H}, \quad \mathbf{j} = \sigma(\mathbf{E} + \mathbf{E}^*), \qquad (10.17)$$

where ε, μ, and σ are the well-known constants characterizing the electric and magnetic properties of the medium (permittivity, permeability, and electroconductivity), and $\mathbf{E}^*$ is the strength of the extraneous force field due to chemical or thermal processes

10.3. Properties of Maxwell's Equations

Maxwell's Equations Are Linear. They contain only the first derivatives of fields **E** and **B** with respect to time and spatial coordinates, and the first powers of densities ρ and **j**

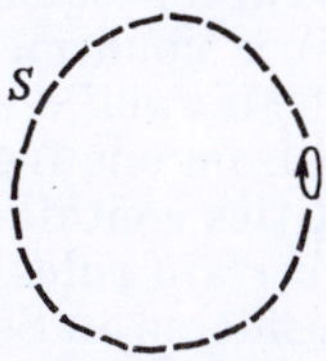

Fig. 10.3

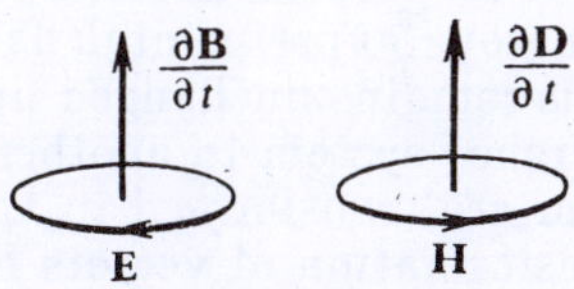

Fig. 10.4

of electric charges and currents. The linearity of Maxwell's equations is directly connected with the principle of superposition: if any two fields satisfy Maxwell's equations, the sum of these fields will also satisfy these equations.

Maxwell's Equations Include the Continuity Equation. This expresses the law of conservation of electric charge. In order to make sure of this, let us take an infinitely small contour Γ, stretch over it an arbitrary finite surface S (Fig. 10.3), and then contract this contour to a point, so that surface S remains finite. In the limit, $\oint \mathbf{H}\, d\mathbf{l}$ vanishes, surface S becomes closed, and the first of Eqs. (10.11) becomes

$$\oint \left(\mathbf{j}+\frac{\partial \mathbf{D}}{\partial t}\right) d\mathbf{S}=0.$$

Hence it follows that

$$\oint \mathbf{j}\, d\mathbf{S}=-\frac{\partial}{\partial t}\oint \mathbf{D}\, d\mathbf{S}=-\frac{\partial q}{\partial t},$$

which is just the continuity equation (5.4). This equation states that the current flowing from a volume V through a closed surface S is equal to the decrease in charge in this volume per unit time.

The same law (viz. the continuity equation) can also be obtained from Maxwell's differential equations. For this purpose, it is sufficient to take the divergence of both sides of the first of Eqs. (10.14) and make use of the second of Eqs. (10.13). This gives $\nabla \cdot \mathbf{j} = -\partial\rho/\partial t$.

Maxwel's Equations Are Satisfied in All Inertial Systems of Reference. They are relativistic invariants. This is a consequence of the relativity principle, according to which all inertial systems of reference are equivalent from the physical point of view. The invariance of Maxwell's equations (with respect to Lorentz' transformations) is confirmed by numerous experimental data. The form of Maxwell's equations remains unchanged upon a transition from one inertial reference system to another, but the quantities contained in them are transformed in accordance with certain rules. The transformation of vectors **E** and **B** was considered in Sec. 8.

Thus, unlike, for example, Newton's equations in mechanics, Maxwell's equations are correct relativistic equations.

On the Symmetry of Maxwell's Equations. Maxwell's equations are not symmetric relative to electric and magnetic fields. This is again due to the fact that there are electric charges in nature, but as far as it is known at present, magnetic charges do not exist. However, for a neutral homogeneous nonconducting medium, where $\rho = 0$ and $\mathbf{j} = 0$, Maxwell's equations acquire a symmetric form, since **E** is related to $\partial\mathbf{B}/\partial t$ in the same way as **B** to $\partial\mathbf{E}/\partial t$:

$$\begin{aligned} \nabla \times \mathbf{E} &= -\partial\mathbf{B}/\partial t, \quad \nabla \cdot \mathbf{D} = 0, \\ \nabla \times \mathbf{H} &= \partial\mathbf{D}/\partial t, \quad \nabla \cdot \mathbf{B} = 0. \end{aligned} \tag{10.18}$$

The symmetry of equations relative to electric and magnetic fields does not involve only the sign of the derivatives $\partial\mathbf{B}/\partial t$ and $\partial\mathbf{D}/\partial t$. The difference in signs of these derivatives indicates that the lines of the vortex electric field induced by a variation of field **B** form a left-hand screw system with vector $\partial\mathbf{B}/\partial t$, while the lines of the magnetic field induced by a variation of **D** form a right-hand screw system with vector $\partial\mathbf{D}/\partial t$ (Fig. 10.4).

On Electromagnetic Waves. Maxwell's equations lead to the following important conclusion about the existence of a new in principle physical phenomenon: electromagnetic field may exist independently, without electric charges and currents. A change in its state in this case necessarily has a wave nature. Such fields are called *electromagnetic waves*. In a vacuum, these waves always propagate at a velocity equal to the velocity of light c.

It also turned out that displacement current ($\partial \mathbf{D}/\partial t$) plays in this phenomenon a primary role. It is its presence,

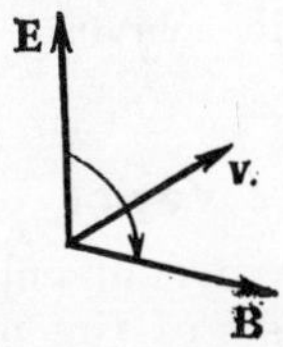

Fig. 10.5

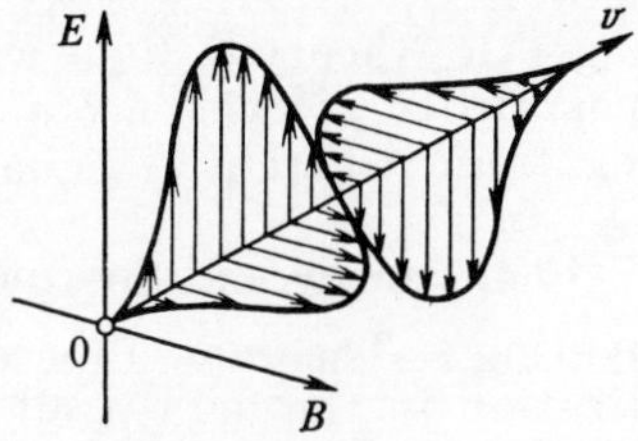

Fig. 10.6

along with the quantity $\partial \mathbf{B}/\partial t$, that makes the appearance of electromagnetic waves possible. Any time variation of a magnetic field induces an electric field, while a variation of an electric field, in its turn, induces a magnetic field. This continuous interconversion or interaction of the fields preserves them, and an electromagnetic perturbation propagates in space.

Maxwell's theory not only predicted the existence of electromagnetic waves, but also made it possible to establish all their basic properties. Any electromagnetic wave, regardless of its specific form (it can be a harmonic wave or an electromagnetic perturbation of any form) is characterized by the following properties:

(1) the velocity of its propagation in a nonconducting neutral nonferromagnetic medium is equal to

$$v = c/\sqrt{\varepsilon\mu}, \text{ where } c = 1/\sqrt{\varepsilon_0\mu_0}; \qquad (10.19)$$

(2) vectors **E**, **B**, and **v** (wave velocity) are *mutually perpendicular* and form a *right-hand screw system* (Fig. 10.5) This right-handed relation is an *intrinsic property* of an electromagnetic wave, independent of the choice of coordinate system;

(3) vectors **E** and **B** in an electromagnetic wave always oscillate in phase (see Fig. 10.6 showing a "photograph" of a wave), the instantaneous values E and B at any point being connected through the relation $E = vB$, or

$$\sqrt{\varepsilon\varepsilon_0}\,E=\sqrt{\mu\mu_0}\,H. \tag{10.20}$$

This means that E and H (or B) simultaneously attain their maximum values, vanish, etc.

Maxwell's brilliant success in the development of the electromagnetic theory of light was due to his understanding of the possibility of existence of electromagnetic waves, following from differential equations (10.18).

10.4. Energy and Energy Flux. Poynting's Vector

Poynting's Theorem. Proceeding from the idea of energy localization in the field itself and on the basis of the principle of energy conservation, we should conclude that if energy decreases in a certain region, this may only occur due to its "flowing out" through the boundaries of the region under consideration (the medium is assumed to be stationary).

In this respect, there is a formal analogy with the law of conservation of charge expressed by Eq. (5.4). The meaning of this law consists in that a decrease in charge per unit time in a given volume is equal to the flux of vector **j** through the surface enclosing this volume.

In a similar manner, for the law of conservation of energy we should assume that in addition to the energy density w in a given region, there exists a certain vector **S** characterizing the *energy flux density*.

If we speak only of the energy of an electromagnetic field, its total energy in a given volume will change due to its flowing out of this volume as well as due to the fact that the field transfers its energy to a substance (charged particles), i.e. accomplishes work over the substance. In macroscopic form, this statement can be written as follows:

$$-\frac{dW}{dt}=\oint \mathbf{S}\,d\mathbf{A}+P, \tag{10.21}$$

where $d\mathbf{A}$ is a surface element.

This equation expresses *Poynting's theorem*: *a decrease in energy per unit time in a given volume is equal to the energy flux through the surface bounding this volume plus the work per unit time* (i.e. *power P*) *which is accomplished by the field over the charges of the substance inside the given volume.*

In Eq. (10.21), $W=\int w\,dV$, where w is the field energy density, and $P=\int \mathbf{j}\cdot\mathbf{E}\,dV$, where $\mathbf{j}$ is the current density and $\mathbf{E}$ is the electric field strength. The above expression for P can be obtained as follows. During the time dt, field $\mathbf{E}$ will do the work $\delta A = q\mathbf{E}\cdot\mathbf{u}\,dt$ over a point charge q, where $\mathbf{u}$ is the charge velocity. Hence the power of force $q\mathbf{E}$ is $P = q\mathbf{u}\cdot\mathbf{E}$. Going over to the volume distribution of charges, we replace q by $\rho\,dV$, where ρ is the volume charge density. Then $dP = \rho\mathbf{u}\cdot\mathbf{E}\,dV = \mathbf{j}\cdot\mathbf{E}\,dV$. It remains for us to integrate dP over the volume under consideration.

It should be noted that power P in (10.21) can be either positive or negative. The latter takes place when positive charges of the substance move against field $\mathbf{E}$, or negative charges move in the opposite direction. For example, such a situation is observed at the points of a medium where in addition to an electric field $\mathbf{E}$, the field $\mathbf{E}^*$ of extraneous forces is also acting. At such points, $\mathbf{j} = \sigma(\mathbf{E}+\mathbf{E}^*)$, and if $\mathbf{E}^* \uparrow\downarrow \mathbf{E}$, and $E^* > E$ in magnitude, the product $\mathbf{j}\cdot\mathbf{E}$ in the expression for P turns out to be negative.

Poynting obtained the expressions for the energy density w and vector $\mathbf{S}$ by using Maxwell's equations (we shall not present this derivation here). If a medium contains no ferroelectrics and ferromagnetics (i.e. in the absence of hysteresis), the energy density of an electromagnetic field is given by

$$w=\frac{\mathbf{E}\cdot\mathbf{D}}{2}+\frac{\mathbf{B}\cdot\mathbf{H}}{2}. \qquad (10.22)$$

It should be noted that individual terms of this expression have been obtained by us earlier [see (4.10) and (9.32)].

The density of the electromagnetic energy flux, which is called *Poynting vector*, is defined as

$$\mathbf{S}=[\mathbf{E}\times\mathbf{H}]. \qquad (10.23)$$

Strictly speaking, Maxwell's equations cannot give unambiguous expressions for the two quantities w and $\mathbf{S}$. The expressions presented above are the simplest from an infinite number of possible equations. Therefore, we should treat these expressions as postulates whose validity should be confirmed by the agreement with experiments of their corollaries.

Several following examples will demonstrate that although the results obtained with the help of formulas (10.22) and (10.23) may sometimes seem strange, we will not be able to find in them anything incredible, any disagreement with experiment. And this is just the evidence that these expressions are correct.

Example 1. Energy flux in an electromagnetic wave (in vacuum). Let us calculate energy dW passing during time dt through a unit area perpendicular to the direction of propagation of the wave.

If we know the values of $\mathbf{E}$ and $\mathbf{B}$ in the region of location of the unit area, then

$$dW = wc\, dt,$$

where w is the energy density, $w = \varepsilon_0 E^2/2 + \mu_0 H^2/2$. In accordance with (10.20), for an electromagnetic wave we have

$$\varepsilon_0 E^2 = \mu_0 H^2.$$

This means that the electric energy density in the electromagnetic wave at any instant is equal to the magnetic energy density at the same point, so we can write for the energy density

$$w = \varepsilon_0 E^2.$$

Then

$$dW = \varepsilon_0 E^2 c\, dt = \sqrt{\varepsilon_0/\mu_0}\, E^2\, dt.$$

Let us now see what we shall obtain with the help of Poynting vector. The same quantity dW can be represented in terms of the magnitude of vector $\mathbf{S}$ as follows:

$$dW = S\, dt = EH\, dt = \sqrt{\varepsilon_0/\mu_0}\, E^2\, dt.$$

Thus, both expressions (for w and for $\mathbf{S}$) lead to the same result (the last two formulas).

Example 2. Evolution of heat in a conductor. Let a current I flow through a straight circular wire of radius a (Fig. 10.7). Since the wire has a resistance, a certain electric field $\mathbf{E}$ is acting along it. The same value of $\mathbf{E}$ will be at the wire surface in a vacuum. Besides, the presence of current generates a magnetic field. According to the theorem on circulation of vector $\mathbf{H}$, near the surface of the wire we have $2\pi a H = I$, $H = I/2\pi a$. Vectors $\mathbf{E}$ and $\mathbf{H}$ are arranged so that the

Poynting vector is directed inside the wire normally to its lateral surface (Fig. 10.7). Consequently, the electromagnetic energy flows into the wire from the surrounding space! But does it agree with the amount of heat liberated in the conductor? Let us calculate the flux of electromagnetic energy through the lateral surface of a section of the wire of length l:

$$EH \cdot 2\pi a l = 2\pi a H \cdot E l = I \cdot U = R I^2,$$

where U is the potential difference across the ends of this section and R is its resistance. Thus, we arrive at the conclusion that the electromagnetic energy flows into the wire from outside and is completely converted into Joule's heat. We must agree that this is a rather unexpected conclusion.

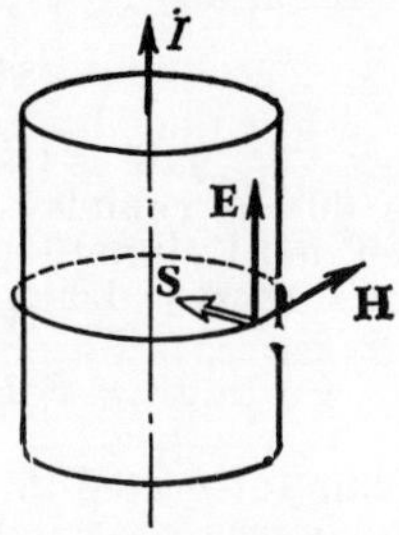

Fig. 10.7

It should be noted that in a power source, vector **E** is directed against current I, and hence in the vicinity of the source the Poynting vector is directed outside: in this region, electromagnetic energy flows into the surrounding space. In other words, it turns out that the energy from the source is transmitted not along the wires but through the space surrounding a conductor in the form of the flux of electromagnetic energy, viz. the flux of vector **S**.

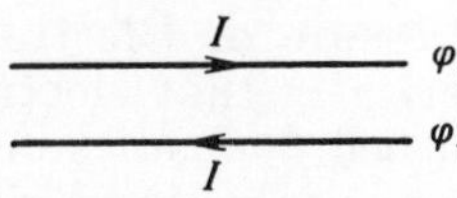

Fig. 10.8

Example 3. Figure 10.8 shows a section of a balanced (twin) line. The directions of currents in the wires are known as well as the fact that the wires' potentials are $\varphi_1 < \varphi_2$. Can we find where (to the right or to the left) the power source (generator) is?

The answer to this question can be obtained with the help of the Poynting vector. In the case under consideration, between the wires, vector **E** is directed downwards while vector **H** is directed behind the plane of the figure. Hence, vector $\mathbf{S} = [\mathbf{E} \times \mathbf{H}]$ is directed to the right, i.e. the power source is on the left and a consumer is on the right.

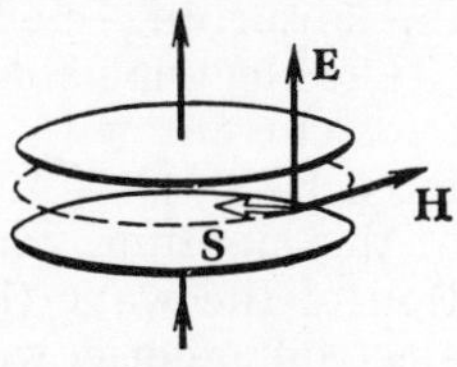

Fig. 10.9

Example 4. Charging of a capacitor. Let us take a parallel-plate capacitor with circular plates of radius a. Ignoring edge effects (field dissipation), find the electromagnetic energy flux through the lateral "surface" of the capacitor, since only in this region the Poynting vector is directed inside the capacitor (Fig. 10.9).

On this surface, there is a varying electric field **E** and a magnetic field **H** generated by its variation. According to the theorem on circulation of vector **H**, it follows that $2\pi a H = \pi a^2\, \partial D/\partial t$, where the right-hand side contains the displacement current through the contour shown in Fig. 10.9 by the dashed line. Hence $H = (1/2)\, a\, \partial D/\partial t$. If the distance between the plates is h, the flux of vector **S** through the lateral surface is

$$EH2\pi ah = E\frac{a}{2}\frac{\partial D}{\partial t}2\pi ah = E\frac{\partial D}{\partial t}V, \tag{1}$$

where $V = \pi a^2 h$ is the volume of the capacitor. We shall assume that this flux is completely spent for increasing the capacitor's energy. Then, multiplying Eq. (1) by dt, we obtain the increment of the capacitor's energy during the time dt:

$$dW = E\, dD \cdot V = d\left(\frac{\varepsilon\varepsilon_0 E^2}{2}V\right) = d\left(\frac{ED}{2}V\right)$$

Having integrated this equation, we find the formula for the energy W of a charged capacitor. Thus, everything is all right in this case too.

10.5. Electromagnetic Field Momentum

Pressure of Electromagnetic Wave. Maxwell showed theoretically that electromagnetic waves, being reflected or absorbed by bodies on which they are incident, exert *pressure* on these bodies. This pressure appears as a result of the action of the magnetic field of a wave on the electric currents induced by the electric field of the same wave.

Suppose an electromagnetic wave propagates in a homogeneous medium capable of absorption. The presence of absorption means that Joule's heat with the volume density σE^2 will be liberated in the medium. Hence $\sigma \neq 0$, i.e. the absorbing medium is conductive.

The electric field of the wave excites in this medium an electric current with a density $\mathbf{j} = \sigma\mathbf{E}$. As a result, Ampère's force $\mathbf{F}_u = [\mathbf{j} \times \mathbf{B}] = \sigma\,[\mathbf{E} \times \mathbf{B}]$ will act on a unit volume of the medium. This force is directed towards the propagation of the wave (Fig. 10.10) and causes the pressure of the electromagnetic wave.

In the absence of absorption, the conductivity $\sigma = 0$ and $\mathbf{F}_u = 0$, i.e. in this case the electromagnetic wave does not exert any pressure on the medium.

Electromagnetic Field Momentum. Since an electromagnetic wave exerts a pressure on a substance, the latter acquires a certain momentum. However, if in a closed system consisting of a substance and an electromagnetic wave only the substance possessed a momentum, the law of conservation of momentum would be violated.

The momentum of such a system can be conserved only provided that the electromagnetic field (wave) also possesses a momentum: the substance acquires the momentum due to that transferred to it by the electromagnetic field.

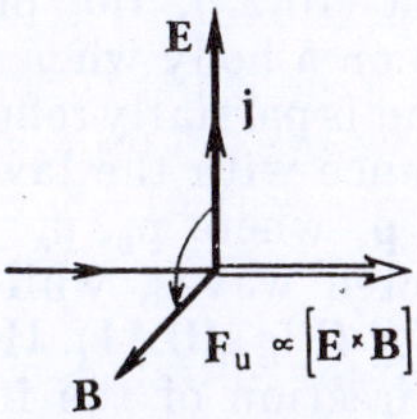

Fig. 10.10

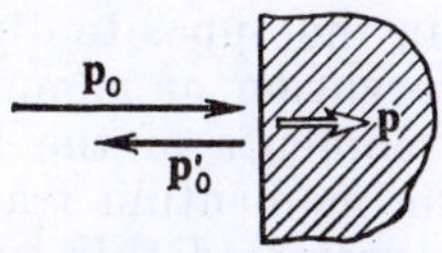

Fig. 10.11

Let us introduce the concept of *momentum density* **G** of electromagnetic field as the quantity numerically equal to the momentum of the field in a unit volume. Calculations which will be omitted here show that the momentum density is given by

$$\boxed{\mathbf{G} = \mathbf{S}/c^2,} \tag{10.24}$$

where $\mathbf{S} = [\mathbf{E} \times \mathbf{H}]$ is Poynting's vector. Just as vector **S**, momentum density **G** is generally a function of time and coordinates.

In accordance with (10.20), for an electromagnetic wave in a vacuum we have $\sqrt{\varepsilon_0}E = \sqrt{\mu_0}H$. Hence the energy density w and the magnitude S of Poynting's vector are respectively equal to

$$w = \varepsilon_0 E^2/2 + \mu_0 H^2/2 = \varepsilon_0 E^2, \quad S = EH = \sqrt{\varepsilon_0/\mu_0}\, E^2.$$

It follows that $S = w/\sqrt{\varepsilon_0 \mu_0}$. And since $\sqrt{\varepsilon_0 \mu_0} = 1/c$, where c is the velocity of light in vacuum, $S = wc$. Taking into account (10.24), we obtain that for an electromagnetic wave in a vacuum

$$\boxed{G = w/c.} \tag{10.25}$$

The same relation between energy and momentum is inherent (as is shown in the theory of relativity) in particles with zero rest mass. This is quite natural since, according to quantum-mechanical notions, an electromagnetic wave is equivalent to the flow of photons, viz. the particles with zero rest mass.

Some More Remarks on the Pressure of Electromagnetic Waves. Let us calculate, using formula (10.25), the pressure exerted by an electromagnetic wave on a body when the wave is incident normal to its surface and is partially reflected in the opposite direction. In accordance with the law of conservation of momentum, $\mathbf{p}_0 = \mathbf{p}_0' + \mathbf{p}$, where $\mathbf{p}_0$, $\mathbf{p}_0'$ are the momenta of the incident and reflected waves, while $\mathbf{p}$ is the momentum transferred to the body (Fig. 10.11). Having projected this equality onto the direction of the incident wave and referring all the quantities to unit time and unit cross-sectional area, we obtain

$$p = p_0 + p_0' = \langle G \rangle c + \langle G' \rangle c,$$

where $\langle G \rangle$ and $\langle G' \rangle$ are the average values of the momentum density for the incident and reflected waves. It remains for us to take into account relation (10.25) between $\langle G \rangle$ and $\langle w \rangle$ as well as the fact that $\langle w' \rangle = \rho \langle w \rangle$, where ρ is the *reflection coefficient*. As a result, the previous expression becomes

$$p = (1 + \rho) \langle w \rangle. \tag{10.26}$$

Here the quantity p is just the pressure of the electromagnetic wave on the body. In the case of total reflection, $\rho = 1$ and pressure $p = 2 \langle w \rangle$ while for total absorption, $\rho = 0$ and $p = \langle w \rangle$.

It should be added that the pressure of electromagnetic radiation is usually very low (the exception is the pressure of high-power laser radiation beams, especially after beam focusing, and the pressure of radiation inside hot stars). For example, the pressure of solar radiation on the Earth amounts to about 10^{-6} Pa, which is smaller than the atmospheric pressure by a factor of 10^{10}. In spite of insignificant values of these quantities, experimental proof of the existence of electromagnetic waves, viz. the pressure of light, was obtained by Lebedev. The results of these experiments were in agreement with the electromagnetic theory of light.

Problems

● **10.1. Displacement current.** A point charge q moves uniformly along a straight line with a nonrelativistic velocity $\mathbf{v}$. Find the vector of the displacement current density at a point P lying at a distance r from the charge on a straight line (1) coinciding with its trajectory; (2) perpendicular to its trajectory and passing through the charge.

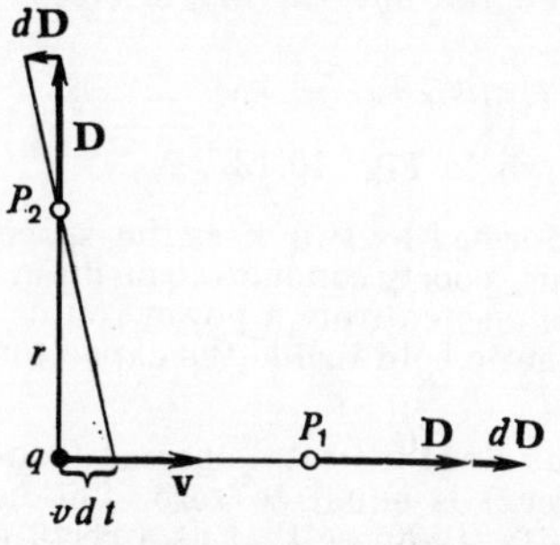

Fig. 10.12

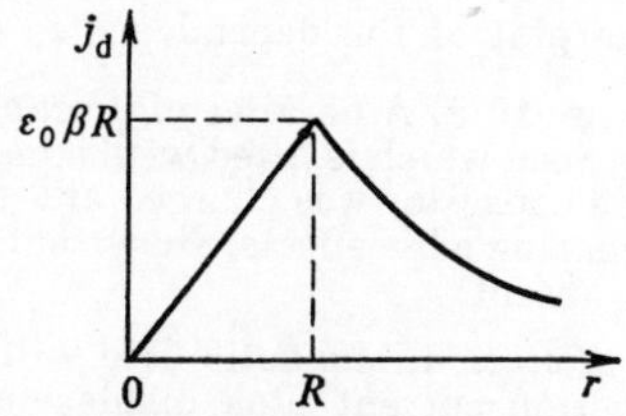

Fig. 10.13

Solution. The displacement current density $\mathbf{j}_d = \partial\mathbf{D}/\partial t$. Hence, the solution of the problem is reduced to determining vector $\mathbf{D}$ at the indicated points and to finding its time derivative. In both cases, $\mathbf{D} = q\mathbf{e}_r/4\pi r^2$, where $\mathbf{e}_r$ is the unit vector of $\mathbf{r}$. Let us find the derivative $\partial\mathbf{D}/\partial t$

(1) At point P_1 (Fig. 10.12, where it is assumed that $q > 0$), we have

$$\frac{\partial\mathbf{D}}{\partial t} = -\frac{2q}{4\pi r^3}\frac{\partial r}{\partial t}\mathbf{e}_r = \frac{2q\mathbf{v}}{4\pi r^3},$$

where we took into account that for point P_1 the derivative $\partial r/\partial t = -v$. If point P_1 were not in front of the charge (in the direction of its motion) but behind it, vector $\mathbf{j}_d$ would be directed in the same way and would have the same magnitude.

Thus, for $q > 0$, vector $\mathbf{j}_d \uparrow\uparrow \mathbf{v}$, and vice versa.

(2) At point P_2 (Fig. 10.12), $|d\mathbf{D}|/D = v\,dt/r$, and hence we can write

$$\partial\mathbf{D}/\partial t = -q\mathbf{v}/4\pi r^3.$$

If $q > 0$, $j_d \uparrow\downarrow \mathbf{v}$, and vice versa.

● **10.2.** A current flowing in a long straight solenoid with the radius R of cross section is varied so that the magnetic field inside the solenoid increases with time according to the law $B = \beta t^2$, where β is a constant. Find the displacement current density as a function of the distance r from the solenoid axis.

Solution. In order to find the displacement current density, we must, in accordance with (10.5), first find the electric field strength (here it will be a vortex field). Using Maxwell's equation for circulation of vector **E**, we write

$$2\pi rE = \pi r^2\, \partial B/\partial t, \quad E = r\beta t \quad (r < R);$$

$$2\pi rE = \pi R^2\, \partial B/\partial t, \quad E = R^2\beta t/r \quad (r > R).$$

Now, using the formula $j_d = \varepsilon_0\, \partial E/\partial t$, we can find the displacement current density:

$$j_d = \varepsilon_0\beta r \quad (r < R); \qquad j_d = \varepsilon_0\beta R^2/r \quad (r > R).$$

The plot of the dependence $j_d(r)$ is shown in Fig. 10.13.

● **10.3.** A parallel-plate capacitor is formed by two discs the space between which is filled with a homogeneous, poorly conducting medium. The capacitor was charged and then disconnected from a power source. Ignoring edge effects, show that the magnetic field inside the capacitor is absent.

Solution. Magnetic field will be absent since the total current (conduction current plus displacement current) is equal to zero. Let us prove this. We consider the current density. Suppose that at a certain instant the density of conduction current is **j**. Obviously, $\mathbf{j} \propto \mathbf{D}$ and $\mathbf{D} = \sigma\mathbf{n}$, where σ is the surface charge density on the positively charged plate and **n** is the normal (Fig. 10.14).

The presence of conduction current leads to a decrease in the surface charge density σ, and hence in **D** as well. This means that conduction current will be accompanied by the displacement current whose density is

$$\mathbf{j}_d = \partial\mathbf{D}/\partial t = (\partial\sigma/\partial t)\,\mathbf{n} = -j\mathbf{n} = -\mathbf{j}.$$

Hence it follows that, indeed

$$\mathbf{j}_t = \mathbf{j} + \mathbf{j}_d = 0.$$

● **10.4.** The space between the plates of a parallel-plate capacitor in the form of circular discs is filled with a homogeneous poorly conducting medium with a conductivity σ and permittivity ε. Ignoring edge effects, find the magnitude of vector **H** between the plates at a distance r from their axes, if the electric field strength between the plates varies with time in accordance with the law $E = E_m \cos \omega t$.

Solution. From Maxwell's equation for circulation of vector **H**, it follows that

$$2\pi rH = \left(j_n + \varepsilon\varepsilon_0 \frac{\partial E_n}{\partial t}\right)\pi r^2.$$

Taking into account Ohm's law $j_n = \sigma E_n(t)$, we obtain

$$H = \frac{r}{2}\left(E_n + \frac{\varepsilon\varepsilon_0}{\sigma}\frac{\partial E_n}{\partial t}\right) = \frac{rE_m}{2}(\sigma \cos \omega t - \varepsilon\varepsilon_0\omega \sin \omega t).$$

Let us transform the expression in the parentheses to cosine. For this purpose, we multiply and divide this expression by $f = \sqrt{\sigma^2 + (\varepsilon\varepsilon_0\omega)^2}$ and then introduce angle δ through the formulas $\sigma/f = \cos\delta$, $\varepsilon\varepsilon_0\omega/f = \sin\delta$. This gives

$$H = \frac{1}{2} rE_m \sqrt{\sigma^2 + (\varepsilon\varepsilon_0\omega)^2} \cos(\omega t + \delta).$$

● **10.5.** A point charge q moves in a vacuum uniformly and rectilinearly with a nonrelativistic velocity **v**. Using Maxwell's

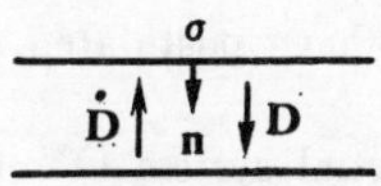

Fig. 10.14

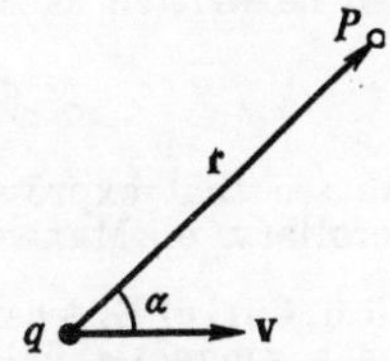

Fig. 10.15

equation for the circulation of vector **H**, obtain the expression for **H** at a point P whose position relative to the charge is characterized by radius vector **r** (Fig. 10.15).

Solution. It is clear from symmetry considerations that for a contour around which the circulation of vector **H** should be taken, we must choose a circle with centre O (its trace is shown in Fig. 10.16 by the dashed line). Then

$$2\pi RH = \frac{\partial}{\partial t} \int D_n \, dS, \tag{1}$$

where R is the radius of the circle.

Let us find the flux of vector **D** through a surface bounded by this circle. For the sake of simplicity, we shall take a spherical surface with the radius of curvature r (Fig. 10.16). Then the flux of **D** through an elementary ring of this spherical surface is

$$D\,dS = \frac{q}{4\pi r^2} 2\pi r \sin\alpha' \cdot r\,d\alpha' = \frac{q}{2} \sin\alpha'\,d\alpha',$$

while the total flux through the selected surface is

$$\int D\,dS = \frac{q}{2}(1 - \cos\alpha). \tag{2}$$

Now, in accordance with (1), we differentiate (2) with respect to time:

$$\frac{\partial}{\partial t} \int D\,dS = \frac{q}{2} \sin\alpha \frac{d\alpha}{dt}. \tag{3}$$

For the displacement of the charge from point *1* to point *2* (Fig. 10.17) over the distance $v\,dt$, we have $v\,dt \cdot \sin\alpha = r\,d\alpha$, whence

$$\frac{d\alpha}{dt} = \frac{v \sin\alpha}{r}. \tag{4}$$

Substituting (4) into (3) and then (3) into (1), we obtain

$$H = qvr \sin\alpha / 4\pi r^3, \tag{5}$$

where we took into account that $R = r \sin\alpha$. Relation (5) in vector form can be written as follows:

$$\mathbf{H} = \frac{q}{4\pi} \frac{[\mathbf{v} \times \mathbf{r}]}{r^3}.$$

Thus we see that expression (6.3) which we have postulated earlier is a corollary of Maxwell's equations.

● **10.6. Curl of E.** A certain region of an inertial system of reference contains a magnetic field of magnitude $B = \text{const}$, rotating at an

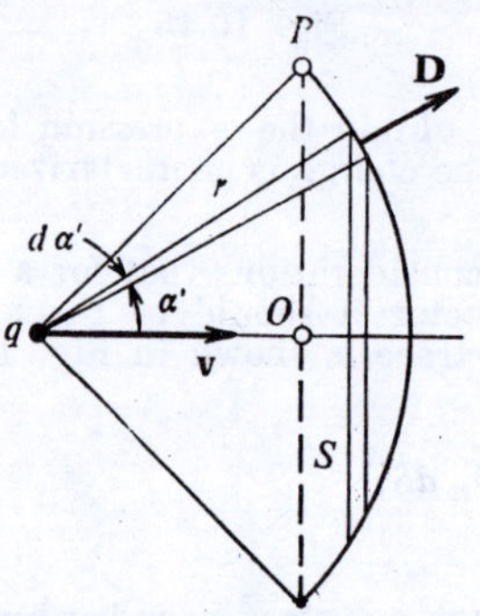

Fig. 10.16

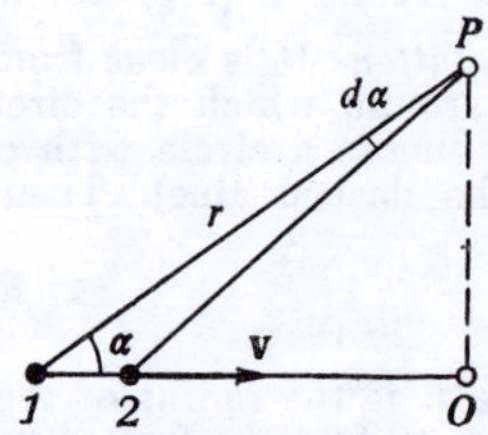

Fig. 10.17

angular velocity $\boldsymbol{\omega}$. Find $\nabla \times \mathbf{E}$ in this region as a function of vectors $\boldsymbol{\omega}$ and $\mathbf{B}$.

Solution. It follows from the equation $\nabla \times \mathbf{E} = -\partial \mathbf{B}/\partial t$ that vector $\nabla \times \mathbf{E}$ is directed oppositely to vector $d\mathbf{B}$. The magnitude of this vector can be calculated with the help of Fig. 10.18:

$$|\,d\mathbf{B}\,| = B\omega\,dt, \qquad |\,d\mathbf{B}/dt\,| = B\omega.$$

Hence

$$\nabla \times \mathbf{E} = -[\boldsymbol{\omega} \times \mathbf{B}].$$

● **10.7. Poynting's vector.** Protons having the same velocity $\mathbf{v}$ form a beam of a circular cross section with current I. Find the direction and magnitude of Poynting's vector $\mathbf{S}$ outside the beam at a distance r from its axis.

Solution. Figure 10.19 shows that S ↑↑ v. Let us find the magnitude of S: $S = EH$, where E and H depend on r. According to the Gauss theorem, we have

$$2\pi rE = \lambda/\varepsilon_0,$$

where λ is the charge per unit length of the beam. Besides, it follows from the theorem on circulation of vector H that

$$2\pi rH = I.$$

Having determined E and H from the last two equations and taking into account that $I = \lambda v$, we obtain

$$S = EH = I^2/4\pi^2\varepsilon_0 vr^2.$$

● **10.8.** A current flowing through the winding of a long straight

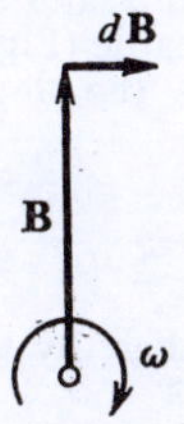

Fig. 10.18

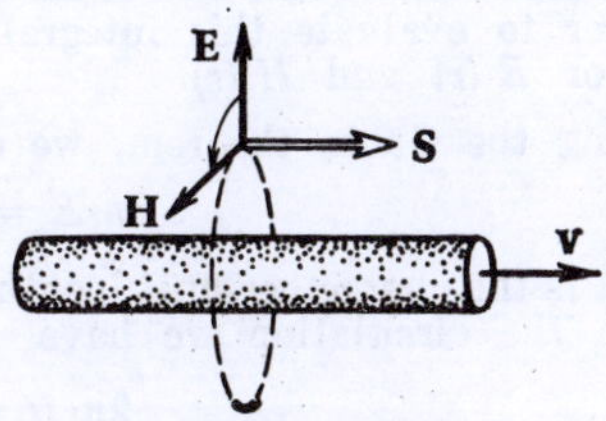

Fig. 10.19

solenoid is being increased. Show that the rate of increase in the energy of the magnetic field in the solenoid is equal to the flux of Poynting's vector through its lateral surface.

Solution. As the current increases, the magnetic field in the solenoid also increases, and hence a vortex electric field appears. Suppose that the radius of the solenoid cross section is equal to a. Then the strength of the vortex electric field near the lateral surface of the solenoid can be determined with the help of Maxwell's equation that expresses the law of electromagnetic induction:

$$2\pi aE = \pi a^2 \frac{\partial B}{\partial t}, \quad E = \frac{a}{2}\frac{\partial B}{\partial t}.$$

The energy flux through the lateral surface of the solenoid can be represented as follows:

$$\Phi = EH \cdot 2\pi al = \pi a^2 l \frac{\partial}{\partial t}\left(\frac{B^2}{2\mu_0}\right),$$

where l is the solenoid length and $\pi a^2 l$ is its volume.

Thus, we see that the energy flux through the lateral surface of the solenoid (the flux of vector S) is equal to the rate of variation of the magnetic energy inside the solenoid:

$$\Phi = S \cdot 2\pi al = \partial W/\partial t.$$

● **10.9.** The energy from a source of constant voltage U is transmitted to a consumer via a long coaxial cable with a negligibly small resistance. The current in the cable is I. Find the energy flux through the cable cross section, assuming that the outer conducting shell of the cable has thin walls.

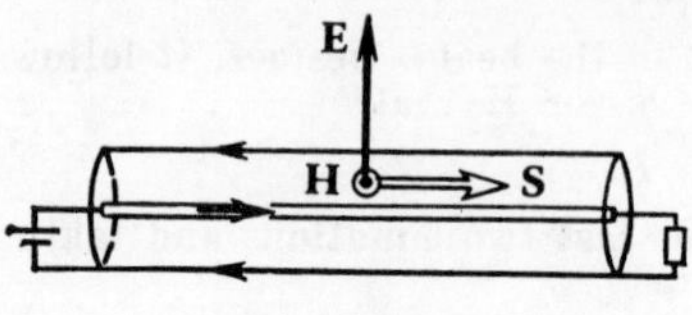

Fig. 10.20

Solution. The required energy flux is defined by the formula

$$\Phi = \int_a^b S \cdot 2\pi r\, dr, \tag{1}$$

where $S = EH$ is the flux density, $2\pi r\, dr$ is the elementary ring of width dr within which the value of S is constant, and a and b are the radii of the internal wire and of the outer shell of the cable (Fig. 10.20). In order to evaluate this integral, we must know the dependence $S(r)$, or $E(r)$ and $H(r)$.

Using the Gauss theorem, we obtain

$$2\pi r E = \lambda/\varepsilon_0, \tag{2}$$

where λ is the charge per unit length of the wire. Further, by the theorem on the circulation we have

$$2\pi r H = I. \tag{3}$$

After substituting E and H from formulas (2) and (3) into expression (1) and integrating, we get

$$\Phi = \frac{\lambda I}{2\pi\varepsilon_0} \ln \frac{b}{a}. \tag{4}$$

The values of λ, a, and b are not given in the problem. Instead, we know U. Let us find the relation between these quantities:

$$U = \int_a^b E\, dr = \frac{\lambda}{2\pi\varepsilon_0} \ln \frac{b}{a}. \tag{5}$$

A comparison of (4) and (5) gives

$$\Phi = UI.$$

This coincides with the value of power liberated in the load.

● **10.10.** A parallel-plate air capacitor whose plates are made in the form of disks of radius a are connected to a source of varying harmonic voltage of frequency ω. Find the ratio of the maximum values of magnetic and electric energy inside the capacitor.

Solution. Let the voltage across the capacitor vary in accordance with the law $U = U_m \cos \omega t$ and the distance between the capacitor plates be h. Then the electric energy of the capacitor is equal to

$$W_e = \frac{\varepsilon_0 E^2}{2} \pi a^2 h = \frac{\varepsilon_0 \pi a^2}{2h} U_m^2 \cos^2 \omega t. \quad (1)$$

The magnetic energy can be determined through the formula

$$W_m = \int \frac{B^2}{2\mu_0} dV. \quad (2)$$

The quantity B required for evaluating this integral can be found from the theorem on the circulation of vector **H**: $2\pi r H = \pi r^2 \, \partial D/\partial t$. Hence, considering that $H = B/\mu_0$ and $\partial D/\partial t = -\varepsilon_0 (U_m/h) \omega \sin \omega t$, we obtain

$$B = \frac{1}{2} \varepsilon_0 \mu_0 \frac{r \omega U_m}{h} |\sin \omega t|. \quad (3)$$

It remains for us to substitute (3) into (2), where for dV we must take an elementary volume in the form of a ring for which $dV = 2\pi r \, dr \cdot h$. As a result of integration, we obtain

$$W_m = \frac{\pi}{16} \frac{\mu_0 \varepsilon_0^2 \omega^2 a^4 U_m^2}{h} \sin^2 \omega t. \quad (4)$$

The ratio of the maximum values of magnetic energy (4) and electric energy (1) is given by

$$\frac{W_{m\,max}}{W_{e\,max}} = \frac{1}{8} \mu_0 \varepsilon_0 a^2 \omega^2.$$

For example, for $a = 6$ cm and $\omega = 1000 \text{ s}^{-1}$, this ratio is equal to 5×10^{-15}.

11. Electric Oscillations

11.1. Equation of an Oscillatory Circuit

Quasi-steady Conditions. When electric oscillations occur, the current in a circuit varies with time and, generally speaking, turns out to be different at each instant of time in different sections of the circuit (due to the fact that electromagnetic perturbations propagate although at a very high but still finite velocity). There are, however, many cases when instantaneous values of current prove to be practically the same in all sections of the circuit (such a current is called *quasistationary*). In this case, all time variations should occur so slowly that the propagation of electromagnetic perturbations could be considered *instantaneous*. If l is the length of a circuit, the time required for an electromagnetic perturbation to cover the distance l is of the order of $\tau = l/c$. For a periodically varying current, quasi-steady condition will be observed if

$$\tau = l/c \ll T,$$

where T is the period of variations.

For example, for a circuit of length $l = 3$ m, the time $\tau = 10^{-8}$ s, and the current can be assumed to be quasista-

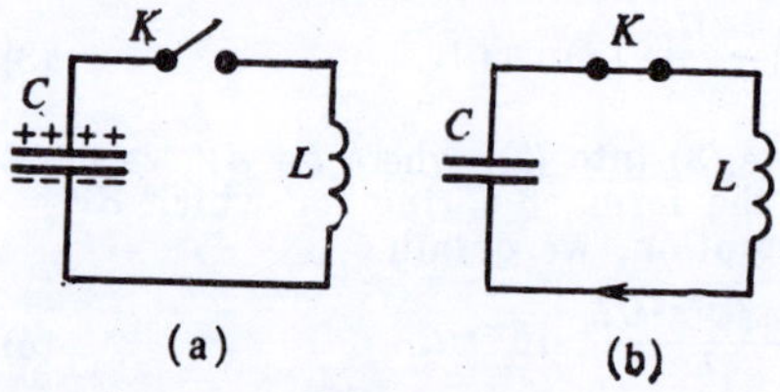

Fig. 11.1

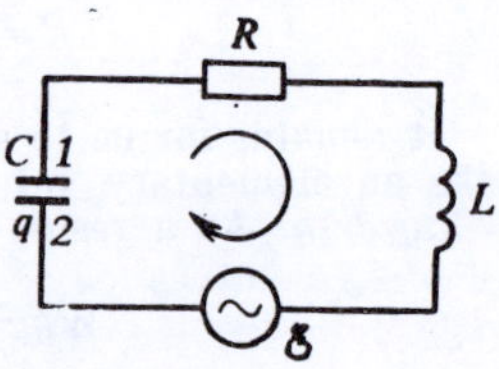

Fig. 11.2

tionary down to frequencies of 10^6 Hz (which corresponds to $T = 10^{-6}$ s).

In this chapter, we shall assume everywhere that in the cases under consideration quasi-steady conditions are observed and currents are *quasistationary*. This will allow us to use formulas obtained for static fields. In particular, we shall be using the fact that instantaneous values of quasistationary currents obey Ohm's law.

Oscillatory Circuit. In a circuit including a coil of inductance L and a capacitor of capacitance C, electric oscillations may appear. For this reason, such a circuit is called an *oscillatory circuit*. Let us find out how electric oscillations emerge and are sustained in an oscillatory circuit.

Suppose that initially, the upper plate of the capacitor is charged positively and the lower plate, negatively (Fig. 11.1*a*). In this case, the entire energy of the oscillatory circuit is concentrated in the capacitor. Let us close key K. The capacitor starts to discharge, and a current flows through coil L. The electric energy of the capacitor is converted into the magnetic energy of the coil. This process terminates when the capacitor is discharged completely, while current in the circuit attains its maximum value (Fig. 11.1*b*). Starting from this moment, the current begins to decrease, retaining its direction. It will not, however, cease immediately since it will be sustained by self-induced e.m.f. The current recharges the capacitor, and the appearing electric field will tend to reduce the current. Finally, the current ceases, while the charge on the capacitor attains its maximum value. From this moment, the capacitor starts to discharge again, the current flows in the opposite direction, and the process is repeated.

If the conductors constituting the oscillatory circuit have no resistance, strictly periodic oscillations will be observed in the circuit. In the course of the process, the charge on the capacitor plates, the voltage across the capacitor and the current in the induction coil vary periodically. The oscillations are accompanied by mutual conversion of the energy of electric and magnetic fields.

If, however, the resistance of conductors $R \neq 0$, then, in addition to the process described above, electromagnetic energy will be transformed into Joule's heat.

Equation of an Oscillatory Circuit. Let us derive the equation describing oscillations in a circuit containing series-connected capacitor C, induction coil L, resistor R, and varying external e.m.f. $\mathscr{E}$ (Fig. 11.2).

First, we choose the positive direction of circumvention, e.g. clockwise. We denote by q the charge on the capacitor plate the direction from which to the other plate coincides with the chosen direction of circumvention. Then current in the circuit is defined as

$$I = dq/dt. \qquad (11.1)$$

Consequently, if $I > 0$, then $dq > 0$ as well, and vice versa (the sign of I coincides with that of dq).

In accordance with Ohm's law for section *1RL2* of the circuit, we have

$$RI = \varphi_1 - \varphi_2 + \mathscr{E}_s + \mathscr{E}, \qquad (11.2)$$

where $\mathscr{E}_s$ is the self-induced e.m.f. In the case under consideration,

$$\mathscr{E}_s = -L\, dI/dt \text{ and } \varphi_2 - \varphi_1 = q/C$$

(the sign of q must coincide with the sign of the potential difference $\varphi_2 - \varphi_1$ since $C > 0$). Hence Eq. (11.2) can be written in the form

$$L\frac{dI}{dt}+RI+\frac{q}{C}=\mathscr{E}, \tag{11.3}$$

or, taking into account (11.1),

$$\boxed{L\frac{d^2q}{dt^2}+R\frac{dq}{dt}+\frac{1}{C}q=\mathscr{E}.} \tag{11.4}$$

This is the *equation of an oscillatory circuit*, which is a linear second-order nonhomogeneous differential equation with constant coefficients. Using this equation for calculating $q(t)$, we can easily obtain the voltage across the capacitor as $U_C = \varphi_2 - \varphi_1 = q/C$ and current I by formula (11.1).

The equation of an oscillatory circuit can be given a different form:

$$\boxed{\ddot{q}+2\beta\dot{q}+\omega_0^2q=\mathscr{E}/L,} \tag{11.5}$$

where the following notation is introduced:

$$2\beta = R/L, \quad \omega_0^2 = 1/LC. \tag{11.6}$$

The quantity ω_0 is called the *natural frequency* of the circuit and β is the *damping factor*. The meaning of these quantities will be explained below.

If $\mathscr{E} = 0$, the oscillations are usually called *free oscillations*. They will be *undamped* for $R = 0$ and *damped* for $R \neq 0$. Let us consider consecutively all these cases.

11.2. Free Electric Oscillations

Free Undamped Oscillations. If a circuit contains no external e.m.f. $\mathscr{E}$ and if its resistance $R = 0$, the oscillations in such a circuit will be *free* and *undamped*. The equation describing these oscillations is a particular case of Eq. (11.5) when $\mathscr{E} = 0$ and $R = 0$:

$$\ddot{q}+\omega_0^2q=0. \tag{11.7}$$

The solution of this equation is the function

$$q = q_m \cos(\omega_0 t + \alpha), \quad (11.8)$$

where q_m is the maximum value of the charge on capacitor plates, ω_0 is the natural frequency of the oscillatory circuit, and α is the initial phase. The value of ω_0 is determined only by the properties of the circuit itself, while the values of q_m and α depend on the initial conditions. For these conditions we can take, for example, the value of charge q and current $I = \dot{q}$ at the moment $t = 0$.

According to (11.6), $\omega_0 = 1/\sqrt{LC}$; hence the period of free undamped oscillations is given by

$$T_0 = 2\pi\sqrt{LC} \quad (11.9)$$

(Thomson's formula).

Having found current I [by differentiating (11.8) with respect to time] and bearing in mind that the voltage across the capacitor plates is in phase with charge q, we can easily see that in free undamped oscillations current I leads in phase the voltage across the capacitor plates by $\pi/2$.

While solving certain problems, energy approach can also be used.

Example. In an oscillatory circuit, free undamped oscillations occur with energy W. The capacitor plates are slowly moved apart so that the frequency of oscillations decreases by a factor of η. Which work against electric forces is done in this case?

The required work can be represented as an increment of the energy of the circuit:

$$A = W' - W = \frac{q_m^2}{2}\left(\frac{1}{C'} - \frac{1}{C}\right) = W\left(\frac{C}{C'} - 1\right).$$

On the other hand, $\omega_0 \propto 1/\sqrt{C}$, and hence $\eta = \omega_0'/\omega_0 = \sqrt{C/C'}$, and consequently

$$A = W(\eta^2 - 1).$$

Free Damped Oscillations. Every real oscillatory circuit has a resistance, and the energy stored in the circuit is gradually spent on heating. Free oscillations will be damped.

We can obtain the equation for a given oscillatory circuit by putting $\mathscr{E} = 0$ in (11.5). This gives

$$\ddot{q} + 2\beta\dot{q} + \omega_0^2 q = 0. \quad (11.10)$$

It can be shown (we shall not do it here since we are interested in another aspect of the problem) that for $\beta^2 < \omega_0^2$, the solution of this homogeneous differential equation has the form

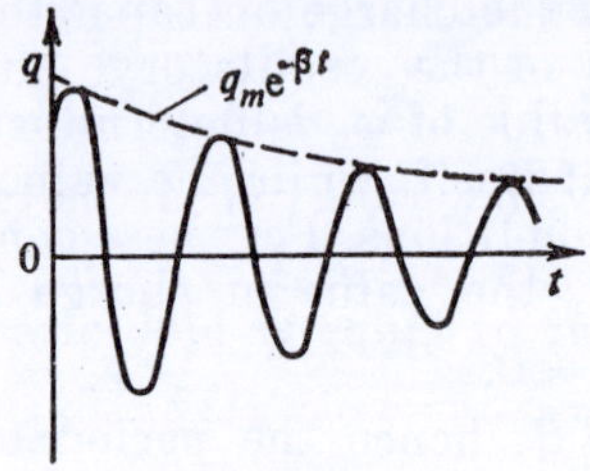

Fig. 11.3

$$q = q_m e^{-\beta t} \cos(\omega t + \alpha), \tag{11.11}$$

where

$$\omega = \sqrt{\omega_0^2 - \beta^2} = \sqrt{\frac{1}{LC} - \left(\frac{R}{2L}\right)^2}, \tag{11.12}$$

and q_m and α are arbitrary constants determined from the initial conditions. The plot of the function (11.11) is shown in Fig. 11.3. It can be seen that this is not a periodic function since it determines damped oscillations.

Nevertheless, the quantity $T = 2\pi/\omega$ is called the *period of damped oscillations*:

$$T = \frac{2\pi}{\sqrt{\omega_0^2 - \beta^2}} = \frac{T_0}{\sqrt{1 - (\beta/\omega_0)^2}}, \tag{11.13}$$

where T_0 is the period of free undamped oscillations.

The factor $q_m e^{-\beta t}$ in (11.11) is called the *amplitude of damped oscillations*. Its dependence on time is shown in Fig. 11.3 by the dashed line.

Voltage across a Capacitor and Current in an Oscillatory Circuit. Knowing $q(t)$, we can find the voltage across a capacitor and the current in a circuit. The voltage across a capacitor is given by

$$U_C = \frac{q}{C} = \frac{q_m}{C} e^{-\beta t} \cos(\omega t + \alpha). \tag{11.14}$$

The current in a circuit is

$$I = \frac{dq}{dt} = q_m e^{-\beta t} [-\beta \cos(\omega t + \alpha) - \omega \sin(\omega t + \alpha)].$$

We transform the expression in the brackets to cosine. For this purpose, we multiply and divide this expression by $\sqrt{\omega^2 + \beta^2} = \omega_0$ and then introduce angle δ by the formulas

$$-\beta/\omega_0 = \cos\delta, \qquad \omega/\omega_0 = \sin\delta. \qquad (11.15)$$

After this, the expression for I takes the following form:

$$I = \omega q_m e^{-\beta t} \cos(\omega t + \alpha + \delta). \qquad (11.16)$$

It follows from (11.15) that angle δ lies in the second quadrant ($\pi/2 < \delta < \pi$). This means that in the case of a non-zero resistance, the current in the circuit *leads* voltage in phase (11.14) across the capacitor by more than $\pi/2$. It should be noted that for $R = 0$ this advance is $\delta = \pi/2$.

The plots of the dependences $U_C(t)$ and $I(t)$ have the form similar to that shown in Fig. 11.3 for $q(t)$.

Example. An oscillatory circuit contains a capacitor of capacitance C and a coil with resistance R and inductance L. Find the ratio between the energies of the magnetic and electric fields in the circuit at the moment when the current is at a maximum.

According to equation (11.3) for an oscillatory circuit,

$$L\frac{dI}{dt} + RI + \frac{q}{C} = 0.$$

When the current is at a maximum, $dI/dt = 0$, and $RI = -q/C$. Hence the required ratio is

$$W_m/W_e = L/CR^2.$$

Quantities Characterizing Damping.

1. *Damping factor* β and *relaxation time* τ, viz. the time during which the amplitude of oscillations decreases by a factor of e. It can be easily seen from formula (11.11) that

$$\tau = 1/\beta. \qquad (11.17)$$

2. *Logarithmic decrement* λ *of damping.* It is defined as the Naperian logarithm of two successive values of amplitudes measured in a period T of oscillations:

$$\lambda = \ln\frac{a(t)}{a(t+T)} = \beta T, \qquad (11.18)$$

where a is the amplitude of the corresponding quantity (q, U, or I). In a different form, we can write

$$\lambda = 1/N_e, \qquad (11.19)$$

where N_e is the number of oscillations during the time τ, i.e. the time required for the amplitude of oscillations to decrease to 1/e of its initial value. This expression can be easily obtained from formulas (11.17) and (11.18).

If damping is small ($\beta^2 \ll \omega_0^2$), then $\omega \approx \omega_0 = 1/\sqrt{LC}$ and, in accordance with (11.18), we have

$$\lambda \approx \beta \cdot 2\pi/\omega_0 = \pi R\sqrt{C/L}. \tag{11.20}$$

3. *Q-factor* of an oscillatory circuit. By definition,

$$Q = \pi/\lambda = \pi N_e, \tag{11.21}$$

where λ is the logarithmic decrement. The smaller the damping, the higher the Q-factor. For a weak damping ($\beta^2 \ll \ll \omega_0^2$), in accordance with (11.20), we have

$$Q \approx \frac{1}{R}\sqrt{\frac{L}{C}} \tag{11.22}$$

Here is one more useful formula for the Q-factor in the case of a weak damping:

$$Q \approx 2\pi \frac{W}{\delta W}, \tag{11.23}$$

where W is the energy stored in the circuit and δW is the decrease in this energy during the period T of oscillations. Indeed, the energy W is proportional to the square of the amplitude value of the capacitor charge, i.e. $W \propto e^{-2\beta t}$. Hence the relative decrease in the energy over the period is $\delta W/W = 2\beta T = 2\lambda$. It remains for us to take into account the fact that according to (11.21), $\lambda = \pi/Q$.

Concluding the section, it should be noted that when $\beta^2 \geqslant \omega_0^2$, an *aperiodic* discharge of the capacitor will occur instead of oscillations. The resistance of the circuit for which the aperiodic process sets in is called the *critical* resistance:

$$R_{cr} = 2\sqrt{L/C}. \tag{11.24}$$

Let us consider two examples.

Example 1. An oscillatory circuit has a capacitance C, inductance L, and resistance R. Find the number of oscillations after which the amplitude of current decreases to 1/e of its initial value.

The current amplitude ($I_m \propto e^{-\beta t}$) decreases to 1/e of its initial value during the time $\tau = 1/\beta$. During this time, N_e oscillations will occur. If T is the period of damped oscillations, then

$$N_e = \frac{\tau}{T} = \frac{1/\beta}{2\pi/\sqrt{\omega_0^2-\beta^2}} = \frac{1}{2\pi}\sqrt{\left(\frac{\omega_0}{\beta}\right)^2-1}.$$

Considering that $\omega_0^2 = 1/LC$ and $\beta = R/2L$, we obtain

$$N_e = \frac{1}{2\pi}\sqrt{\frac{4L}{CR^2}-1}.$$

Example 2. Find the time during which the amplitude of current oscillations in a circuit with a given Q-factor will decrease to $1/\eta$ of its initial value, if the frequency of damped oscillations is equal to ω.

Since the current amplitude $I_m \propto e^{-\beta t}$, the time t_0 during which the amplitude decreases by a factor of η is determined by the equation $\eta = e^{\beta t_0}$. Hence

$$t_0 = (\ln \eta)/\beta.$$

On the other hand, the Q-factor is also related to β:

$$Q = \pi/\lambda = \pi/\beta T = \omega/2\beta.$$

Eliminating β from the last two equations, we obtain

$$t_0 = \frac{2Q}{\omega}\ln \eta.$$

11.3. Forced Electric Oscillations

Steady-state Oscillations. Let us return to equations (11.3) and (11.4) for an oscillatory circuit and consider the case when the circuit includes a varying external e.m.f. $\mathscr{E}$ whose dependence on time is decreased by the harmonic law:

$$\mathscr{E} = \mathscr{E}_m \cos \omega t. \tag{11.25}$$

This law occupies a special place owing to the properties of the oscillatory circuit itself to retain a harmonic form of oscillations under the action of external harmonic e.m.f.

In this case, the equation for an oscillatory circuit is written in the form

$$L\frac{dI}{dt} + RI + \frac{q}{C} = \mathscr{E}_m \cos \omega t, \tag{11.26}$$

or

$$\ddot{q} + 2\beta\dot{q} + \omega_0^2 q = (\mathscr{E}_m/L)\cos \omega t. \tag{11.27}$$

As is known from mathematics, the solution of this equation is the sum of the general solution of the homogeneous equation (with the zero right-hand side) and a particular solution of the nonhomogeneous equation.

We shall be interested only in steady-state oscillations, i.e. in the particular solution of this equation (the general solution of the homogeneous equation exponentially attenuates and after the elapse of a certain time it virtually vanishes). It can be easily seen that this solution has the form

$$q = q_m \cos(\omega t - \psi), \tag{11.28}$$

where q_m is the amplitude of charge on the capacitor and ψ is the phase shift between oscillations of the charge and of the external e.m.f. $\mathscr{E}$ (11.25). It will be shown that q_m and ψ depend only on the properties of the circuit itself and the driving e.m.f. $\mathscr{E}$. It turns out that $\psi > 0$, and hence q always *lags behind* $\mathscr{E}$ in phase.

In order to determine the constants q_m and ψ, we must substitute (11.28) into the initial equation (11.27) and transform the result. However, for the sake of simplicity, we shall proceed in a different way: first find current I and then substitute its expression into (11.26). By the way, we shall solve the problem of determining the constants q_m and ψ.

Differentiation of (11.28) with respect to time t gives

$$I = -\omega q_m \sin(\omega t - \psi) = \omega q_m \cos(\omega t - \psi + \pi/2).$$

Let us write this expression as follows:

$$I = I_m \cos(\omega t - \varphi), \tag{11.29}$$

where I_m is the current amplitude and φ is the phase shift between the current and external e.m.f. $\mathscr{E}$,

$$I_m = \omega q_m, \quad \varphi = \psi - \pi/2. \tag{11.30}$$

We aim at finding I_m and φ. For this purpose, we proceed as follows. Let us represent the initial equation (11.26) in the form

$$U_L + U_R + U_C = \mathscr{E}_m \cos \omega t, \tag{11.31}$$

where the left-hand side is the sum of voltages across induction coil L, capacitor C and resistor R. Thus, we see that at each instant of time, the sum of these voltages is equal to the external e.m.f. $\mathscr{E}$. Taking into account relation (11.30), we write

$$U_R = RI = RI_m \cos(\omega t - \varphi), \qquad (11.32)$$

$$U_C = \frac{q}{C} = \frac{q_m}{C} \cos(\omega t - \psi) = \frac{I_m}{\omega C} \cos\left(\omega t - \varphi - \frac{\pi}{2}\right), \qquad (11.33)$$

$$U_L = L\frac{dI}{dt} = -\omega L I_m \sin(\omega t - \varphi)$$

$$= \omega L I_m \cos\left(\omega t - \varphi + \frac{\pi}{2}\right). \qquad (11.34)$$

Vector Diagram. The last three formulas show that U_R is *in phase* with current I, U_C *lags behind* I by $\pi/2$, and U_L

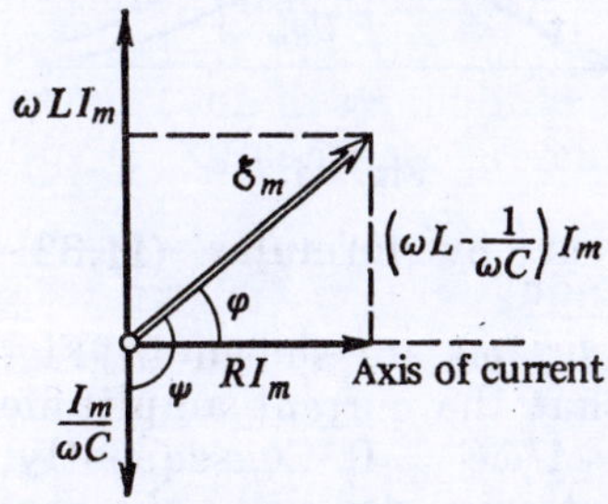

Fig. 11.4

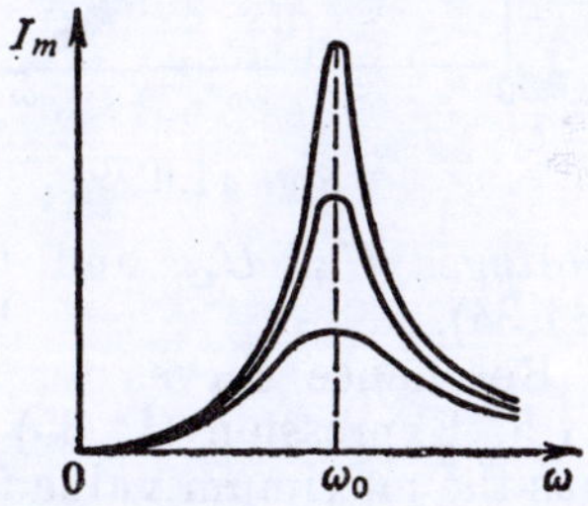

Fig. 11.5

leads I by $\pi/2$. This can be visually represented with the help of a *vector diagram*, depicting the amplitudes of voltages

$$U_{Rm} = RI_m, \quad U_{Cm} = I_m/\omega C, \quad U_{Lm} = \omega L I_m$$

and their vector sum which, according to (11.31), is equal to vector $\mathscr{E}_m$ (Fig. 11.4).

Considering the right triangle of this diagram, we can easily obtain the following expressions for I_m and φ:

$$I_m = \frac{\mathscr{E}_m}{\sqrt{R^2 + (\omega L - 1/\omega C)^2}}, \qquad (11.35)$$

$$\tan\varphi = \frac{\omega L - 1/\omega C}{R}. \qquad (11.36)$$

Thus, the problem is solved.

It should be noted that the vector diagram obtained above

proves to be very convenient for solving many specific problems. It permits a visual, easy, and rapid analysis of various situations.

Resonance Curves. This is the name given to the plots of dependences of the following quantities on the frequency ω of external e.m.f. $\mathscr{E}$: current I, charge q on a capacitor, and

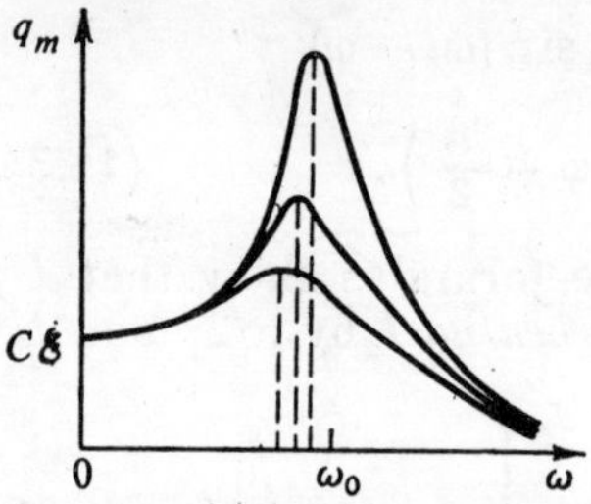

Fig. 11.6

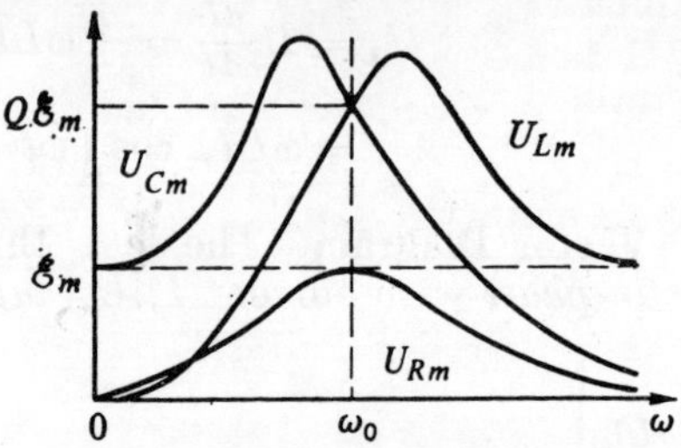

Fig. 11.7

voltages U_R, U_C, and U_L defined by formulas (11.32)-(11.34).

Resonance curves for current $I_m(\omega)$ are shown in Fig. 11.5. Expression (11.35) shows that the current amplitude has the maximum value for $\omega L - 1/\omega C = 0$. Consequently, the resonance frequency for current coincides with the natural frequency of the oscillatory circuit:

$$\omega_{I\,\text{res}} = \omega_0 = 1/\sqrt{LC}. \tag{11.37}$$

The maximum at resonance is the higher and the sharper the smaller the damping factor $\beta = R/2L$.

Resonance curves for the charge $q_m(\omega)$ on a capacitor are shown in Fig. 11.6 (resonance curves for voltage U_{Cm} across the capacitor have the same shape). The maximum of charge amplitude is attained at the resonance frequency

$$\omega_{q\,\text{res}} = \sqrt{\omega_0^2 - 2\beta^2}, \tag{11.38}$$

which comes closer and closer to ω_0 with decreasing β. In order to obtain (11.38), we must represent q_m, in accordance with (11.30), in the form $q_m = I_m/\omega$ where I_m is defined by (11.35). Then

$$q_m = \frac{\mathscr{E}_m/L}{\sqrt{(\omega_0^2-\omega^2)^2+4\beta^2\omega^2}} \tag{11.39}$$

The maximum of this function or, which is the same, the minimum of the radicand can be found by equating to zero the derivative of the radicand with respect to ω. This gives the resonance frequency (11.38).

Let us now see how the amplitudes of voltages U_R, U_C, and U_L are redistributed depending on the frequency ω of the external e.m.f. This pattern is depicted in Fig. 11.7.

The resonance frequencies for U_R, U_C and U_L are determined by the following formulas:

$$\omega_{R\text{ res}} = \omega_0,$$
$$\omega_{C\text{ res}} = \omega_0 \sqrt{1 - 2(\beta/\omega_0)^2}, \tag{11.40}$$
$$\omega_{L\text{ res}} = \omega_0 \sqrt{1 - 2(\beta/\omega_0)^2}.$$

The smaller the value of β, the closer are the resonance frequencies for all quantities to the value ω_0.

Resonance Curves and Q-factor. The shape of resonance curves is connected in a certain way with the Q-factor of an oscillatory circuit. This connection has the simplest form for the case of weak damping, i.e. for $\beta^2 \ll \omega_0^2$. In this case,

$$U_{C\text{res}}/\mathscr{E}_m = Q \tag{11.41}$$

(Fig. 11.7). Indeed, for $\beta^2 \ll \omega_0^2$, the quantity $\omega_{\text{res}} \simeq \omega_0$ and, according to (11.33) and (11.35), $U_{Cm\text{ res}} = I_m/\omega_0 C = \mathscr{E}_m/\omega_0 CR$, or $U_{Cm\text{ res}}/\mathscr{E}_m = \sqrt{LC}/CR = (1/R)\sqrt{L/C}$ which is, in accordance with formula (11.22), just the Q-factor.

Thus, the Q-factor of a circuit (for $\beta^2 \ll \omega_0^2$) shows how many times the maximum value of the voltage amplitude across the capacitor (and induction coil) exceeds the amplitude of the external e.m.f.

The Q-factor of a circuit is also connected with another important characteristic of the resonance curve, viz. its width. It turns out that for $\beta^2 \ll \omega_0^2$

$$Q = \omega_0/\delta\omega, \tag{11.42}$$

where ω_0 is the resonance frequency and $\delta\omega$ is the width of the resonance curve at a "height" equal to 0.7 of the peak height, i.e. at resonance.

Resonance. Resonance in the case under consideration is

the excitation of strong oscillations at the frequency of external e.m.f. or voltage. This frequency is equal to the natural frequency of an oscillatory circuit. Resonance is used for singling out a required component from a composite voltage. The entire radio reception technique is based on resonance. In order to receive with a given radio receiver the station we are interested in, the receiver must be tuned. In other words, by varying C and L of the oscillatory circuit, we must attain the coincidence between its natural frequency and the frequency of radio waves emitted by the radio station.

The phenomenon of resonance is also associated with a certain *danger*: the external e.m.f. or voltage may be small, but the voltages across individual elements of the circuit (the capacitor or induction coil) may attain the values dangerous for people. This should always be remembered.

11.4. Alternating Current

Total Resistance (Impedance). Steady-state forced electric oscillations can be treated as an alternating current flowing in a circuit having a capacitance, inductance, and resistance. Under the action of external voltage (which plays the role of external e.m.f.)

$$U = U_m \cos \omega t, \tag{11.43}$$

the current in the circuit varies according to the law

$$I = I_m \cos (\omega t - \varphi), \tag{11.44}$$

where

$$I_m = \frac{U_m}{R^2 + (\omega L - 1/\omega C)^2}, \qquad \tan \varphi = \frac{\omega L - 1/\omega C}{R}. \tag{11.45}$$

The problem is reduced to determining the current amplitude and the phase shift of the current relative to U.

The obtained expression for the current amplitude $I_m(\omega)$ can be formally interpreted as Ohm's law for the amplitude values of current and voltage. The quantity in the denominator of this expression, which has the dimension of resistance, is denoted by Z and is called the *total resistance*, or *impedance*:

$$Z=\sqrt{R^2+(\omega L-1/\omega C)^2}. \quad (11.46)$$

It can be seen that for $\omega = \omega_0 = 1/\sqrt{LC}$, the impedance has the minimum value and is equal to the resistance R. The quantity appearing in the parentheses in formula (11.46) is denoted by X and is called the *reactance*:

$$X = \omega L - 1/\omega C. \quad (11.47)$$

Here the quantity ωL is called the *inductive reactance*, while the quantity $1/\omega C$ is called the *capacitive reactance*. They are denoted X_L and X_C respectively. Thus,

$$X_L=\omega L, \quad X_C=1/\omega C, \quad X=X_L-X_C, \quad Z=\sqrt{R^2+X^2}. \quad (11.48)$$

It should be noted that the inductive reactance grows with the frequency ω, while the capacitive reactance decreases with increasing ω. When it is said that a circuit has no capacitance, this must be understood so that there is no capacitive reactance which is equal to $1/\omega C$ and hence vanishes if $C \to \infty$ (when a capacitor is replaced by a short-circuited section).

Finally, although the reactance is measured in the same units as the resistance, they differ in principle. This difference consists in that only the resistance determines irreversible processes in a circuit such as, for example, the conversion of electromagnetic energy into Joule's heat.

Power Liberated in an A.C. Circuit. The instantaneous value of power is equal to the product of instantaneous values of voltage and current:

$$P(t) = UI = U_m I_m \cos \omega t \cos (\omega t - \varphi). \quad (11.49)$$

Using the formula $\cos(\omega t - \varphi) = \cos \omega t \cos \varphi + \sin \omega t \sin \varphi$, we transform (11.49) as follows:

$$P(t)=U_m I_m(\cos^2 \omega t \cos \varphi+\sin \omega t \cos \omega t \sin \varphi).$$

Of practical importance is the value of power averaged over a period of oscillation. Considering that $\langle \cos^2 \omega t \rangle = 1/2$ and $\langle \sin \omega t \cos \omega t \rangle = 0$, we obtain

$$\langle P \rangle=\frac{U_m I_m}{2}\cos \varphi. \quad (11.50)$$

This expression can be written in a different form if we take into account that, as follows from the vector diagram (see Fig. 11.4), $U_m \cos\varphi = RI_m$. Hence,

$$\langle P\rangle = \frac{1}{2} RI_m^2. \tag{11.51}$$

This power is equal to that of the direct current $I = I_m/\sqrt{2}$. The quantities

$$I = I_m/\sqrt{2}, \quad U = U_m/\sqrt{2} \tag{11.52}$$

are called *effective* (root-mean-square) values of current and voltage. All ammeters and voltmeters are graduated for r.m.s. values of current and voltage.

The expression for the average power (11.50) in terms of r.m.s. values of current and voltage has the form

$$\langle P\rangle = UI\cos\varphi, \tag{11.53}$$

where the factor $\cos\varphi$ is usually called the *power factor*. Thus, the power developed in an a.c. circuit depends not only on the voltage and current but also on the phase angle between them.

When $\varphi = \pi/2$, the value $\langle P\rangle = 0$ regardless of the values of U and I. In this case, the energy transmitted over a quarter of a period from a generator to an external circuit is exactly equal to the energy transmitted from the external circuit to the generator during the next quarter of the period so that the entire energy "oscillates" uselessly between the generator and the external circuit.

The dependence of the power on $\cos\varphi$ should be taken into account while designing transmission lines for alternating current. If loads being fed have a high reactance X, $\cos\varphi$ may be considerably less than one. In such cases, in order to transmit the required power to a consumer (for a given voltage of the generator), it is necessary to increase current I, which leads to an increase of useless energy losses in feeding wires. Hence, loads, inductances and capacitances should be always distributed so as to make $\cos\varphi$ as close to one as possible. For this purpose, it is sufficient to make reactance X as small as possible, i.e. to ensure the equality of inductive and capacitive reactances ($X_L = X_C$).

When the conductors forming an a.c. circuit are in motion,

the concept of electric resistance becomes wider since in addition to the conversion of electric energy into Joule's heat, other types of energy transformations are also possible. For example, a part of electric energy can be converted into mechanical work (electric motors).

Problems

● **11.1. Free undamped oscillations.** Free undamped oscillations occur in an oscillatory circuit consisting of a capacitor with capacitance C and an induction coil with inductance L. The voltage amplitude on the capacitor is U_m. Find the e.m.f. of self-induction in the coil at the moments when its magnetic energy is equal to the electric energy of the capacitor.

Solution. According to Ohm's law,

$$RI = U + \mathscr{E}_s$$

where U is the voltage across the capacitor ($U = \varphi_1 - \varphi_2$). In our case, $R = 0$ and hence $\mathscr{E}_s = -U$.

It remains for us to find voltage U at the moments when the electric energy of the capacitor is equal to the magnetic energy of the coil. Under this condition, we can write

$$\frac{CU_m^2}{2} = \frac{CU^2}{2} + \frac{LI^2}{2} = 2\,\frac{CU^2}{2},$$

whence $|U| = U_m/\sqrt{2}$.

As a result, we have $|\mathscr{E}_s| = U_m/\sqrt{2}$.

● **11.2.** An oscillatory circuit consists of an induction coil with inductance L and an uncharged capacitor of capacitance C. The resistance of the circuit $R = 0$. The coil is in a permanent magnetic field. The total magnetic flux piercing all the turns of the coil is Φ. At the moment $t = 0$, the magnetic field was abruptly switched off. Find the current in the circuit as a function of time t.

Solution. Upon an abrupt switching off of the external magnetic field at the moment $t = 0$, an induced current appears, but the capacitor still remains uncharged. In accordance with Ohm's law, we have

$$RI = -\frac{d\Phi}{dt} - L\,\frac{dI}{dt}.$$

In the given case $R = 0$, and hence $\dot{\Phi} + L\dot{I} = 0$. This gives $\Phi = LI_0$, where I_0 is the initial current (immediately after switching off the field).

After the external field has been switched off, the process is described by the following equation:

$$0 = -\frac{q}{C} - L\,\frac{dI}{dt}. \tag{1}$$

Differentiation of this equation with respect to time gives

$$\ddot{I}+\frac{1}{LC}I=0.$$

This is the equation of harmonic oscillations. We seek its solution in the form

$$I=I_m\cos(\omega_0 t+\alpha).$$

The constants I_m and α can be found from the initial conditions

$$I(0)=I_0,\quad \dot{I}(0)=0$$

(the second condition follows from Eq. (1), since at the initial moment $t=0$ the capacitor was uncharged). From these conditions we find $\alpha=0$ and $I_m=I_0$. As a result, we obtain

$$I=I_0\cos\omega_0 t=(\Phi/L)\cos\omega_0 t,$$

where $\omega_0=1/\sqrt{LC}$.

● **11.3.** *Q*-**factor of a circuit.** An oscillatory circuit with a low damping has a capacitance C and inductance L. In order to sustain in it undamped harmonic oscillations with the voltage amplitude U_m across the capacitor, it is necessary to supply the average power $\langle P\rangle$. Find the Q-factor of the circuit.

Solution. Since damping is low, we can make use of formula (11.23):

$$Q=2\pi W/\delta W,\tag{1}$$

where $W=CU_m^2/2$ and $\delta W=\langle P\rangle T$, T being the period of damped oscillations. In our case, $T\simeq T_0=2\pi\sqrt{LC}$. Having substituted these expressions into (1), we obtain

$$Q=\frac{U_m^2}{2\langle P\rangle}\sqrt{\frac{C}{L}}.$$

● **11.4. Damped oscillations.** An oscillatory circuit includes a capacitor of capacitance C, an induction coil of inductance L, a resistor of resistance R, and a key. The capacitor was charged with the key open. When the key was closed, a current began to flow. Find the ratio of the voltage across the capacitor at time t to the voltage at the initial moment (immediately after closing the key).

Solution. The voltage across the capacitor depends on time in the same way as the charge does. Hence we can write

$$U=U_m e^{-\beta t}\cos(\omega t+\alpha).\tag{1}$$

At the initial moment $t=0$, the voltage $U(0)=U_m\cos\alpha$, where U_m is the amplitude at this moment. We must find $U(0)/U_m$, i.e. $\cos\alpha$.

For this purpose, we shall use another initial condition: at the moment $t=0$, current $I=\dot{q}=0$. Since $q=CU$, it is sufficient to differentiate (1) with respect to time and equate the obtained expression to zero at $t = 0$. We obtain $-\beta \cos \alpha - \omega \sin \alpha = 0$, whence $\tan \alpha = -\beta/\omega$. The required ratio is

$$\frac{U(0)}{U_m} = \cos \alpha = \frac{1}{\sqrt{1+\tan^2 \alpha}} = \frac{1}{\sqrt{1+(\beta/\omega)^2}}. \qquad (2)$$

The quantities U_m and $U(0)$ are shown in Fig. 11.8.

Considering that $\omega^2 = \omega_0^2 - \beta^2$, we transform (2) as follows:

$$U(0)/U_m = \sqrt{1-(\beta/\omega_0)^2} = \sqrt{1-R^2C/4L},$$

where we took into account that $\beta = R/2L$ and $\omega_0^2 = 1/LC$.

● **11.5.** In an oscillatory circuit with a capacitance C and induc-

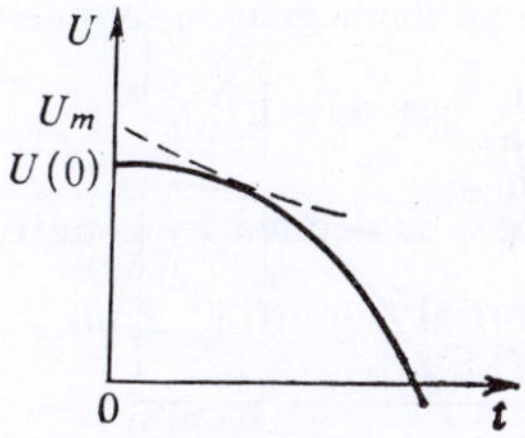

Fig. 11.8

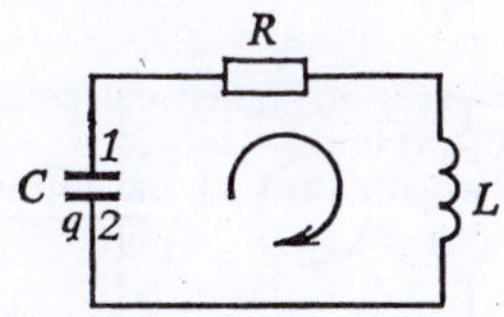

Fig. 11.9

tance L damped oscillations occur, in which the current varies with time in accordance with the law $I(t) = I_m e^{-\beta t} \sin \omega t$. Find the voltage across the capacitor as a function of time.

Solution. Let us choose the clockwise direction as the positive direction of circumvention (Fig. 11.9). According to Ohm's law for section *1RL2* of the circuit, we have $RI = \varphi_1 - \varphi_2 + \mathscr{E}_s$. In our case, $\mathscr{E}_s = -L\dot{I}$ and $\varphi_2 - \varphi_1 = q/C = U_C$, where q is the charge on plate *2*. Hence Ohm's law can be written as

$$U_C = -RI - L\dot{I}.$$

Having substituted into this formula the expression for $I(t)$ and its derivative, we obtain

$$U_C = \frac{RI_m e^{-\beta t}}{2\beta}(-\beta \sin \omega t - \omega \cos \omega t).$$

Let us transform the expression in the parentheses to sine. For this purpose, we multiply and divide it by $\sqrt{\omega^2+\beta^2} = \omega_0$ and introduce angle δ through the formulas

$$-\beta/\omega_0 = \cos \delta, \quad \omega/\omega_0 = \sin \delta. \qquad (1)$$

Then

$$U_C = \frac{RI_m\omega_0}{2\beta} e^{-\beta t} \sin(\omega t - \delta) = I_m \sqrt{L/C} e^{-\beta t} \sin(\omega t - \delta),$$

where angle δ is, in accordance with (1), in the second quadrant, i.e. assumes the values $\pi/2 < \delta < \pi$. Thus, the voltage across the capacitor lags behind the current in phase.

● **11.6. Steady-state oscillations.** An induction coil of inductance L and resistance R was connected at the moment $t = 0$ to a source of external voltage $U = U_m \cos \omega t$. Find the current in the circuit as a function of time.

Solution. In our case, $RI = U - L\dot{I}$, or

$$\dot{I} + (R/L) I = U_m \cos \omega t.$$

The solution of this equation is the general solution of the homogeneous equation plus the particular solution of the nonhomogeneous equation:

$$I(t) = Ae^{-(R/L)t} = \frac{U_m}{\sqrt{R^2 + \omega^2 L^2}} \cos(\omega t - \varphi),$$

where A is an arbitrary constant and angle φ is defined by condition (11.36): $\tan \varphi = \omega L/R$.

The constant A is found from the initial condition $I(0) = 0$. Hence $A = -(U_m/\sqrt{R^2 + \omega^2 L^2}) \cos \varphi$. This gives

$$I(t) = \frac{U_m}{\sqrt{R^2 + \omega^2 L^2}} [\cos(\omega t - \varphi) - e^{-(R/L)t} \cos \varphi].$$

For a sufficiently large t, the second term in the brackets becomes negligibly small, and we obtain the steady-state solution $I(t) \propto \cos(\omega t - \varphi)$.

● **11.7. Forced oscillations.** A section of a circuit consisting of a series-connected capacitor and resistor R, is connected to a source of varying voltage with the amplitude U_m. The amplitude of the steady-state current turned out to be I_m. Find the phase angle between the current and external voltage.

Solution. In the case under consideration,

$$U = U_m \cos \omega t, \quad I = I_m \cos(\omega t - \varphi),$$

where φ is defined by formula (11.36): $\tan \varphi = -1/\omega CR$.

The unknown value of capacitance C can be found from the expression for the current amplitude: $I_m = U_m/\sqrt{R^2 + (1/\omega C)^2}$, whence

$$C = 1/\omega \sqrt{(U_m/I_m)^2 - R^2}.$$

Having substituted this expression into formula for $\tan \varphi$, we obtain

$$\tan \varphi = -\sqrt{(U_m/RI_m)^2 - 1}.$$

In our case $\varphi < 0$, which means that the current leads the external voltage (Fig. 11.10).

● **11.8.** An a.c. circuit, containing a series-connected capacitor and induction coil with a certain resistance, is connected to the source

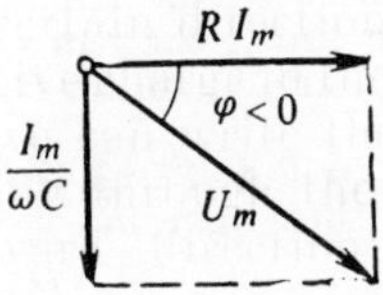

Fig. 11.10

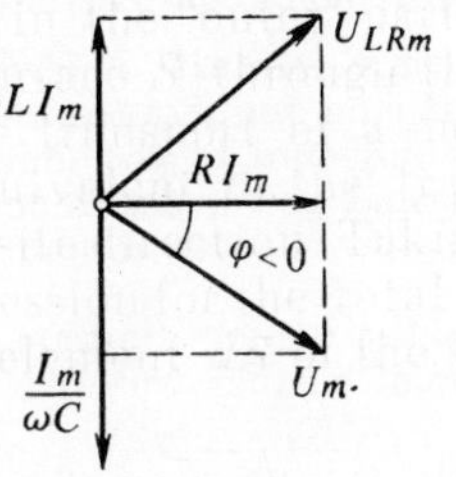

Fig. 11.11

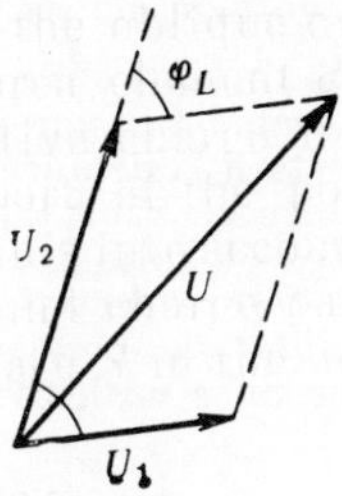

Fig. 11.12

of external alternating voltage whose frequency can be altered without changing its amplitude. At frequencies ω_1 and ω_2 the current amplitudes in the circuit proved to be the same. Find the resonance frequency of the current.

Solution. According to (11.35), the amplitudes are equal under the condition

$$\left| \omega_1 L - \frac{1}{\omega_1 C} \right| = \left| \omega_2 L - \frac{1}{\omega_2 C} \right|. \tag{1}$$

The maximum of the resonance curve for current corresponds to the frequency equal to the natural frequency $\omega_0 = 1/\sqrt{LC}$. Further, let $\omega_1 < \omega_0 < \omega_2$ (the opposite inequality is also possible, it will not affect the final result). Equality (1) can then be written in a more general form: $\omega_0^2/\omega_1 - \omega_1 = \omega_2 - \omega_0^2/\omega_2$, or

$$\omega_2 - \omega_1 = \omega_0^2 \left(\frac{1}{\omega_1} - \frac{1}{\omega_2} \right).$$

Cancelling out $\omega_2 - \omega_1$ on both sides of this equality, we obtain $1 = \omega_0^2/\omega_1\omega_2$, whence

$$\omega_0 = \sqrt{\omega_1 \omega_2}.$$

● **11.9. Vector diagram.** A circuit consisting of a series-connected capacitor of capacitance C and induction coil with resistance R and inductance L is connected to an external voltage with the amplitude U_m and frequency ω. Assuming that current in the circuit leads in phase the external voltage, construct the vector diagram and use it for determining the amplitude of voltage across the coil.

Solution. The vector diagram for the case under consideration is shown in Fig. 11.11. It is readily seen from this diagram that the amplitude of voltage across the coil is

$$U_{LRm}=I_m\sqrt{R^2+\omega^2L^2},$$

where $I_m = U_m/\sqrt{R^2+(\omega L-1/\omega C)^2}$. In the presence of resistance the voltage across the coil leads the current by less than $\pi/2$.

●**11.10. Power in an a.c. circuit.** A circuit consisting of a series-connected resistor R with no inductance and an induction coil with a certain resistance is connected to the main system with the r.m.s. voltage U. Find the thermal power developed in the coil if the r.m.s. values of voltage across the resistor R and the coil are equal to U_1 and U_2 respectively.

Solution. Let us use the vector diagram shown in Fig. 11.12. In accordance with the law of cosines we obtain from this diagram

$$U^2=U_1^2+U_2^2-2U_1U_2\cos\varphi_L. \tag{1}$$

The power developed in the coil is

$$P_2=IU_2\cos\varphi_L, \tag{2}$$

where $I=U_1/R$.

Combining Eqs. (1) and (2), we obtain

$$P_2=(U^2-U_1^2-U_2^2)/2R.$$

Appendices

1. Notations for Units of Measurement

A	ampere	min	minute
C	coulomb	Mx	maxwell
eV	electronvolt	N	newton
F	farad	Oe	oersted
G	gauss	Ω	ohm
g	gram	rad	radian
H	henry	s	second
hr	hour	S	siemens
Hz	hertz	T	tesla
J	joule	V	volt
K	kelvin	W	watt
m	meter	Wb	weber

2. Decimal Prefixes for Units of Measurement

T,	thera-,	10^{12}	d,	deci-,	10^{-1}	n,	nano,	10^{-9}
G,	giga-,	10^{9}	c,	centi-,	10^{-2}	p,	pico-,	10^{-12}
M,	mega-,	10^{6}	m,	milli-,	10^{-3}			
k,	kilo-,	10^{3}	μ,	micro-,	10^{-6}			
h,	hecto-,	10^{2}						
da,	deca-,	10^{1}						

3. Units of Measurement of Electric and Magnetic Quantities in SI and Gaussian Systems

Quantity	Notation	Unit of measurement		Ratio $\frac{\text{SI unit}}{\text{CGS unit}}$
		SI	CGS	
Force	F	N	dyne	10^{5}
Work, energy	A, W	J	erg	10^{7}
Charge	q	C	CGSE unit	3×10^{9}

Table 3. Cont.

Quantity	Notation	Unit of measurement		Ratio $\frac{\text{SI unit}}{\text{CGS unit}}$
		SI	CGS	
Electric field strength	E	V/m	CGSE unit	$1/(3\times10^4)$
Potential, voltage	φ, U	V	CGSE unit	1/300
Electric moment	p	C·m	CGSE unit	3×10^{11}
Polarization	P	C/m^2	CGSE unit	3×10^5
Vector **D**	D	C/m^2	CGSE unit	$12\pi\times10^5$
Capacitance	C	F	cm	9×10^{11}
Current	I	A	CGSE unit	3×10^9
Current density	j	A/m^2	CGSE unit	3×10^5
Resistance	R	Ω	CGSE unit	$1/(9\times10^{11})$
Resistivity	ρ	Ω·m	CGSE unit	$1/(9\times10^9)$
Conductance	Σ	S	CGSE unit	9×10^{11}
Conductivity	σ	S/m	CGSE unit	9×10^9
Magnetic induction	B	T	G	10^4
Magnetic flux, magnetic-flux linkage	Φ, Ψ	Wb	Mx	10^8
Magnetic moment	p_m	$A\cdot m^2$	CGSE unit	10^3
Magnetization	J	A/m	CGSE unit	10^{-3}
Vector **H**	H	A/m	Oe	$4\pi\times10^{-3}$
Inductance	L	H	cm	10^9

4. Basic Formulas of Electricity and Magnetism in SI and Gaussian Systems

Relation	SI	Gaussian system
Field E of a point charge	$E=\frac{1}{4\pi\varepsilon_0}\frac{q}{r^2}$	$E=\frac{q}{r^2}$
Field E in a parallel-plate capacitor and at the surface of a conductor	$E=\frac{\sigma}{\varepsilon_0\varepsilon}$	$E=\frac{4\pi\sigma}{\varepsilon}$
Potential of the field of a point charge	$\varphi=\frac{1}{4\pi\varepsilon_0}\frac{q}{r}$	$\varphi=\frac{q}{r}$
Relation between **E** and φ	$\mathbf{E}=-\nabla\cdot\varphi,\ \varphi_1-\varphi_2=\int_1^2\mathbf{E}\,d\mathbf{l}$	

Table 4. Cont.

Relation	SI	Gaussian system
Circulation of vector E in an electrostatic field	$\oint \mathbf{E}\, d\mathbf{l}=0$	$\oint \mathbf{E}\, d\mathbf{l}=0$
Electric moment of a dipole	$\mathbf{p}=q\mathbf{l}$	$\mathbf{p}=q\mathbf{l}$
Electric dipole p in field E	$\mathbf{F}=p\dfrac{\partial \mathbf{E}}{\partial l}$, $\mathbf{M}=[\mathbf{p}\times\mathbf{E}]$, $W=-\mathbf{p}\cdot\mathbf{E}$	$\mathbf{F}=p\dfrac{\partial \mathbf{E}}{\partial l}$, $\mathbf{M}=[\mathbf{p}\times\mathbf{E}]$, $W=-\mathbf{p}\cdot\mathbf{E}$
Relation between polarization and field strength	$\mathbf{P}=\varkappa\varepsilon_0\mathbf{E}$	$\mathbf{P}=\varkappa\mathbf{E}$
Relation between σ′, P and E	$\sigma'=P_n=\varkappa\varepsilon_0 E_n$	$\sigma'=P_n=\varkappa E_n$
Definition of vector D	$\mathbf{D}=\varepsilon_0\mathbf{E}+\mathbf{P}$	$\mathbf{D}=\mathbf{E}+4\pi\mathbf{P}$
Relation between ε and ϰ	$\varepsilon=1+\varkappa$	$\varepsilon=1+4\pi\varkappa$
Relation between D and E	$\mathbf{D}=\varepsilon\varepsilon_0\mathbf{E}$	$\mathbf{D}=\varepsilon\mathbf{E}$
Gauss theorem for vector D	$\oint \mathbf{D}\, d\mathbf{S}=q$	$\oint \mathbf{D}\, d\mathbf{S}=4\pi q$
Capacitance of a capacitor	$C=q/U$	$C=q/U$
Capacitance of a parallel-plate capacitor	$C=\dfrac{\varepsilon\varepsilon_0 S}{h}$	$C=\dfrac{\varepsilon S}{4\pi h}$
Energy of system of charges	$W=\dfrac{1}{2}\sum q_i\varphi_i$	$W=\dfrac{1}{2}\sum q_i\varphi_i$
Total energy of interaction	$W=\dfrac{1}{2}\int \rho\varphi\, dV$	$W=\dfrac{1}{2}\int \rho\varphi\, dV$
Energy of capacitor	$W=qU/2=CU^2/2=q^2/2C$	$W=qU/2=CU^2/2=q^2/2C$
Electric field energy density	$w=\dfrac{\mathbf{E}\cdot\mathbf{D}}{2}$	$w=\dfrac{\mathbf{E}\cdot\mathbf{D}}{8\pi}$
Continuity equation	$\int \mathbf{j}\, d\mathbf{S}=-\dfrac{dq}{dt}$	$\nabla\cdot\mathbf{j}=-\dfrac{\partial\rho}{\partial t}$
Ohm's law	$RI=\varphi_1-\varphi_2+\mathscr{E}_{12}$	$\mathbf{j}=\sigma(\mathbf{E}+\mathbf{E}^*)$
Joule-Lenz law	$\dot{Q}=RI^2$	$\dot{Q}_{sp}=\rho j^2$
Lorentz force	$\mathbf{F}=q\mathbf{E}+q[\mathbf{v}\times\mathbf{B}]$	$\mathbf{F}=q\mathbf{E}+\dfrac{q}{c}[\mathbf{v}\times\mathbf{B}]$

Table 4. Cont.

Relation	SI	Gaussian system
Field **B** of a moving charge	$\mathbf{B}=\frac{\mu_0}{4\pi}\frac{q\,[\mathbf{v}\times\mathbf{r}]}{r^3}$	$\mathbf{B}=\frac{1}{c}\frac{q\,[\mathbf{v}\times\mathbf{r}]}{r^3}$
Biot-Savart law	$d\mathbf{B}=\frac{\mu_0}{4\pi}\frac{[\mathbf{j}\times\mathbf{r}]\,dV}{r^3}$	$d\mathbf{B}=\frac{1}{c}\frac{[\mathbf{j}\times\mathbf{r}]\,dV}{r^3}$
	$d\mathbf{B}=\frac{\mu_0}{4\pi}\frac{I\,[d\mathbf{l}\times\mathbf{r}]}{r^3}$	$d\mathbf{B}=\frac{1}{c}\frac{I[d\mathbf{l}\times\mathbf{r}]}{r^3}$
Field **B**		
(a) of straight current	$B=\frac{\mu_0}{4\pi}\frac{2I}{b}$	$B=\frac{1}{c}\frac{2I}{b}$
(b) at the centre of a loop	$B=\frac{\mu_0}{4\pi}\frac{2\pi I}{R}$	$B=\frac{1}{c}\frac{2\pi I}{R}$
(c) in solenoid	$B=\mu_0 nI$	$B=\frac{4\pi}{c}nI$
Ampère's law	$d\mathbf{F}=I\,[d\mathbf{l}\times\mathbf{B}]$	$d\mathbf{F}=\frac{I}{c}[d\mathbf{l}\times\mathbf{B}]$
	$d\mathbf{F}=[\mathbf{j}\times\mathbf{B}]\,dV$	$d\mathbf{F}=\frac{1}{c}[\mathbf{j}\times\mathbf{B}]\,dV$
Force of interaction between parallel currents	$F_{\text{u}}=\frac{\mu_0}{4\pi}\frac{2I_1I_2}{b}$	$F_{\text{u}}=\frac{2I_1I_2}{b}$
Magnetic moment of a current loop	$p_{\text{m}}=IS$	$p_{\text{m}}=\frac{1}{c}IS$
Magnetic dipole $\mathbf{p}_{\text{m}}$ in field **B**	$F=p_{\text{m}}\frac{\partial B}{\partial n}$,	$\mathbf{M}=[\mathbf{p}_{\text{m}}\times\mathbf{B}]$
Work done in displacement of a current loop	$A=I\,(\Phi_2-\Phi_1)$	$A=\frac{1}{c}I\,(\Phi_2-\Phi_1)$
Circulation of magnetization	$\oint \mathbf{J}\,d\mathbf{l}=I'$	$\oint \mathbf{J}\,d\mathbf{l}=\frac{1}{c}I'$
Definition of vector **H**	$\mathbf{H}=\mathbf{B}/\mu_0-\mathbf{J}$	$\mathbf{H}=\mathbf{B}-4\pi\mathbf{J}$
Circulation of vector **H** in a stationary field	$\oint \mathbf{H}\,d\mathbf{l}=I$	$\oint \mathbf{H}\,d\mathbf{l}=\frac{4\pi}{c}I$
Relations between		
J and **H**	$\mathbf{J}=\chi\mathbf{H}$	
μ and χ	$\mu=1+\chi$	$\mu=1+4\pi\chi$
B and **H**	$\mathbf{B}=\mu\mu_0\mathbf{H}$	$\mathbf{B}=\mu\mathbf{H}$

Table 4. Cont.

Relation	SI	Gaussian system
Laws of transformation of fields **E** and **B** for $v_0 \ll c$	$\mathbf{E}' = \mathbf{E} + [\mathbf{v}_0 \times \mathbf{B}]$	$\mathbf{E}' = \mathbf{E} + \frac{1}{c}[\mathbf{v}_0 \times \mathbf{B}]$
	$\mathbf{B}' = \mathbf{B} - \frac{1}{c^2}[\mathbf{v}_0 \times \mathbf{E}]$	$\mathbf{B}' = \mathbf{B} - \frac{1}{c}[\mathbf{v}_0 \times \mathbf{E}]$
Electromagnetic field invariants	$\mathbf{E}\cdot\mathbf{B} = \mathrm{inv}$	
	$E^2 - c^2B^2 = \mathrm{inv}$	$E^2 - B^2 = \mathrm{inv}$
Induced e.m.f.	$\mathscr{E}_i = -\frac{d\Phi}{dt}$	$\mathscr{E}_i = -\frac{1}{c}\frac{d\Phi}{dt}$
Inductance	$L = \Phi/I$	$L = c\Phi/I$
Inductance of a solenoid	$L = \mu\mu_0 n^2 V$	$L = 4\pi\mu n^2 V$
E.m.f. of self-induction	$\mathscr{E}_s = -L\frac{dI}{dt}$	$\mathscr{E}_s = -\frac{1}{c^2}L\frac{dI}{dt}$
Energy of the magnetic field of a current	$W = \frac{LI^2}{2}$	$W = \frac{1}{c}\frac{LI^2}{2}$
Magnetic field energy density	$w = \frac{\mathbf{B}\cdot\mathbf{H}}{2}$	$w = \frac{\mathbf{B}\cdot\mathbf{H}}{8\pi}$
Displacement current density	$\mathbf{j}_d = \frac{\partial\mathbf{D}}{\partial t}$	$\mathbf{j}_d = \frac{1}{4\pi}\frac{\partial\mathbf{D}}{\partial t}$
Maxwell's equations in integral form	$\oint \mathbf{E}\,d\mathbf{l} = -\int \dot{\mathbf{B}}\,d\mathbf{S}$	$\oint \mathbf{E}\,d\mathbf{l} = -\frac{1}{c}\int \dot{\mathbf{B}}\,d\mathbf{S}$
	$\oint \mathbf{D}\,d\mathbf{S} = \int \rho\,dV$	$\oint \mathbf{D}\,\partial\mathbf{S} = 4\pi\int \rho\,dV$
	$\oint \mathbf{H}\,d\mathbf{l} = \int (\mathbf{j} + \dot{\mathbf{D}})\,d\mathbf{S}$	$\oint \mathbf{H}\,d\mathbf{l} = \frac{4\pi}{c}\int\left(\mathbf{j} + \frac{\dot{\mathbf{D}}}{4\pi}\right)d\mathbf{S}$
	$\oint \mathbf{B}\,d\mathbf{S} = 0$	$\oint \mathbf{B}\,d\mathbf{S} = 0$
Maxwell's equations in differential form	$\nabla \times \mathbf{E} = -\dot{\mathbf{B}}$	$\nabla \times \mathbf{E} = -\frac{1}{c}\dot{\mathbf{B}}$
	$\nabla \cdot \mathbf{D} = \rho$	$\nabla \cdot \mathbf{D} = 4\pi\rho$
	$\nabla \times \mathbf{H} = \mathbf{j} + \dot{\mathbf{D}}$	$\nabla \times \mathbf{H} = \frac{4\pi}{c}\left(\mathbf{j} + \frac{\dot{\mathbf{D}}}{4\pi}\right)$
	$\nabla \cdot \mathbf{B} = 0$	$\nabla \cdot \mathbf{B} = 0$

Tabl. 4. Cont.

Relation	SI	Gaussian system
Relation between E and H in electromagnetic wave	$E\sqrt{\varepsilon_0\varepsilon}=H\sqrt{\mu\mu_0}$	$E\sqrt{\varepsilon}=H\sqrt{\mu}$
Poynting's vector	$\mathbf{S}=[\mathbf{E}\times\mathbf{H}]$	$\mathbf{S}=\frac{c}{4\pi}[\mathbf{E}\times\mathbf{H}]$
Density of electromagnetic field momentum	$\mathbf{G}=\frac{1}{c^2}[\mathbf{E}\times\mathbf{H}]$	$\mathbf{G}=\frac{1}{4\pi c}[\mathbf{E}\times\mathbf{H}$

5. Some Physical Constants

Velocity of light in vacuum	$c=2.998\times10^8$ m/s
Gravitational constant	$G=\begin{cases}6.67\times10^{-11}\ \text{m}^3/(\text{kg}\cdot\text{s}^2)\\ 6.67\times10^{-8}\ \text{cm}^3/(\text{g}\cdot\text{s}^2)\end{cases}$
Acceleration of free fall	$g=9.807$ m/s^2
Avogadro constant	$N_A=6.022\times10^{23}$ mole^{-1}
Charge of electron or proton	$e=\begin{cases}1.602\times10^{-19}\ \text{C}\\ 4.80\times10^{-10}\ \text{CGSE units}\end{cases}$
Rest mass of electron	$m_e=0.911\times10^{-30}$ kg
Specific charge of electron	$\frac{e}{m_e}=\begin{cases}1.76\times10^{11}\ \text{C/kg}\\ 5.27\times10^{17}\ \text{CGSE units/g}\end{cases}$
Rest mass of proton	$m_p=1.672\times10^{-27}$ kg
Electric constant	$\varepsilon_0=0.885\times10^{-11}$ F/m $1/4\pi\varepsilon_0=9\times10^9$ m/F
Magnetic constant	$\mu_0=1.257\times10^{-6}$ H/m $\mu_0/4\pi=10^{-7}$ H/m
Relation between velocity of light and ε_0 and μ_0	$c=1/\sqrt{\varepsilon_0\mu_0}$

Subject Index

TO THE READER

Mir Publishers would be grateful for your comments on the content, translation and design of this book. We would also be pleased to receive any other suggestions you may wish to make.
Our address is:

Mir Publishers
2 Pervy Rizhsky Pereulok
I-110, GSP, Moscow, 129820
USSR

Handbook of Physics

B. Yavorsky and A. Detlaf

A companion volume to Vygodsky's *Handbook of Higher Mathematics*, designed for use by engineers, technicians, research workers, students, and teachers of physics. Includes definitions of basic physical concepts, brief formulations of physical laws, concise descriptions of phenomena, tables of physical quantities in various systems of units, universal physical constants, etc.